SELECTIVE REDUCTION OF CARBON-CARBON MULTIPLE BONDS CATALYZED BY 3D METALS AND REGIOSELECTIVE C-H FUNCTIONALIZATION OF HETEROARENES

Félix Roi Berger

SELECTIVE REDUCTION OF CARBON-CARBON MULTIPLE BONDS CATALYZED BY 3D METALS AND REGIOSELECTIVE C-H FUNCTIONALIZATION OF HETEROARENES

First Edition November 2024

Written by Félix Roi Berger

Chapter 3. Hydrogenation of *α,β*-Epoxy Ketones to *α*-Hydroxy Epoxides by a Well-Defined Manganese Catalyst

Chapter 4. Manganese-Catalyzed Chemoselective Direct Hydrogenation of *α*-Ketoamides

Chapter 5. Copper-Catalyzed Regioselective C(2)–H Bond Arylation of Indoles and Related (Hetero)arenes

Chapter 6. Summary and Outlook

Chapter 1

Introduction

Hydrogenation is a cornerstone of synthetic chemistry that offers a powerful means to selectively modify functional groups within complex molecular structures. Since the early 20th century, catalytic hydrogenative processes have grown and advanced significantly with the help of many transition metal catalysts. For example, well-known hydrogenation catalysis such as Pd or Pt on carbon in the presence of H_2 are employed for the catalytic hydrogenation of multiple bonds. However, the lack of selectivity in such a heterogeneous hydrogenation process resulted in multiple product formation. Hence researchers moved toward homogeneous hydrogenation processes that are much milder and more product-selective.[1] Over the time, ruthenium, rhodium, palladium, and iridium, the 4d and 5d transition metals have played a significant role in the homogeneous hydrogenation process with molecular hydrogen. While homogeneous hydrogenation reactions resulted in improved product selectivity and were achieved under milder conditions, the heavy 4d/5d transition metals were expensive, naturally less abundant, and toxic.[2–4] As a result, more affordable, naturally abundant and relatively non-hazardous 3d transition metals were employed instead for these changes.[5]

In recent years, the pursuit of chemoselective hydrogenation methods has gained significant attention. Chemoselective hydrogenation refers to the chemoselective hydrogenation of one unsaturated moiety in the presence of others within a molecule. Driven by the imperative to streamline synthetic routes while minimizing waste and maximizing atom economy, 3d-metal catalysis emerges as a promising frontier, offering unparalleled precision and efficiency in the selective reduction of multiple functional groups. Traditional hydrogenation processes often suffer from a lack of selectivity, necessitating laborious protection and deprotection steps or leading to undesired side reactions.[6–8] The advent of 3d-metal catalysts has revolutionized this landscape, enabling direct and chemoselective hydrogenation of specific functional groups within complex organic frameworks. By harnessing the unique coordination environments and electronic properties of 3d-metal complexes, various research groups have unlocked a new realm of catalytic selective hydrogenation, paving the way for more sustainable and efficient synthetic pathways.[5] In this chapter, we delve into the growing field of 3d-metal catalyzed chemoselective direct hydrogenation, exploring the underlying principles, synthetic methodologies, and transformative applications. We highlight the key advances in catalyst design, and mechanistic insights, illustrating how these catalysts enable precise control over reaction outcomes in diverse chemical contexts.

Besides the hydrogenation, the regioselective C−H arylation of indoles has garnered considerable interest in recent years, owing in part to the pivotal role of 3d-metal catalysis in enabling this synthetic transformation.[9,10] Leveraging the unique reactivity and coordination properties of 3d transition metals including cobalt, nickel, and copper, researchers developed a plethora of innovative catalytic systems that not only facilitate the direct arylation of indoles but also afford remarkable levels of regiocontrol. By harnessing the redox versatility, ligand design strategies, and tunable reactivity profiles of 3d-metal catalysts, synthetic chemists have achieved exquisite selectivity in accessing diverse regioisomeric products, thereby expanding the practical applicability of this powerful method. In this context, this chapter, also explores the promising field of 3d-metal catalyzed selective C−H bond arylation of indoles, highlighting key advancements, and mechanistic insights in the pursuit of efficient, sustainable, and versatile synthetic methodologies.

1.1 3d-METAL CATALYZED CHEMOSELECTIVE DIRECT HYDROGENATION REACTIONS

Chemoselective hydrogenation using noble metal catalysts under hydrogenation conditions is well developed.[11−15] Along with this, the heterogeneous chemoselective hydrogenation of various unsaturated moieties is explored using noble and non-noble metal catalysts.[14,16−19] However, the limited abundance of noble metals found in the Earth's crust and the harsh reaction conditions associated with heterogeneous catalysis force chemists to find alternative methods for sustainable development. In the last twenty years, remarkable progress has been achieved in this field, driven by a persistent commitment to innovation and advancement. Moreover, chemoselective hydrogenation using molecular hydrogen under 3d-metal catalysis stands as a pivotal advancement in synthetic organic chemistry.[20,21] While traditional methods rely on various hydrogenation sources such as $NaBH_4$, $LiAlH_4$, and isopropanol, they often encounter limitations. These limitations include the formation of multiple products, lack of functional group tolerance, harsh reaction conditions, and the generation of hazardous stoichiometric waste. Addressing these drawbacks, recent research has focused on the development of 3d-metal catalyzed chemoselective hydrogenation utilizing molecular hydrogen as a clean source. This approach offers several advantages, including the use of molecular H_2, milder reaction conditions, and improved chemoselectivity. By harnessing the catalytic properties of 3d-transition metals, such as cobalt, iron, or nickel, chemists can selectively reduce specific functional groups within the complex molecules while leaving others untouched. This advancement not only enhances the efficiency of hydrogenation but also aligns with the principles of green chemistry by minimizing waste generation and utilizing a renewable and environmentally benign hydrogen source. As a result, 3d-metal catalyzed chemoselective hydrogenation holds promise for enabling more sustainable and efficient synthetic routes in organic chemistry.

1.1.1 Manganese-Catalyzed Chemoselective Hydrogenation

In 2016, Beller's groundbreaking discovery unveiled manganese's ability to activate molecular hydrogen, thus catalyzing the hydrogenation of unsaturated moieties.[22] This pivotal breakthrough marked a significant turning point, sparking new avenues of exploration and innovation within the field of hydrogenation. After this many researchers developed numerous manganese-centered catalysts for the reduction of various polar and non-polar multiple bonds including, imine, aldehyde, ketone, amide, ester, alkene and alkyne.[23–29] In these reports, a little modification in the ligand backbone resulted in a drastic change in the reactivity of the catalyst.

In the last decade, manganese has emerged as an attractive metal for chemoselective hydrogenation due to its abundance, low toxicity, and relatively low cost compared to precious metals like platinum, palladium, or ruthenium often used in hydrogenation reactions.[30] Moreover, manganese catalysts can exhibit high activity and selectivity under appropriate reaction conditions. One of the key challenges in chemoselective hydrogenation is designing catalysts that can distinguish between different functional groups based on their electronic and steric properties. Manganese catalysts, often in coordination with suitable ligands, can be tailored to selectively hydrogenate certain functional groups while leaving others untouched.

The pioneering work on the application of manganese-based catalyst for the hydrogenation of polar double bonds via activation of H_2 is typically attributed to the research conducted by Beller and co-workers (Scheme 1.1).[22] By carefully designing ligands around the manganese center, they were able to control the selectivity in the hydrogenation reaction, enabling the reduction of polar functional groups such as ketones, aldehydes, and nitriles while leaving other functionalities intact. This breakthrough opened up new avenues for the development of manganese-catalyzed hydrogenation protocols, expanding the scope of chemoselective hydrogenation reactions and contributing to the advancement of sustainable and efficient catalytic processes in organic synthesis. The synthesized **Mn-1** catalyst showed excellent selectively for C=O bond hydrogenation in *α,β*-unsaturated aldehydes, however formation of two products was observed when *α,β*-unsaturated nitrile was subjected to the hydrogenation.

Soon after, Kempe developed a (PNP)Mn(I) catalyst for the reduction of ketones to secondary alcohols (Scheme 1.2).[31] This catalyst was more active compared to Beller's catalyst. In this report, authors have shown two examples of chemoselective C=O bond hydrogenation of ketone having isolated double bonds. The hydrogenation of even simple alkyl and aromatic ketone was achieved in excellent yields. The developed method shows the chemoselective C=O bond hydrogenation in presence of highly reducible functionalities such as nitro and nitrile groups.

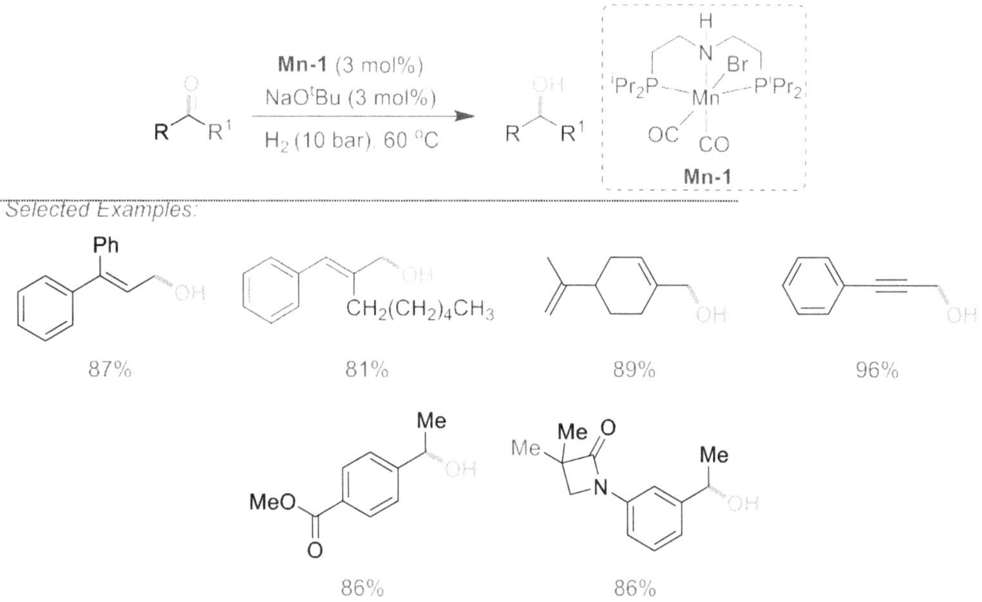

Mn-1 (3 mol%)
NaO^{t}Bu (3 mol%)
H_2 (10 bar), 60 °C

Mn-1

Selected Examples:

87% 81% 89% 96%

86% 86%

Scheme 1.1 Mn-Catalyzed Chemoselective Hydrogenation of Carbonyl Compounds.

Mn-2 (0.1 mol%)
KO^{t}Bu, H_2 (20 bar)
80 °C

2 exmples
up to 95% yield

Mn-2

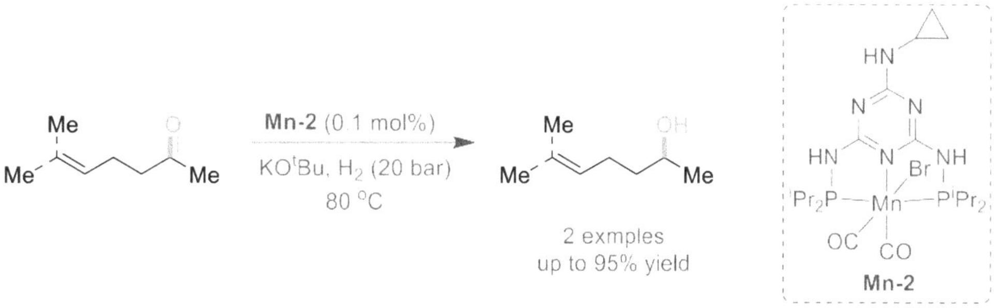

Scheme 1.2 Mn-Catalyzed Chemoselective Hydrogenation of Unconjugated Ketones.

In 2017, Beller synthesized a chiral version of the **Mn-1** catalyst for asymmetric hydrogenation of carbonyl compounds (Scheme 1.3).[32] In this report, they have shown three examples of chemoselective chiral C=O bond hydrogenation in α,β-unsaturated ketones. The developed catalyst is efficient for the hydrogenation of unconjugated aliphatic as well as aromatic ketone moieties with good enantioselectivity.

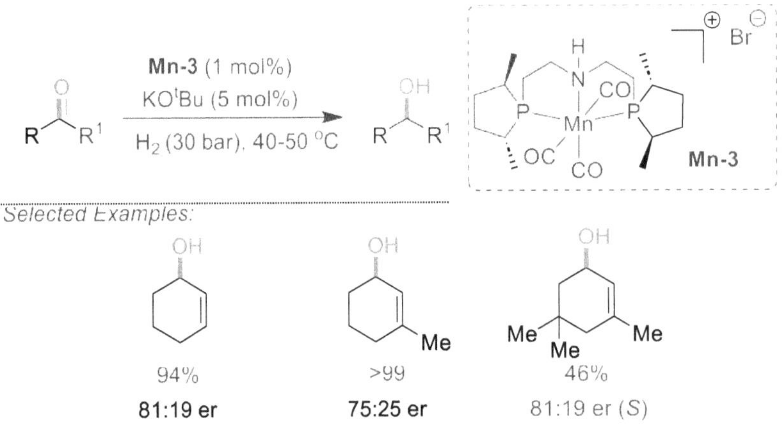

Mn-3 (1 mol%)
KO^{t}Bu (5 mol%)
H_2 (30 bar), 40-50 °C

Mn-3

Selected Examples:

94%
81:19 er

>99
75:25 er

46%
81:19 er (*S*)

Scheme 1.3 Mn-3 Catalyzed Chemoselective Hydrogenation of Conjugated Ketones.

Furthermore, Kirchner has developed a sequence of (PNP)Mn(I) hydride complexes for the chemoselective reduction of aldehydes to corresponding alcohols (Scheme 1.4).[33] Among all these complexes, **Mn-4** with *NH* group in the ligand backbone for metal-ligand cooperation has shown excellent selectivity for C=O bond hydrogenation of aldehyde in the presence of other reducible groups. In this report, authors have compared the reactivity of manganese and rhenium pincer complexes for hydrogenation where the synthesized manganese catalyst showed better activity than the rhenium catalyst. This reaction is highly selective for aldehyde hydrogenation in the presence of other sensitive functional groups such as ketones, esters, alkynes, alkene, cyano, and conjugated double bonds. The reaction proceeded even at 0.05 mol% catalyst loading with excellent chemoselectivity at room temperature (Scheme 1.4). Similar to the previously discussed manganese catalyst, the hydrogenation catalyzed by **Mn-4** was also proceeding via metal-ligand cooperative H_2 activation. The plausible catalytic pathway starts from the manganese hydride (**Mn-4**), which upon coordination with the aldehyde followed by hydride transfer (via **A**) will result in the formation of manganese alkoxy complex **B**. This intermediate upon deprotonation of the *NH* group will result in the dearomatized species coordinated to the molecular hydrogen (**C**). This intermediate upon H_2 activation will regenerate the active catalyst **Mn-4** (Figure 1.1).

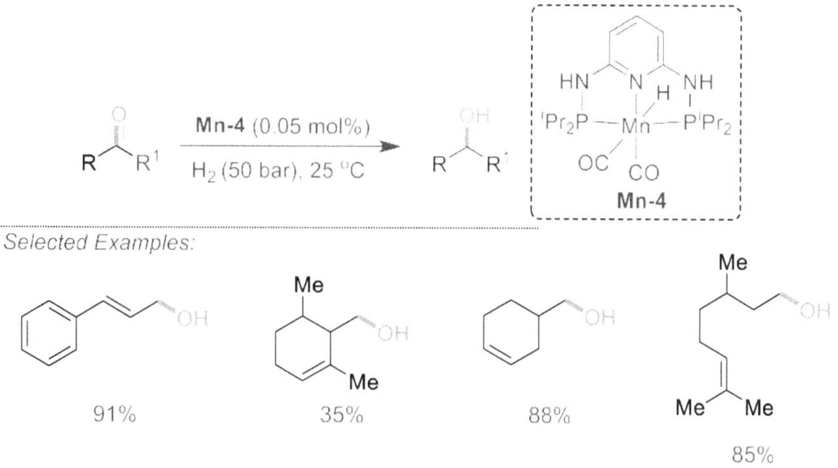

Scheme 1.4 **Mn-4** Catalyzed Chemoselective Hydrogenation of Aldehydes.

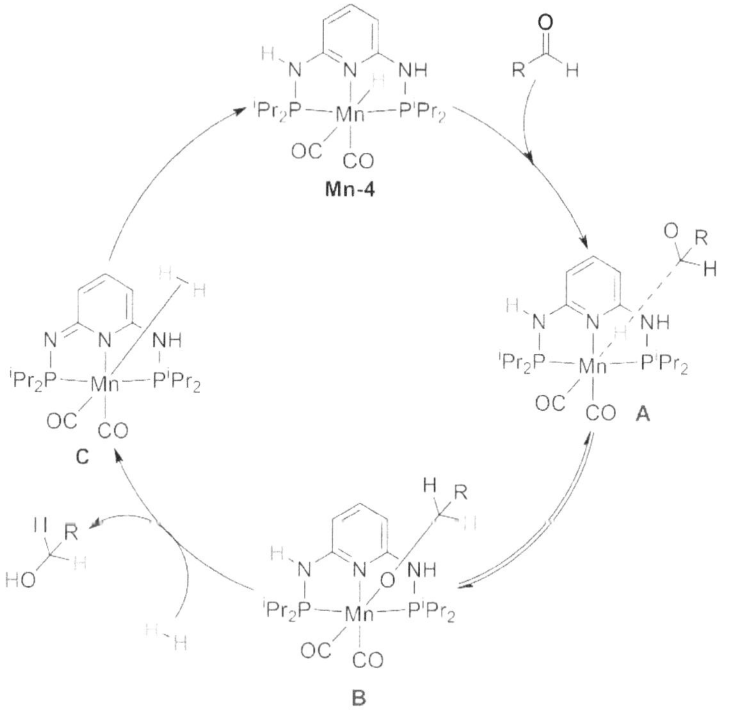

Figure 1.1 Plausible Catalytic Pathway for the Mn-Catalyzed Chemoselective Hydrogenation of Aldehyde.

With significant advancement, the Huang group developed air-stable Mn(II) pincer PNN complexes for the selective C=O bond reduction in conjugated aldehydes, and ynals (Scheme 1.5).[34] The developed protocols exhibit selectivity solely for the C=O bond of aldehydes, leaving the C=O bonds of ketone and esters unreacted. In addition to α,β-unsaturated aldehydes, and ynals simple alkyl and aryl aldehydes were also hydrogenated efficiently. The affordability of Mn(II) in comparison with Mn(I) and the robustness of the developed catalysts opens doors for large-scale applications.

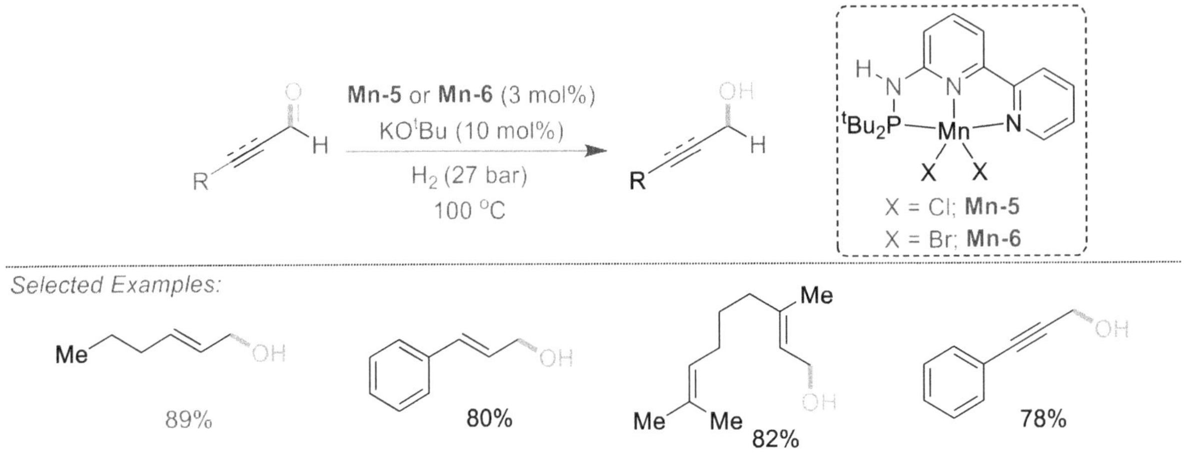

Scheme 1.5 Manganese-Catalyzed Chemoselective C=O Bond Hydrogenation of α,β-Unsaturated Aldehydes, and Ynals.

In a remarkable development, Wang, Han, and Ding reported the chemo- and diastereoselective C=O bond hydrogenation of α-substituted β-ketoamides (Scheme 1.6).[35] This reaction proceeds through dynamic kinetic resolution for the first time under manganese catalysis. This method was suitable for the reduction of various substituted β- including heteroaryl, alkyl, aryl, and halogen groups. The developed **Mn-7** catalyst showed the highest TON of 10000 for the hydrogenation of α-substituted β-ketoamides. DFT calculations unveiled that the distinct anti-stereoselectivity observed in the catalysis stems from an appealing interaction involving the $\pi\cdots\pi$ stacking of the substrate's phenyl ring with the ligand's pyridinyl ring in the preferred transition state.

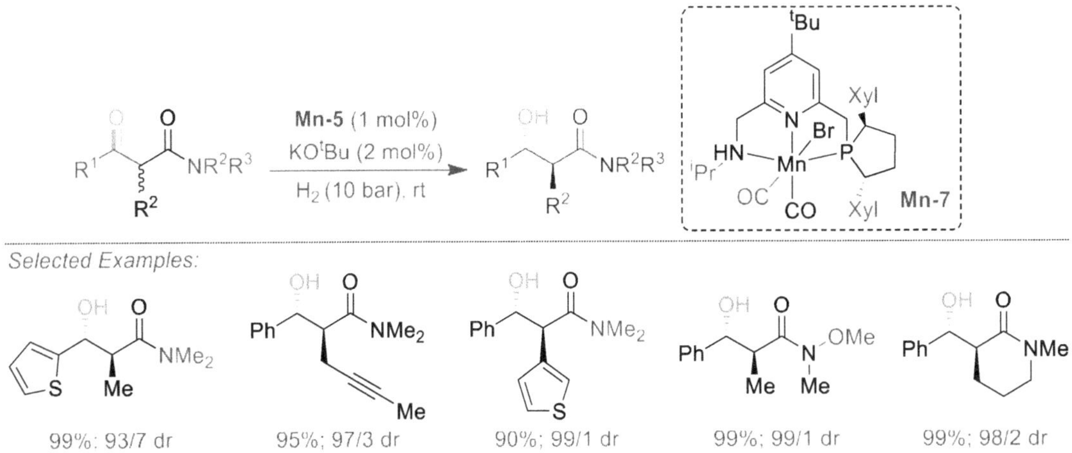

Scheme 1.6 Mn-7 Catalyzed Chemo- and Diastereoselective C=O Bond Hydrogenation of α-

Substituted β-Ketoamides.

Sortais has shown the chemoselective C=O bond hydrogenation in unconjugated ketones using a PNP-based manganese pincer catalyst.[36] With significant modification later on the same group synthesized bidentate Mn(I) complex and used it for the chemoselective C=C and complete hydrogenation of conjugated ketone (Scheme 1.7).[37] By using changing bases along with reaction temperatures, they were able to achieve the other chemoselectivity of the hydrogenationreaction.

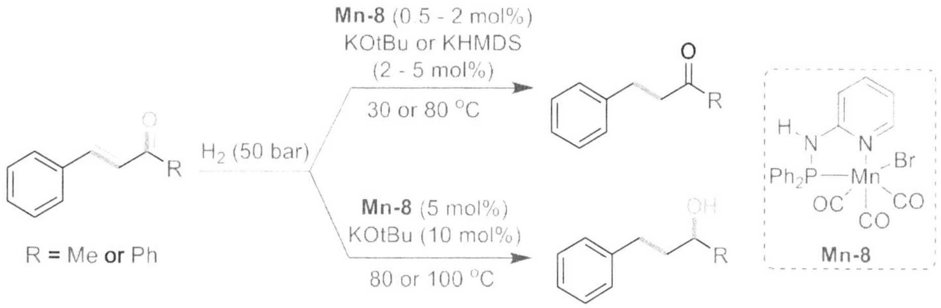
Mn-8 (0.5 - 2 mol%)
KOtBu or KHMDS
(2 - 5 mol%)
30 or 80 °C
H2 (50 bar)
R
R = Me or Ph
Mn-8 (5 mol%)
KOtBu (10 mol%)
80 or 100 °C
O
OH
H
N
N
Br
Ph2P—Mn
OC
CO
CO
Mn-8

Scheme 1.7 Manganese-Catalyzed C=C and Complete Hydrogenation of Conjugated Ketone.

A phosphine-free method for the selective alkene hydrogenation in conjugated esters and nitriles was demonstrated by Topf (Scheme 1.8).[38] This system utilizes the commercially available $Mn(CO)_5Br$ precursor in combination with the 2-picolylamine ligand. Various di-, tri- and tetrasubstituted conjugated alkenes were hydrogenated efficiently under developed protocols. A mixture of products was formed when α,β-unsaturated ketone was reacted under the standard conditions. However, this report is associated with some drawbacks including high reaction temperature and formation of multiple products in the hydrogenation of chalcone, a conjugated ketone.

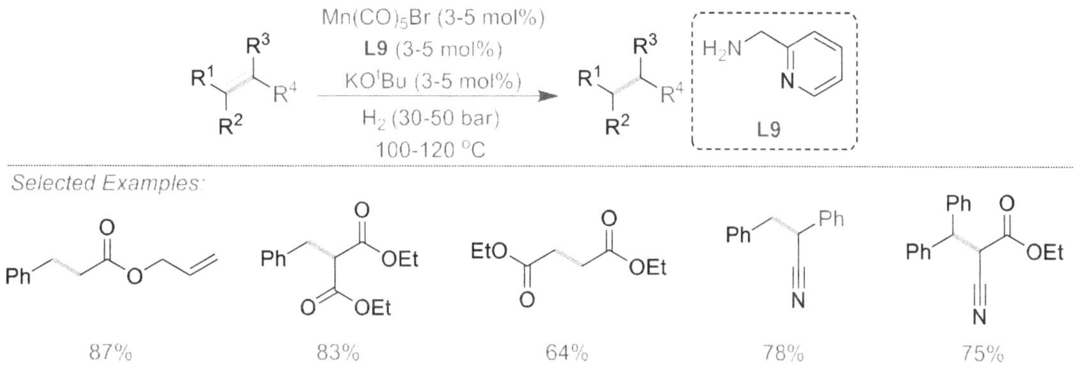
Mn(CO)5Br (3-5 mol%)
L9 (3-5 mol%)
KOtBu (3-5 mol%)
H2 (30-50 bar)
100-120 °C
R1
R2
R3
R4
H2N
N
L9
Selected Examples:
Ph
O
OEt
EtO
Ph
N
87%
83%
64%
78%
75%

Scheme 1.8 Mn-Catalyzed Chemoselective C=C Bond Hydrogenation of Conjugated Double Bonds.

Recently, Rueping and co-workers have demonstrated chemoselective hydrogenation of nitroarenes to the corresponding anilines using (PNP)Mn(I) complex (Scheme 1.9).[39] In this report, they have shown the wide substrate scope where highly reducible functional groups including alkynyl, benzyl, ester and amide groups were tolerated. Based on the experimental outcomes they proposed a plausible catalytic pathway that proceeds via metal-ligand cooperation and direct formation of aniline without observing the accumulation of intermediates including hydroxylamine, azo, hydrazo and azoxy compounds. In addition to the nitroarenes, the advanced system was appropriate for the reduction of azoarene derivatives.

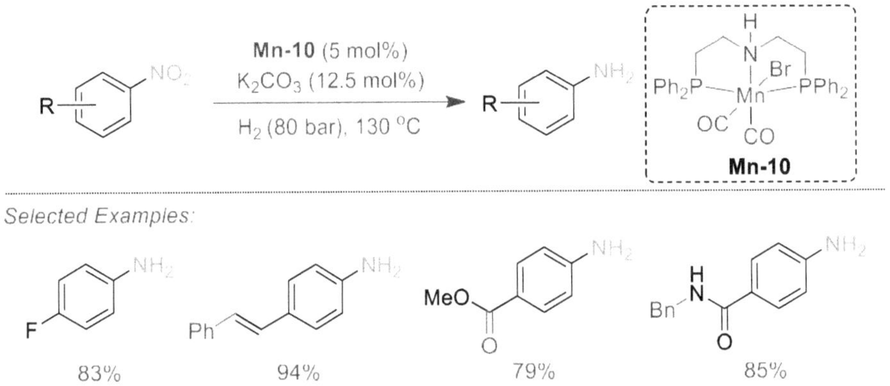

Mn-10 (5 mol%)
K_2CO_3 (12.5 mol%)
H_2 (80 bar), 130 °C

Mn-10

Selected Examples:

83% 94% 79% 85%

Scheme 1.9 Manganese-Catalyzed Chemoselective Hydrogenation of Nitroarenes.

1.1.2 Iron-Catalyzed Chemoselective Hydrogenation

Iron is one of the most plentiful, least toxic, and inexpensive transition metal available in the lithosphere. Iron-based catalysts were well explored in the various organic transformations in synthetic organic chemistry.[40–44] With its convincing chemical properties, iron has found significant applications in the pharmaceutical and agrochemical industries. These properties of iron, make it an attractive option for large-scale synthesis, aligning with the economic demands of these sectors.

With these aspects, Milstein and co-workers synthesized lutidine-based PNP iron complex (**Fe-1**) for direct hydrogenation of ketone (Figure 1.2a).[45] However, in the case of hydrogenation of *α,β*-unsaturated ketones they end up getting multiple products in a moderate ratio. In addition, the borohydride derivative of **Fe-1** (**Fe-2**) developed by the same group also resulted in multiple product formation on the reduction of conjugated (Figure 1.2b).[46] Moreover, these catalysts were efficient for the reduction of alkyl and aryl ketones in good yields with the tolerance of various sensitive functional groups including halides and trifluoromethyl substituents.

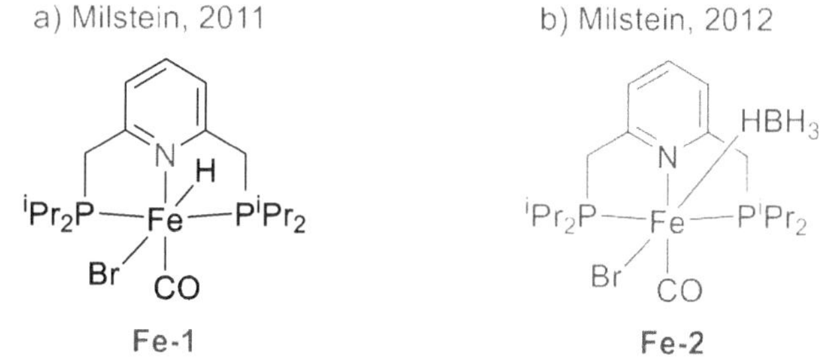

a) Milstein, 2011 b) Milstein, 2012

Fe-1 Fe-2

Figure 1.2 Developed (PNP)Fe(II) Complexes by Milstein.

Beller showed selective C=O bond reduction of conjugated aldehydes using a robust, non-pincer iron catalyst (Scheme 1.10).[47] The developed catalyst was stable under air and water. Using catalyst **Fe-3**, they have shown 8 examples of selective C=O bond hydrogenation of conjugated aldehydes, up to >99% yield using catalyst loading up to 0.1 mol%. In addition, this catalyst hydrogenated simple aryl and alkyl aldehydes, and ketones hydrogenated efficiently. The aromatic ketones containing the other reducible functional groups including nitro, ester, and amide were hydrogenated selectively to the corresponding alcohol.

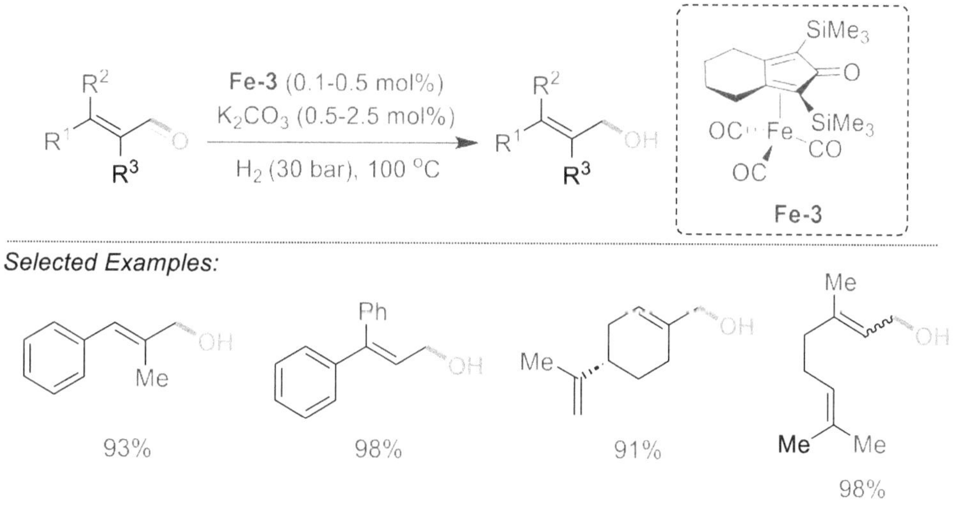

Fe-3 (0.1-0.5 mol%), K_2CO_3 (0.5-2.5 mol%), H_2 (30 bar), 100 °C

Selected Examples: 93%, 98%, 91%, 98%

Scheme 1.10 Iron-Catalyzed Chemoselective C=O Bond Hydrogenation in α,β-Unsaturated Aldehydes.

Later a catalytic system for iron-catalyzed chemoselective hydrogenation of α,β-unsaturated aldehydes under acidic conditions was reported by the same group (Scheme 1.11).[48] In contrast to the majority of other catalytic systems, that work under basic conditions, this system is active under acidic conditions (using catalytic TFA). Various α,β-unsaturated aldehydes with alkyl and aryl backbone were tested under the optimized reaction conditions which provided good yield for corresponding allyl alcohols. The chemoselective C=O bond hydrogenation of aldehyde was observed in the case of substituted benzaldehyde with reducible functional groups including alkene, ester, and ketone.

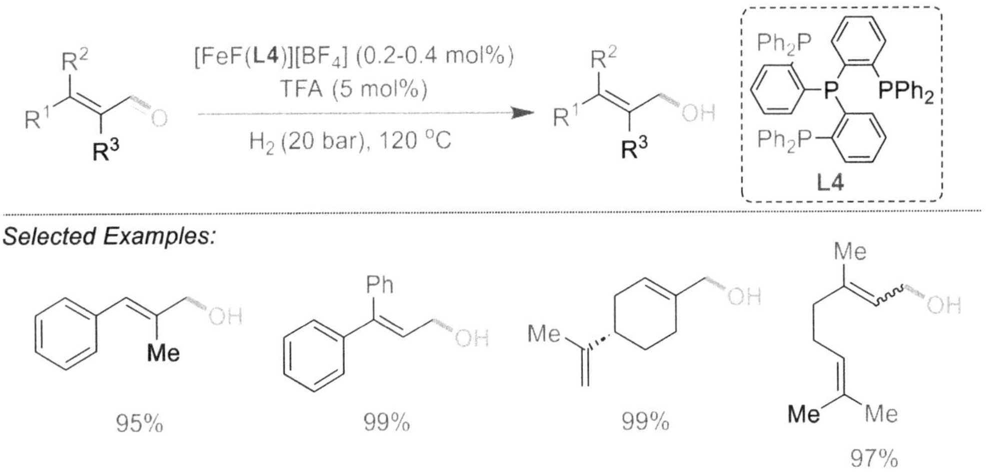

Scheme 1.11 Iron-Catalyzed Chemoselective C=O Bond Hydrogenation of α,β-Unsaturated Aldehydes.

Later, Kirchner developed PNP-based iron complexes with *NH* (**Fe-5**) and NMe (**Fe-6**) groups in the ligand backbone (Figure 1.3).[49] Catalysts having *NH* group were active for the hydrogenation of ketones and aldehydes; however, catalysts having NMe group in the ligand backbone were selective for only aldehyde hydrogenation. When α,β-unsaturated ketone was reacted under the reaction conditions, they resulted in formation of C=C, C=O, and fully hydrogenated products. The reaction using **Fe-5** proceeds through metal-ligand cooperation (MLC) however the mechanism with **Fe-6** does not involve MLC.

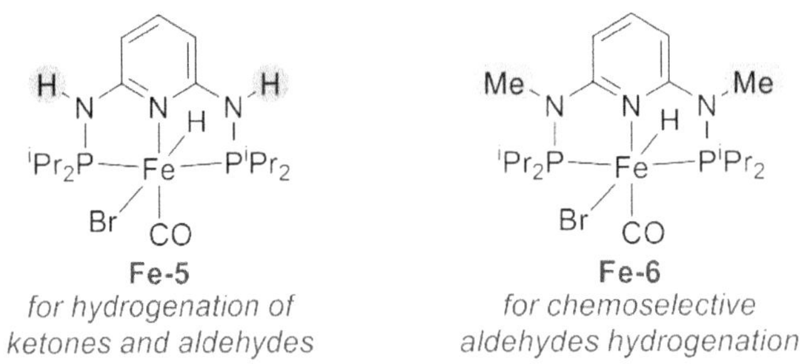

Figure 1.3 Iron-Catalyst for Chemoselective Hydrogenation of C=O Bonds.

Based on the experimental results and theoretical calculations the catalytic cycle for **Fe-6** catalyzed aldehyde hydrogenation was proposed (Figure 1.3).[50] This reaction proceeds with the activation of precatalyst **Fe-6** in the presence of a base and molecular hydrogen to form an iron dihydride intermediate either *trans* **A** or *cis* **A**. Upon coordination with aldehyde followed by hydride migration led to the intermediate **B**. The intermediate **B** will undergo ligand exchange directly with H_2 or, in the presence of ethanol, in two steps via **C** to form intermediate **D**. Then the alkoxide ion on protonolysis will result in the product formation and regenerate the active catalyst **A** (Figure 1.4).

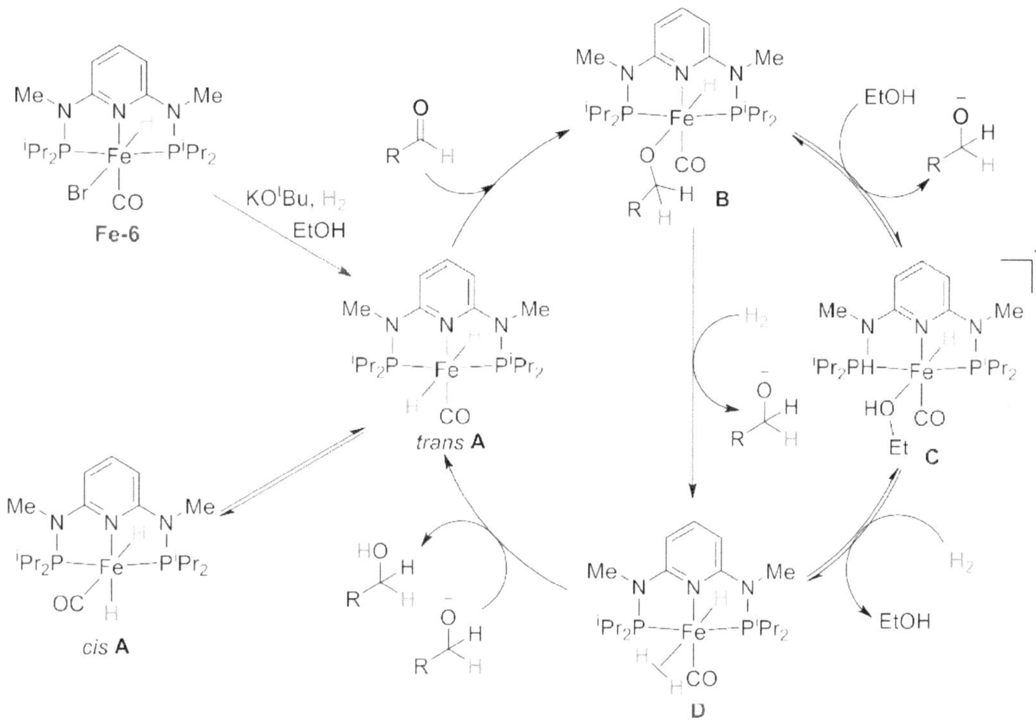

Figure 1.4 Plausible Catalytic Pathway of Iron-Catalyzed Chemoselective Aldehyde Hydrogenation.

With significant modifications, Hu and co-workers developed (PONOP)Fe(II) complexes for chemoselective C=O bond hydrogenation of aldehydes at room temperature (Scheme 1.12).[51] The exceptional selectivity shown towards aldehyde groups over alkene, and particularly ketone groups, in chemical reactions is highly noteworthy and valuable in synthetic applications. The authors showed the potential of **Fe-7** and **Fe-8** in the chemoselective reduction of the aldehyde's C=O bond under direct as well as transfer hydrogenation conditions. In addition, simple aromatic aldehydes can also be hydrogenated using these catalysts.

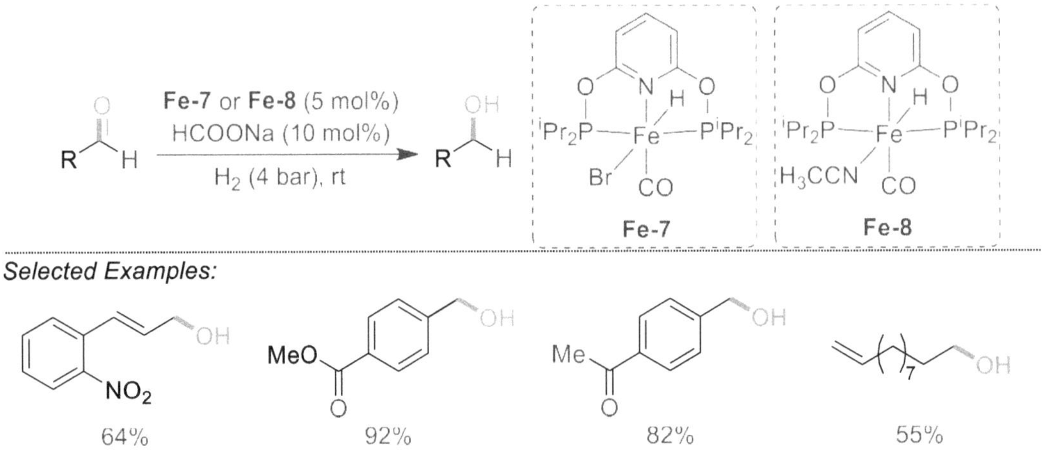

Fe-7 or Fe-8 (5 mol%)
HCOONa (10 mol%)
H2 (4 bar), rt
Fe-7
Fe-8
Selected Examples:
64%
92%
82%
55%

Scheme 1.12 (PONOP)Fe(II) Catalyzed Selective Hydrogenation of Aldehydes.

On the other side, Duan, Wen, and Zhang reported the single catalytic system for the selective reduction of C=O, C=C, and NO_2 functionalities using commercially available iron precursor/tetraphos ligand (**L9**).[52] This method tolerated various reducible functional groups including isolated alkene, alkyne, acetyl group, amide, and sulphonyl group (Scheme 1.13).

Fe(acac)2 (0.1 mol%)
L9 (0.11 mol%)
K2CO3 (1 mol%)
H2 (30 bar), 80-120 °C
L9
Selected Examples:
93%
95%
87%
95%
99%
99%
98%
95%
80%
98%
96%

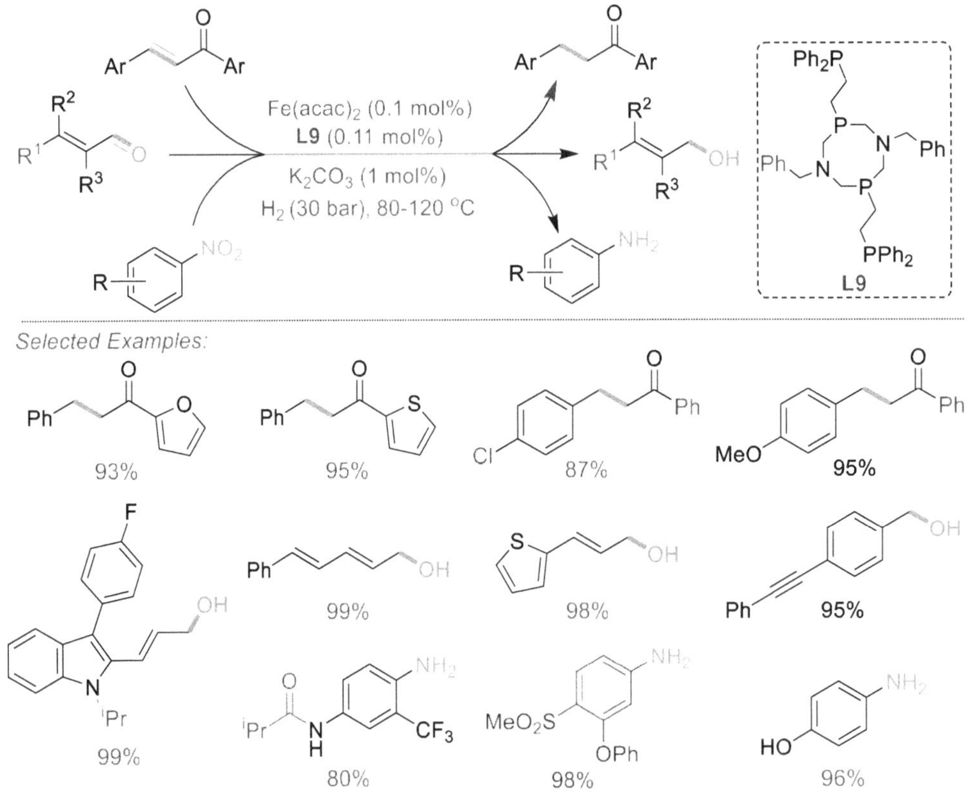

Scheme 1.13 Iron-Catalyzed Chemoselective C=C, C=O, and NO_2 Hydrogenation.

Furthermore, an efficient method for iron-catalyzed chemoselective hydrogenation of the nitrobenzene derivatives in the presence of other sensitive functionalities including heteroarenes, alkene, esters, and ketone was developed (Scheme 1.14).[53] In this method, a tetraphos-based iron catalyst and 20 bar H_2 pressure were employed. Based on the experimental outcomes, a catalytic pathway was proposed, involving either the formation of hydroxylamine or azobenzene as intermediates (Figure 1.5).

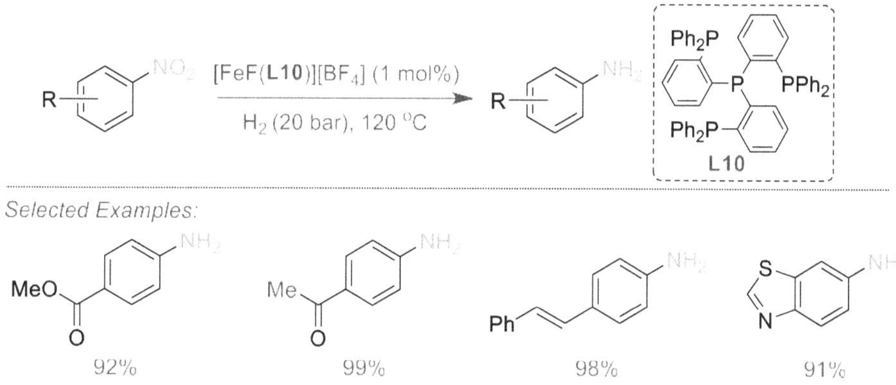

[FeF(L10)][BF$_4$] (1 mol%), H_2 (20 bar), 120 °C; L10

Selected Examples: 92%, 99%, 98%, 91%

Scheme 1.14 Iron-Catalyzed Chemoselective Hydrogenation of Nitroarenes.

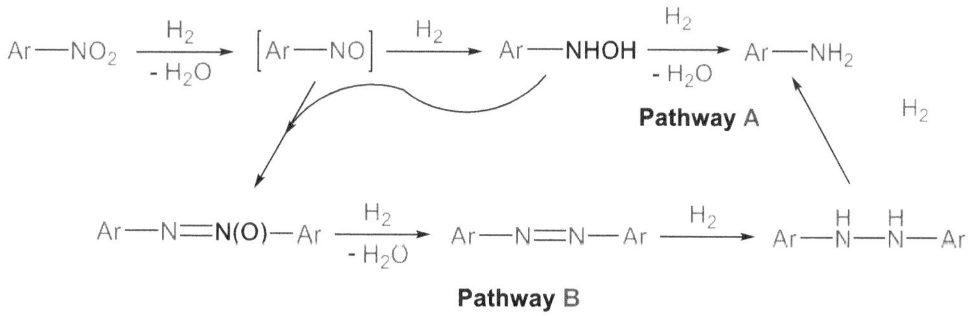

Pathway A; Pathway B

Figure 1.5 Plausible Catalytic Pathways for Hydrogenation of Nitroarene.

1.1.3 Cobalt-Catalyzed Chemoselective Hydrogenation

Cobalt, a relatively inexpensive and low-toxic metal, plays a crucial role in biology, notably through its presence in vitamin B12. Additionally, the distinctive electronic structural properties of this base metal may enable the observation of uncommon activity or selectivity profiles. Thus, in recent times, employing rational ligand design has unveiled several novel cobalt catalysts for homogeneous hydrogenation.[54,55] Cobalt catalysts are well-studied for the independent hydrogenation of various unsaturated moieties including alkene, alkyne, carbonyl, and other groups under both direct and transfer hydrogenation conditions.[56–62] However, the chemoselective hydrogenation using cobalt catalysts is underdeveloped.

In 2015, Kempe developed a $(PN^3P)CoCl_2$ catalyst **Co-1** for the reduction of the carbonyl group (Scheme 1.15).[63] Furthermore, alongside simple ketones, this catalyst demonstrated outstanding efficacy in selectively hydrogenating C=O bonds in *α,β*-unsaturated aldehydes, requiring catalyst loadings as low as 0.5 mol%.

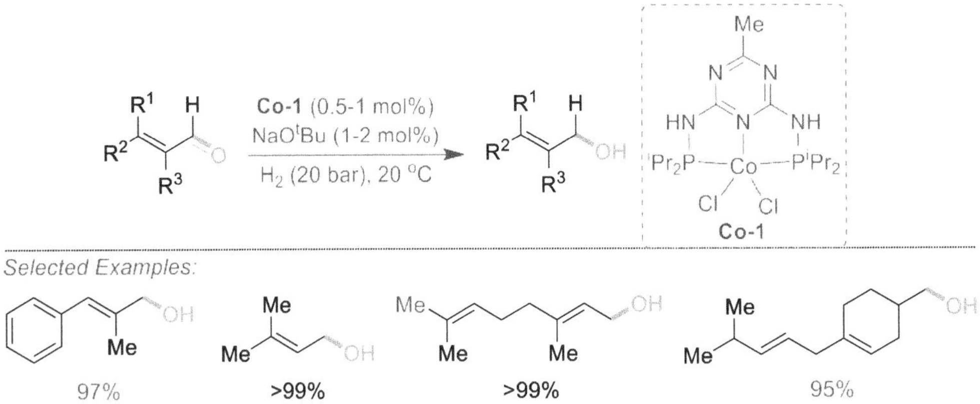

Co-1 (0.5-1 mol%)
NaO^{t}Bu (1-2 mol%)
H_2 (20 bar), 20 °C

Selected Examples:
97% >99% >99% 95%

Scheme 1.15 Cobalt-Catalyzed Chemoselective C=O Bond Hydrogenation in *α,β*-Unsaturated Aldehydes.

Subsequently, Wen and Zhang developed a general catalytic system for cobalt-catalyzed chemoselective hydrogenation of various unsaturated moieties including conjugated aldehydes, ketones, imines as well as *α*-ketoamides, and heteroarenes (Scheme 1.16).[64] The developed tetraphos/Co(II) system was very efficient and showed excellent activity even at 0.002 mol% catalyst loading. This reaction proceeds through the formation of radical species **A** and **B**, followed by cobalt dihydride as an intermediate **C** (Figure 1.6).

Recently, Li and Liu represented an NHC-cobalt catalyzed C=C bond hydrogenation chemoselectively in conjugated carbonyl compounds (Scheme 1.17).[65] The developed catalyst efficiently hydrogenated the challenging hindered, tetrasubstituted C=C bonds selectively by taking advantage of empty coordination sites and less steric hindrance because of the bidentate NHC ligand. The developed system efficiently hydrogenated isolated hindered alkenes to the corresponding alkanes with a tolerance of highly reducible ester group. The outcomes achieved with this in-situ method, utilizing readily available carbene and cobalt precursor, proved to be highly competitive when compared to the results obtained through meticulously developed well-defined catalysis systems.

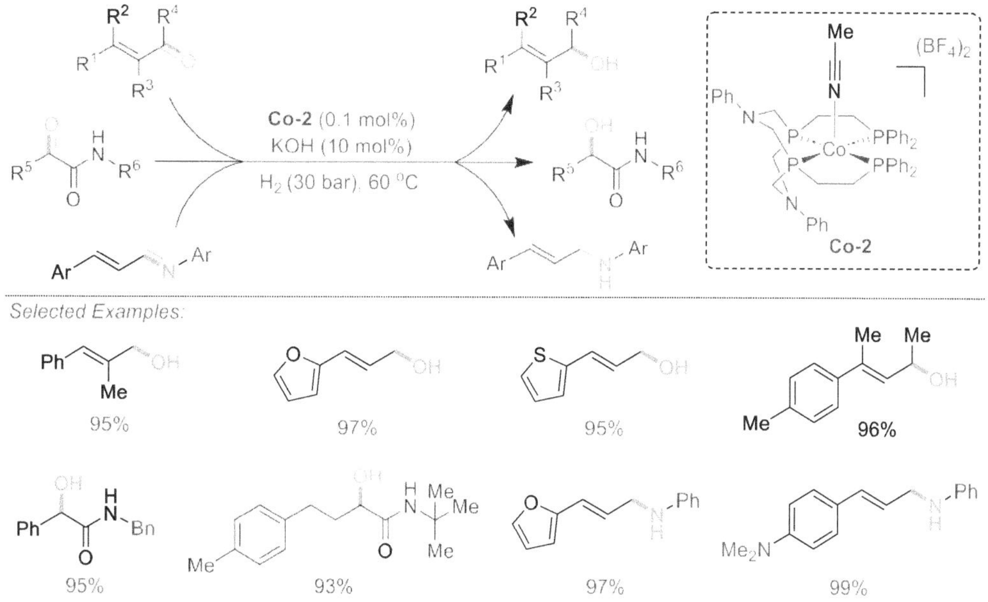

Scheme 1.16 Cobalt-Catalyzed Chemoselective Hydrogenation of α,β-Unsaturated Aldehydes, Ketones, Imines, α-Ketoamides.

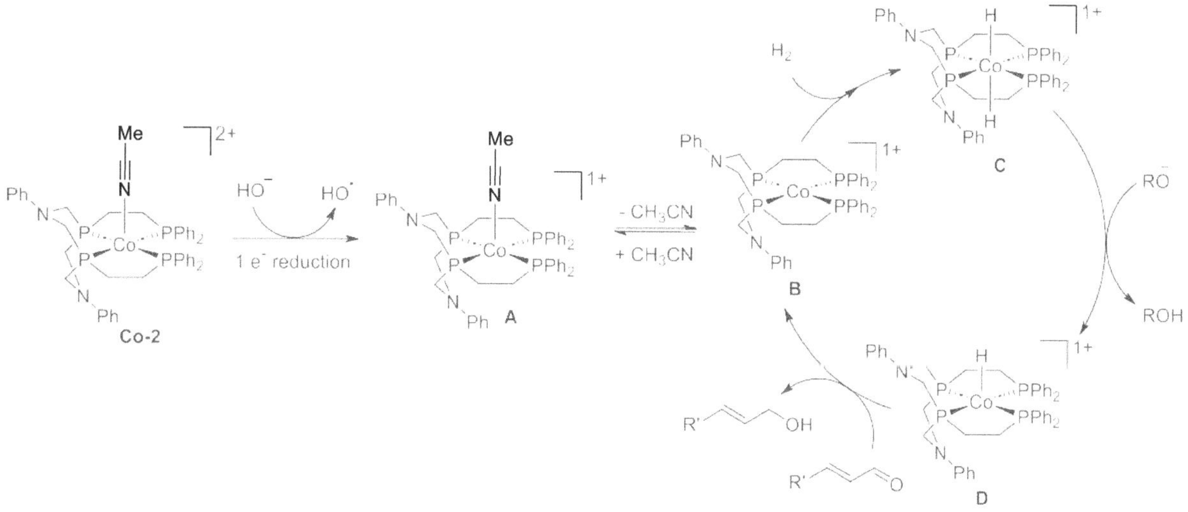

Figure 1.6 Plausible Catalytic Pathway for Cobalt-Catalyzed Chemoselective Hydrogenation.

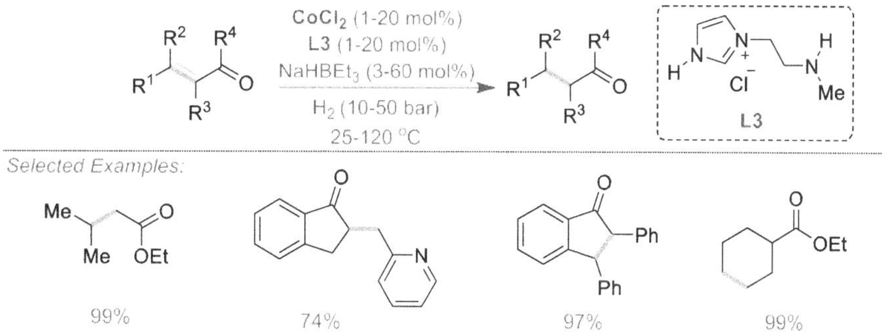

Scheme 1.17. Cobalt-Catalyzed Chemoselective C=C Bond Hydrogenation.

1.2 3d-METAL CATALYZED REGIOSELECTIVE C−H ARYLATION OF INDOLES

C−H bond functionalization stands as a potent technique within organic synthesis, involving the selective substitution or alteration of a C−H bond within an organic compound to introduce a different functional group, including C−O, C−C, C−X, or C−N bonds. (where X represents a halogen).[66–68] This method offers a more direct and efficient approach by utilizing the inherent C−H bonds present in organic molecules as starting materials. Over the years, significant advancement has been achieved in developing methods for C−H bond functionalization, driven by advances in catalysis, organometallic chemistry, and reaction methodology. Transition metal catalysis, particularly using palladium, ruthenium, rhodium, and iron catalysts, has emerged as a dominant strategy for achieving selective C−H bond transformations. Additionally, strategies for controlling regioselectivity and stereoselectivity in C−H bond functionalization reactions have been developed, enabling chemists to selectively target specific C−H bonds within complex molecules. These advancements have greatly expanded the synthetic toolbox available to organic chemists, facilitating the rapid construction of complex molecular architectures via selective C−H arylation, alkylation, alkenylation, alkynylation, and oxygenation. Regioselective C−H arylation is a significant transformation in organic synthesis. It involves the selective transformation of one carbon-hydrogen bond of substrate with an aryl group. Regioselectivity is a crucial aspect of this reaction, ensuring that the aryl group is added to the desired position of the substrate. Recently, arylation of various moieties such as aromatic compounds such as benzamide, anilines, phenols and quinolines, and heteroarenes including indoles, azoles and pyrroles have been studied.[69–76] The arylating agents used for these transformations include organometallic agents such as boronic acids, and Grignard as well as organic coupling agents including phenol derivatives, aryl iodonium salts and aryl halides. In particular, the production of naturally occurring product derivatives and physiologically active chemicals has found use for the regioselective C−H arylation of indoles, thereby contributing to advancements in medicinal chemistry and chemical biology.[9,10] For this reason, many transition metals have been investigated, including ruthenium, iridium, palladium, and rhodium.[77–82] These metals often form complexes with ligands that control the regioselectivity of the C−H activation step. Several research groups have contributed to the development of this methodology, continually improving its efficiency, selectivity, and applicability to various substrates. However, the limited abundance of 4d/5d metals in the Earth's crust has led researchers to consider alternative metals with greater abundance. Exploration into alternative metals aims to address concerns regarding the sustainability and accessibility of transition metal catalysts. Metals such as copper, nickel, and cobalt, which are plentiful in the Earth's crust, have emerged as promising candidates for catalyzing C−H bond arylation reactions. These metals offer potential solutions to the challenges associated with the scarcity and cost of traditional transition metals. Additionally, research efforts focus on developing new ligands and reaction conditions tailored to these alternative metal catalysts, aiming to achieve comparable levels of selectivity and efficiency. By harnessing the catalytic properties of earth-abundant metals, scientists strive to create more sustainable and environmentally friendly synthetic methodologies for organic synthesis.

1.2.1 Cobalt-Catalyzed Regioselective C−H Arylation of Indoles

Cobalt is a relatively inexpensive transition metal, making it an attractive option from a sustainability and cost perspective. By adding a 2-pyridyl ring to the nitrogen of the indole, Ackermann and colleagues reported the first regioselective C2 arylation of indoles enabled by cobalt catalysis (Scheme 1.18). In this method, the pyridine ring directs the arylation to the indole's C2 and couples with the phenol-derived organic electrophiles. With significant development, the same group showed the use of aryl chloride as a coupling for cobalt-catalyzed C2 arylation of indole.[83]

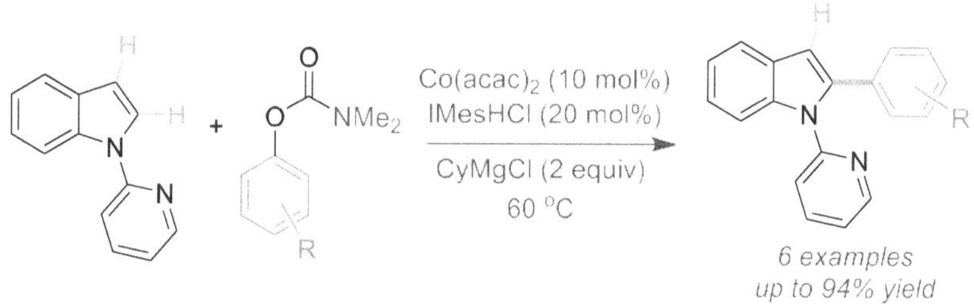

$Co(acac)_2$ (10 mol%)
IMesHCl (20 mol%)
CyMgCl (2 equiv)
60 °C

6 examples
up to 94% yield

Scheme 1.18 Cobalt-Catalyzed Regioselective C2 Arylation of Indoles.

Similarly, aryl boronic acids can be coupled regioselectively to the indole at C2 (Scheme 1.19).[84] In the seminal report, Niu and Song demonstrated a method for cobalt-catalyzed C2 arylation of indoles in the air using stoichiometric amounts of manganese as an oxidant. Various electronically distinct indoles and aryl boronic acids reacted efficiently to provide comparable C2-coupled indoles with respectable yields. The catalytic cycle was proposed based on experimental outcomes and literature, which proceeds from the Co(II) to Co(IV) pathway using manganese as an oxidant (Figure 1.7).

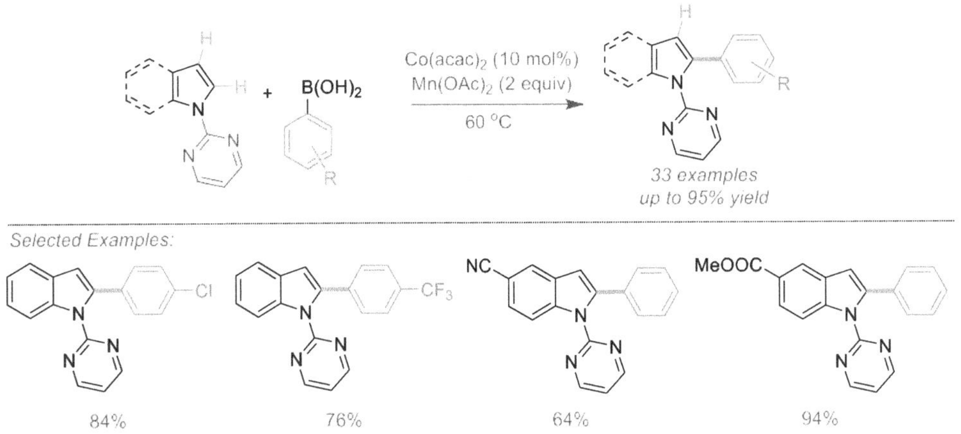

Scheme 1.19 Cobalt-Catalyzed C2 Arylation Using Aryl Boronic Acid.

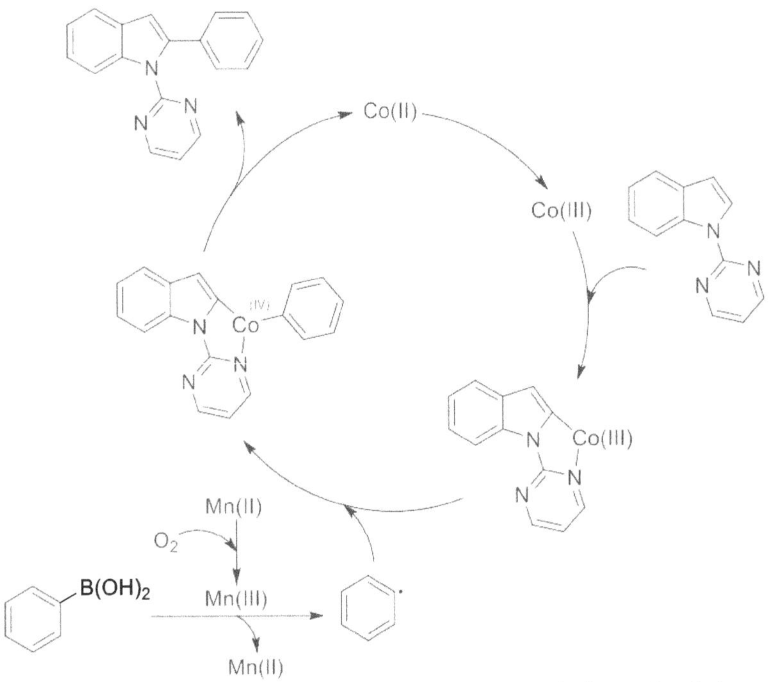

Figure 1.7 Plausible Catalytic Cycle of Cobalt-Catalyzed C2 Arylation of Indoles.

Very recently, Ackermann and Delord developed a method for enantioselective C2 arylation of indole using copper catalysis (Scheme 1.20).[85] This method utilizes chiral carbene ligand in combination with commercially available $CoBr_2$ salt in the presence of stoichiometric Grignard reagent and aryl chloride for the regioselective arylation of indoles. Here, aldimine at the C3 position of indole directs the arylation through weak chelation assistance. After treating with the reaction mixture under the HCl workup, the corresponding aldehyde will be generated from the aldimine. By using this method, they showed more than 30 examples, 96% ee for the arylated indoles and up to 99% yields.

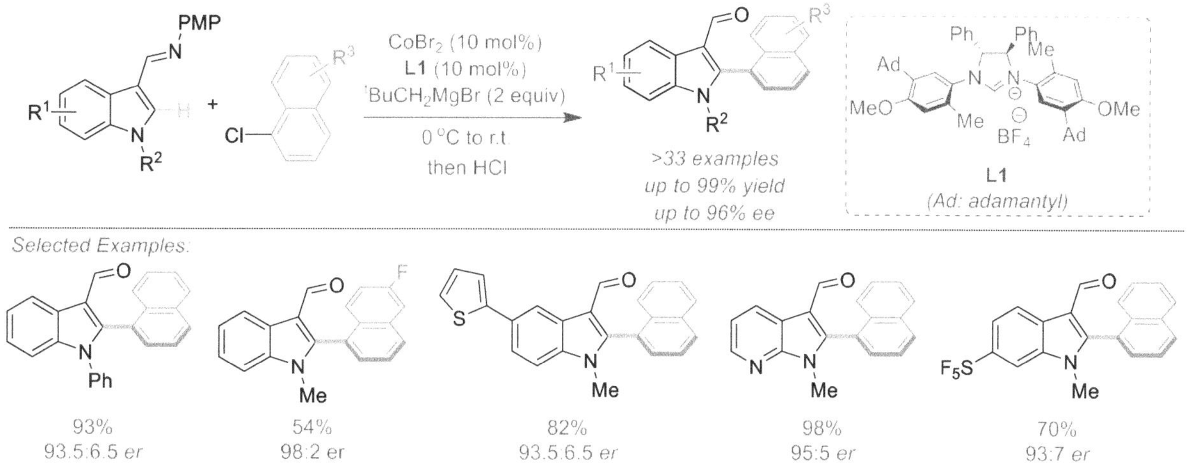

Scheme 1.20 Cobalt-Catalyzed Enantioselective C2 Arylation of Indoles.

1.2.2 Nickel-Catalyzed Regioselective C−H Arylation of Indoles

Nickel is a useful tool for C−H bond arylation reactions in organic synthesis because of its unusual mix of reactivity, adaptability, and affordability. Using a (NNN)Ni(II) catalyst and neat conditions, Punji and colleagues reported the first nickel-catalyzed regioselective C2 arylation of indoles [Scheme 1.21 (a)].[88] Additionally, they used the commercially available $Ni(OAc)_2$/dppf system to demonstrate an easy-to-use procedure for the C2 arylation of indoles after making further modifications [Scheme 1.21 (b)].[89] Based on experimental analysis, they proposed a catalytic cycle for this method, which starts from complex **A** (Figure 1.8). Upon base-mediated reversible C−H activation, it will result in the formation of **B**. From **B**, two pathways are possible. In path 1, either in the presence of an aryl halide, **B** will undergo 1-electron oxidation via **C** and recombination of the aryl radical to form **D**. The intermediate **D** will undergo reductive elimination to provide the desired coupled product and regenerate the active intermediate **A**. Alternatively, in path 2, the aryl chloride will be oxidatively added onto **B** in a bimetallic fashion to give **E**. The reductive elimination will form the desired product and **F**. The intermediate **F** upon comproportionation with **G** regenerated the active catalyst.

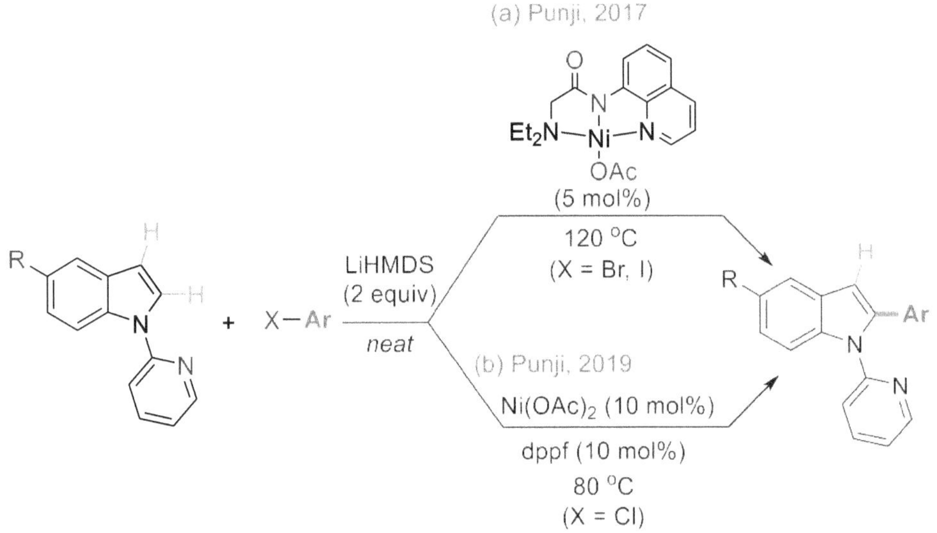

Scheme 1.21 Nickel-Catalyzed Regioselective Arylation of Indoles Using Aryl Halides.

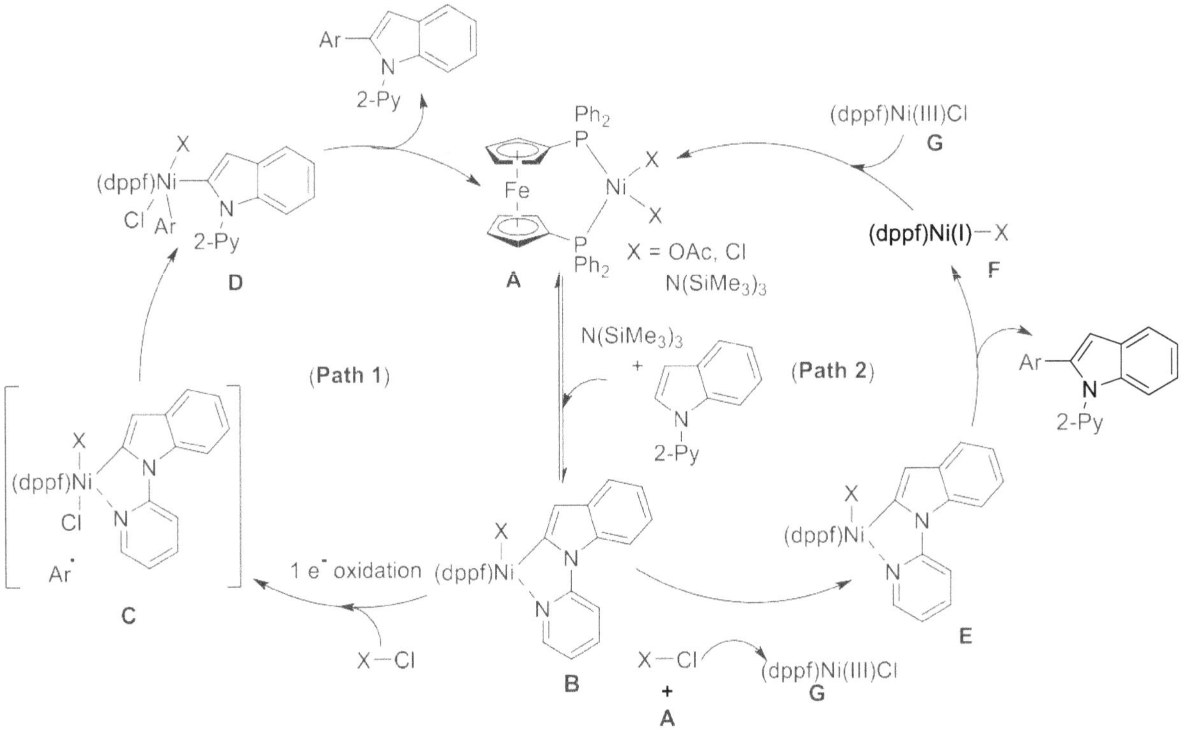

Figure 1.8 Plausible Mechanism of Nickel-Catalyzed C2 arylation of indoles.

1.2.3 Copper-Catalyzed Regioselective C–H Arylation of Indoles

Copper catalysts can undergo oxidative addition with aryl halides or related electrophiles, facilitating the activation of the aryl group for subsequent reaction with the C–H bond. In addition, it can exist in multiple oxidation states, allowing for redox flexibility in catalytic cycles. This versatility enables the activation of both aryl halides and C–H bonds. The first regioselective C–H arylation of indoles was created by Gaunt using copper as a catalyst (Scheme 1.22).[86] This reaction proceeds via the migration of C3 arylated indoles to C2 arylated indoles. This selectivity can be controlled by varying the substituent on the nitrogen of indoles as well as by differing the reaction temperature. However, the limitation associated with this method is that utilizes aryl iodonium salts as an arylating agent, which requires an additional synthetic step for this process. The tandem diarylation (of C−H and N−H bonds) method was developed by Greaney and co-workers in one pot using a copper catalyst (Scheme 1.23).[87] By using an aryl iodonium salt as a coupling partner they were able to transform *NH* indoles to corresponding C1 and C3 arylated products in good yields. In this, they have shown 16 examples with substituted aryl iodonium salts including methyl, methoxy, trifluoromethyl and chloro substitution.

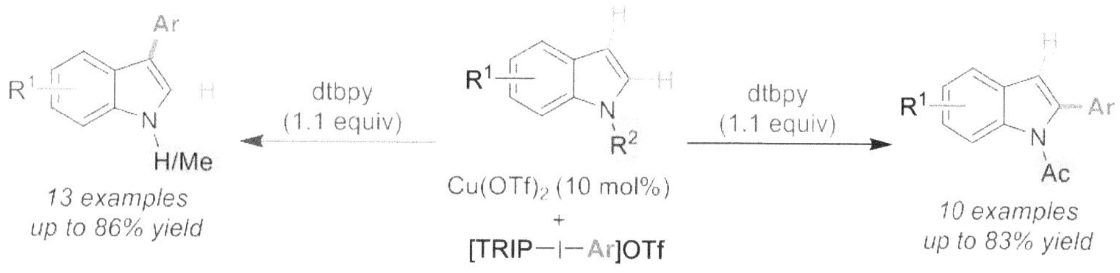
Ar
R1
H
N
H/Me
13 examples
up to 86% yield
dtbpy
(1.1 equiv)
R1
H
H
N
R2
Cu(OTf)2 (10 mol%)
+
[TRIP−I−Ar]OTf
dtbpy
(1.1 equiv)
R1
H
Ar
N
Ac
10 examples
up to 83% yield

Scheme 1.22 Copper-Catalyzed Regioselective C−H Arylation of Indoles.

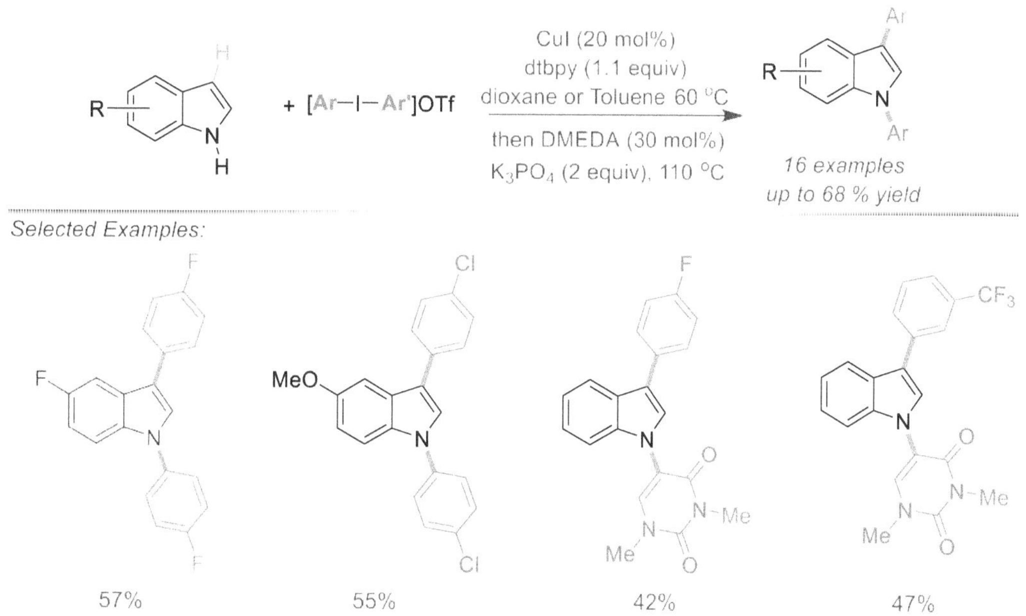
R
H
N
H
+ [Ar−I−Ar']OTf
CuI (20 mol%)
dtbpy (1.1 equiv)
dioxane or Toluene 60 °C
then DMEDA (30 mol%)
K3PO4 (2 equiv), 110 °C
R
Ar
N
Ar
16 examples
up to 68 % yield
Selected Examples:
F
F
N
F
57%
Cl
MeO
N
Cl
55%
F
N
O
N-Me
N
Me
O
42%
CF3
N
O
N-Me
N
Me
O
47%

Scheme 1.23 Copper-Catalyzed Tandem C1 and C3 Diarylation of Indoles.

Soon after, Shi and the group showed the first copper-catalyzed regioselective C6 arylation of indole using sterically hindered removable *N*-substituted phosphine oxide as a directing group

(Scheme 1.24).[90] Authors have shown 31 examples by varying different substitutions on indoles and arylating agents in good to excellent yields. The tolerance of different functional groups including halide, trifluoromethyl, ester, alkene, ketones, and heteroarene moieties was remarkable. However, only a 25% yield for C6 alkenylation was observed when an alkenylating coupling partner was used under standard reaction conditions. Further deprotection of the directing group was demonstrated followed by C2 and C3 arylation using Pd and Cu catalysts respectively was represented.

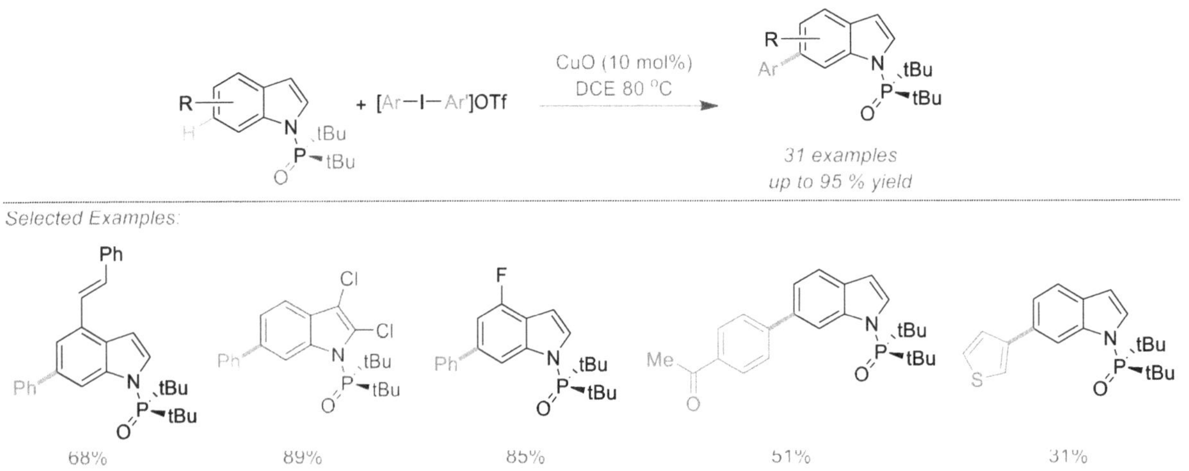

Scheme 1.24 Copper-Catalyzed C6 Arylation of Indoles.

Afterward, the C5 arylation of indoles was developed by substituting the pivaloyl directing group at C3 of indole and using copper-(I)thiophene-2-carboxylate (**Cu-1**) catalyst (Scheme 1.25).[91] Other less hindered directing groups including formyl, acetyl, and isobutyryl resulted in lower yields for the C5 arylation of indole. Good functional group tolerance was observed when arylating agents with various substituents including trifluoromethyl, trifluoromethoxy, halides and esters reacted smoothly. In addition, authors have shown the deprotections of the pivaloyl group in good yields. However, the protection of the *NH* group was necessary for this process.

Very recently, Zhou and Yang has reported the regioselective C2 and C3 arylation of indoles under copper catalysis (Scheme 1.26).[92] In this, the solvent and phosphine oxide substituted at C4 played a very important role in getting particular selectivity by controlling the migration of the aromatic ring. In the presence of toluene as a solvent, the migration of the aryl ring results in the C2 arylated product, whereas in the presence of THF (tetrahydrofuran), only the C3 arylated product is formed. This indicates that the choice of solvent can significantly influence the regioselectivity of the arylation reaction.

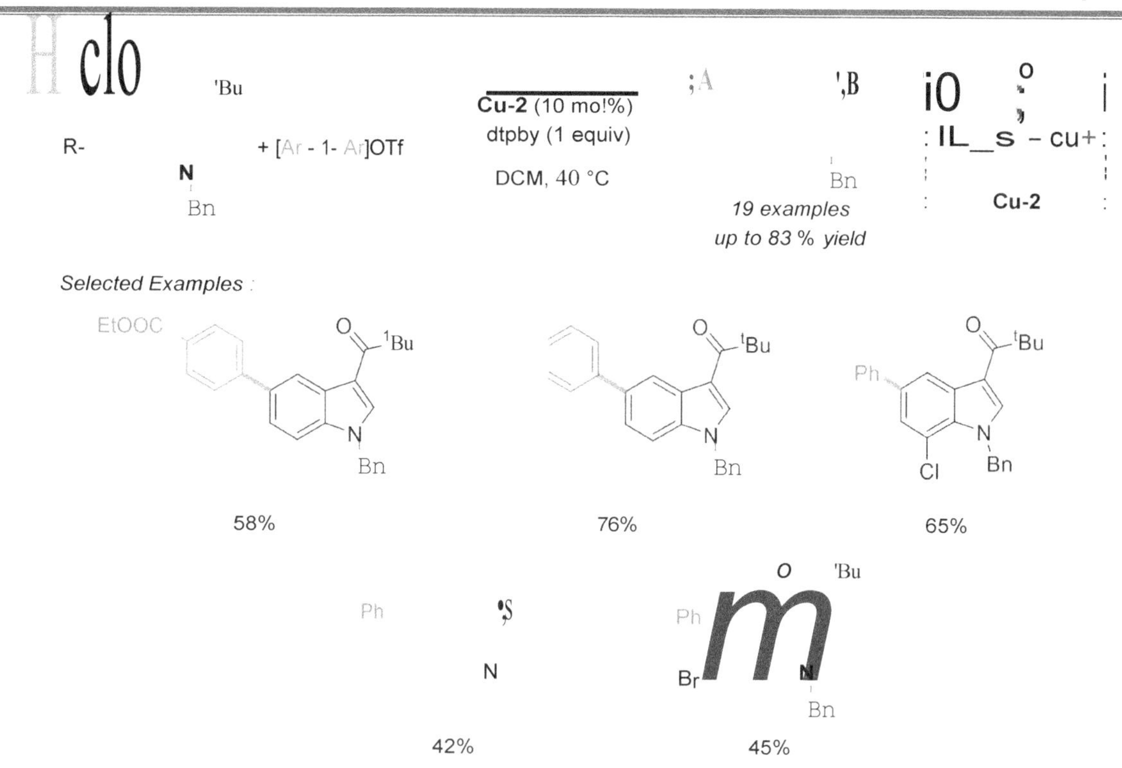

Scheme 1.25 Copper-Catalyzed C5 Arylation of Indoles Directed by Pivaloyl Group.

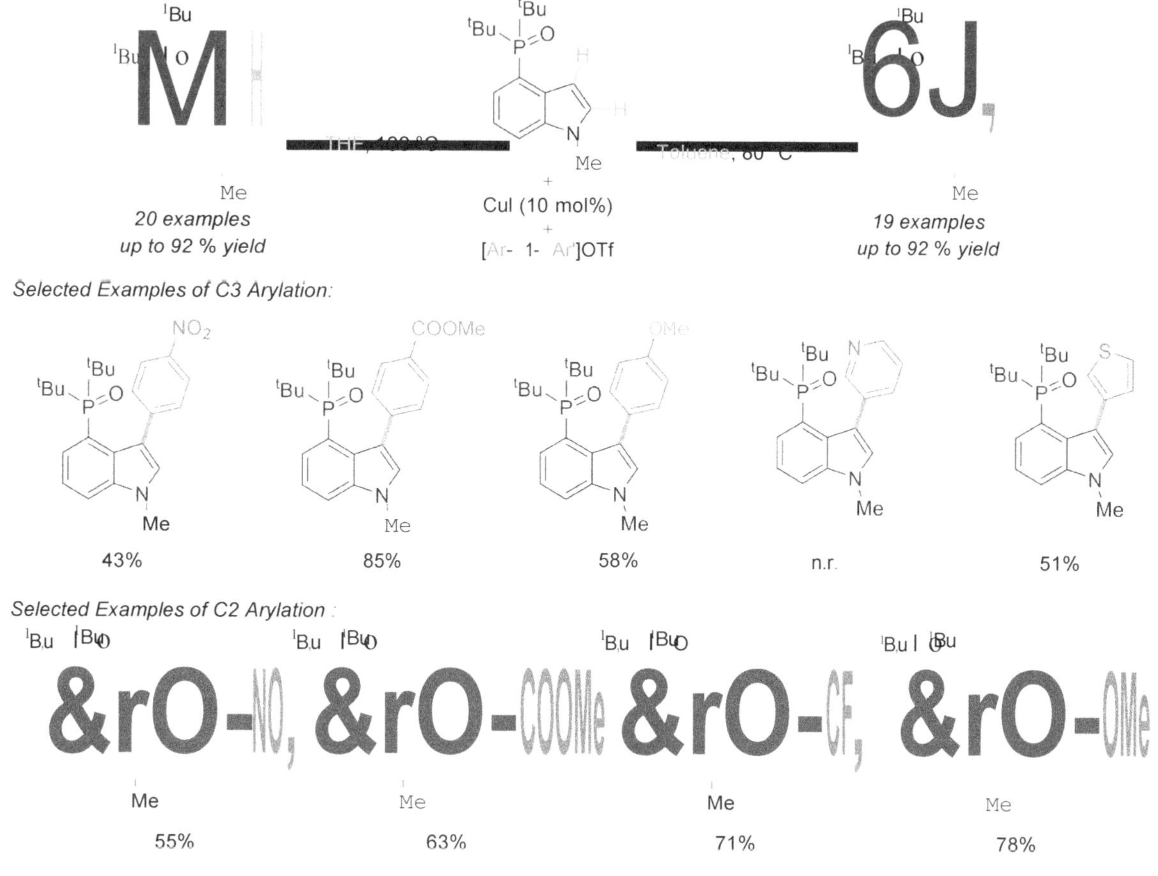

Scheme 1.26 Copper-Catalyzed C2 and C3 Arylation of Indoles.

1.3 CONCLUSION

In recent decades, significant attention has focused on innovating the shift from noble metal-catalyzed direct hydrogenation and C–H bond arylation to the use of more abundant metal catalysts. Particularly, manganese-catalyzed chemoselective hydrogenation has been extensively explored. However, iron and cobalt have been less developed for chemoselective hydrogenation of unsaturated moieties. While 3d-metal-catalyzed chemoselective hydrogenation is advancing, it has notable drawbacks. Many protocols require excess highly reactive additives, which can damage sensitive functional groups. Additionally, the need for high hydrogen pressure and elevated reaction temperatures remains a concern for both industrial and laboratory applications. Moreover, most of these reactions use phosphine ligands, which require specialized handling for their synthesis and are not stable in air. Therefore, there is potential to develop methods for chemoselective hydrogenation that use lower hydrogen pressure, ambient reaction temperatures, and phosphine-free ligands. Additionally, investigating the reactivity of other 3d metal catalysts for these types of hydrogenations could be highly insightful. The regioselective C–H bond arylation of indoles and related heteroarenes has been well-studied with noble metal catalysts. However, regioselective C–H arylation using 3d-metal catalysts remains underdeveloped. Recently, research groups have explored cobalt, nickel, and copper-catalyzed arylation of indoles. Nonetheless, these studies often face limitations due to the use of excessive amounts of pre-functionalized coupling partners, such as functionalized phenols, boronic acids, aryl iodonium salts, and Grignard reagents, which necessitate additional synthetic steps. Furthermore, the use of over-stoichiometric amounts of Grignard reagents can damage sensitive functional groups. Therefore, there is an opportunity to develop more atom-economical 3d-metal-catalyzed reactions using commercially available and less reactive aryl halides as arylating agents.

1.4 REFERENCES

(1) J. G. de Vries; C. J. Elsevier, de Vries, J. G., Elsevier, C. J., Eds.; Wiley, 2006.

(2) Noyori, R.; Ohkuma, T., *Angew. Chem. Int. Ed.* ***2001***, *40*, 40–73

(3) Cabré, A.; Verdaguer, X.; Riera, A., *Chem. Rev.* **2022**, *122*, 269–339.

(4) He, Y.-M.; Song, F.-T.; Fan, Q.-H., 2013; pp 145–190.

(5) Alig, L.; Fritz, M.; Schneider, S., *Chem. Rev.* **2019**, *119*, 2681–2751

(6) Pritchard, J.; Filonenko, G. A.; van Putten, R.; Hensen, E. J. M.; Pidko, E. A., *Chem. Soc. Rev.* **2015**, *44*, 3808–3833.

(7) Tamura, M.; Nakagawa, Y.; Tomishige, K., *J. JPN. Petrol. Inst.* **2019**, *62*, 106–119.

(8) Wang, Y.; Wang, M.; Li, Y.; Liu, Q., *Chem* **2021**, *7*, 1180–1223.

(9) Sharma, V.; Kumar, P.; Pathak, D., *J. Heterocycl. Chem.* **2010**, *47*, 491–502.

(10) Kaushik, N.; Kaushik, N.; Attri, P.; Kumar, N.; Kim, C.; Verma, A.; Choi, E, *Molecules* **2013**, *18*, 6620–6662.

(11) Lindlar, H., *Helv. Chim. Acta.* **1952**, *35*, 446–450.

(12) Noyori, R.; Hashiguchi, S., *Acc. Chem. Res.* **1997**, *30*, 97–102.

(13) Takaya, H.; Ohta, T.; Sayo, N.; Kumobayashi, H.; Akutagawa, S.; Inoue, S.; Kasahara, I.; Noyori, R., *J. Am. Chem. Soc.* **1987**, *109*, 1596–1597.

(14) Knifton, J. F., *J. Org. Chem.* **1976**, *41*, 1200–1206.

(15) Deshmukh, A. A.; Prashar, A. K.; Kinage, A. K.; Kumar, R.; Meijboom, R., *Ind. Eng. Chem. Res.* **2010**, *49*, 12180–12184.

(16) Sun, K.-Q.; Hong, Y.-C.; Zhang, G.-R.; Xu, B.-Q., *ACS Catal.* **2011**, *1*, 1336–1346.

(17) Corma, A.; Serna, P., *Science* **2006**, *313*, 332–334.

(18) Lin, L.; Yao, S.; Gao, R.; Liang, X.; Yu, Q.; Deng, Y.; Liu, J.; Peng, M.; Jiang, Z.; Li, S.; Li, Y.-W.; Wen, X.-D.; Zhou, W.; Ma, D., *Nat. Nanotechnol.* **2019**, *14*, 354–361.

(19) Vielhaber, T.; Topf, C., *Appl. Catal. A Gen.* **2021**, *623*, 118280.

(20) Andersson, P. G., Munslow, I. J., Eds.; Wiley, 2008.

(21) P. N. Rylander, Ed.; Academic Press: New York, 1979.

(22) Elangovan, S.; Topf, C.; Fischer, S.; Jiao, H.; Spannenberg, A.; Baumann, W.; Ludwig, R.; Junge, K.; Beller, M., *J. Am. Chem. Soc.* **2016**, *138*, 8809–8814.

(23) Weber, S.; Stöger, B.; Veiros, L. F.; Kirchner, K., *ACS Catal.* **2019**, *9*, 9715–9720.

(24) Farrar-Tobar, R. A.; Weber, S.; Csendes, Z.; Ammaturo, A.; Fleissner, S.; Hoffmann, H.; Veiros, L. F.; Kirchner, K., *ACS Catal.* **2022**, *12*, 2253–2260.

(25) Wang, Y.; Wang, M.; Li, Y.; Liu, Q., *Chem* **2021**, *7*, 1180–1223.

(26) Espinosa-Jalapa, N. A.; Nerush, A.; Shimon, L. J. W.; Leitus, G.; Avram, L.; Ben-

David, Y.; Milstein, D., *Chem. Eur. J.* **2017**, *23*, 5934–5938.

(27) Zubar, V.; Lichtenberger, N.; Schelwies, M.; Oeser, T.; Hashmi, A. S. K.; Schaub, T., *ChemCatChem* **2022**, *14*, DOI: 10.1002/cctc.202101443.

(28) Weber, S.; Brünig, J.; Veiros, L. F.; Kirchner, K., *Organometallics* **2021**, *40*, 1388–1394.

(29) Luk, J.; Oates, C. L.; Fuentes Garcia, J. A.; Clarke, M. L.; Kumar, A., *Organometallics* **2024**, *43*, 85–93.

(30) Das, K.; Waiba, S.; Jana, A.; Maji, B., *Chem. Soc. Rev.* **2022**, *51*, 4386–4464.

(31) Kallmeier, F.; Irrgang, T.; Dietel, T.; Kempe, R., *Angew. Chem. Int. Ed.* **2016**, *55*, 11806–11809.

(32) Garbe, M.; Junge, K.; Walker, S.; Wei, Z.; Jiao, H.; Spannenberg, A.; Bachmann, S.; Scalone, M.; Beller, M., *Angew. Chem. Int. Ed.* **2017**, *129*, 11389–11393.

(33) Glatz, M.; Stöger, B.; Himmelbauer, D.; Veiros, L. F.; Kirchner, K., *ACS Catal.* **2018**, *8*, 4009–4016.

(34) Gholap, S. S.; Dakhil, A. Al; Chakraborty, P.; Li, H.; Dutta, I.; Das, P. K.; Huang, K. W., *Chem. Commun.* **2021**, *57*, 11815–11818.

(35) Zhang, L.; Wang, Z.; Han, Z.; Ding, K., *Angew. Chem. Int. Ed.* **2020**, *59*, 15565–15569.

(36) Bruneau-Voisine, A.; Wang, D.; Roisnel, T.; Darcel, C.; Sortais, J.-B., *Catal. Commun.* **2017**, *92*, 1–4.

(37) Wei, D.; Bruneau-Voisine, A.; Chauvin, T.; Dorcet, V.; Roisnel, T.; Valyaev, D. A.; Lugan, N.; Sortais, J., *Adv. Synth. Catal.* **2018**, *360*, 676–681.

(38) Vielhaber, T.; Topf, C., *Appl. Catal. A Gen.* **2021**, *623*.

(39) Zubar, V.; Dewanji, A.; Rueping, M., *Org. Lett.* **2021**, *23*, 2742–2747.

(40) Balaraman, E.; Nandakumar, A.; Jaiswal, G.; Sahoo, M. K., *Catal. Sci. Technol.* **2017**, *7*, 3177–3195.

(41) Bolm, C.; Legros, J.; Le Paih, J.; Zani, L., *Chem. Rev.* **2004**, *104*, 6217–6254.

(42) Vaishak, T. B.; Soumya, P. K.; Saranya, P. V.; Anilkumar, G., *Catal. Sci. Technol.* **2021**, *11*, 4690–4701.

(43) Neidig, M. L.; Bhatia, S.; DeMuth, J. C., Wiley, 2022; pp 1–34.

(44) Bhavyesh, D.; Soliya, S.; Konakanchi, R.; Begari, E.; Ashalu, K. C.; Naveen, T., *Chem. Asian J.* **2024**, *19*, DOI: 10.1002/asia.202301056.

(45) Langer, R.; Leitus, G.; Ben-David, Y.; Milstein, D., *Angew. Chem. Int. Ed.* **2011**, *50*, 2120–2124.

(46) Langer, R.; Iron, M. A.; Konstantinovski, L.; Diskin-Posner, Y.; Leitus, G.; Ben-David,

Y.; Milstein, D., *Chem. Eur. J.* **2012**, *18*, 7196–7209.

(47) Fleischer, S.; Zhou, S.; Junge, K.; Beller, M., *Angew. Chem. Int. Ed.* **2013**, *125*, 5224–5228.

(48) Wienhöfer, G.; Westerhaus, F. A.; Junge, K.; Ludwig, R.; Beller, M., *Chem. Eur. J.* **2013**, *19*, 7701–7707.

(49) Gorgas, N.; Stöger, B.; Veiros, L. F.; Pittenauer, E.; Allmaier, G.; Kirchner, K., *Organometallics* **2014**, *33*, 6905–6914.

(50) Gorgas, N.; Stöger, B.; Veiros, L. F.; Kirchner, K., *ACS Catal.* **2016**, *6*, 2664–2672.

(51) Mazza, S.; Scopelliti, R.; Hu, X., *Organometallics* **2015**, *34*, 1538–1545.

(52) Duan, Y. N.; Zeng, Y.; Cui, Z.; Wen, J.; Zhang, X., *J. Catal.* **2023**, *417*, 109–115.

(53) Wienhöfer, G.; Baseda-Krüger, M.; Ziebart, C.; Westerhaus, F. A.; Baumann, W.; Jackstell, R.; Junge, K.; Beller, M., *Chem. Commun.* **2013**, *49*, 9089–9091.

(54) Junge, K.; Papa, V.; Beller, M. Cobalt–Pincer Complexes in Catalysis. *Chem. Eur. J.* **2019**, *25*, 122–143.

(55) Tohidi, M. M.; Paymard, B.; Vasquez-García, S. R.; Fernández-Quiroz, D., *Tetrahedron* **2023**, *136*, 133352.

(56) Viereck, P.; Krautwald, S.; Pabst, T. P.; Chirik, P. J., *J. Am. Chem. Soc.* **2020**, *142*, 3923–3930.

(57) Friedfeld, M. R.; Shevlin, M.; Margulieux, G. W.; Campeau, L.-C.; Chirik, P. J., *J. Am. Chem. Soc.* **2016**, *138*, 3314–3324.

(58) Zhang, G.; Hanson, S. K., *Chem. Commun.* **2013**, *49*, 10151.

(59) Ai, W.; Zhong, R.; Liu, X.; Liu, Q., *Chem. Rev.* **2019**, *119*, 2876–2953.

(60) Fu, S.; Chen, N.-Y.; Liu, X.; Shao, Z.; Luo, S.-P.; Liu, Q., *J. Am. Chem. Soc.* **2016**, *138*, 8588–8594.

(61) Srimani, D.; Mukherjee, A.; Goldberg, A. F. G.; Leitus, G.; Diskin-Posner, Y.; Shimon, L. J. W.; Ben David, Y.; Milstein, D., *Angew. Chem. Int. Ed.* **2015**, *54*, 12357–12360.

(62) Anferov, S. W.; Filatov, A. S.; Anderson, J. S., *ACS Catal.* **2022**, *12*, 9933–9943.

(63) Rösler, S.; Obenauf, J.; Kempe, R., *J. Am. Chem. Soc.* **2015**, *137*, 7998–8001.

(64) Duan, Y. N.; Du, X.; Cui, Z.; Zeng, Y.; Liu, Y.; Yang, T.; Wen, J.; Zhang, X., *J. Am. Chem. Soc.* **2019**, *141*, 20424–20433.

(65) Wei, Z.; Wang, Y.; Li, Y.; Ferraccioli, R.; Liu, Q., *Organometallics* **2020**, *39*, 3082–3087.

(66) Li, B.; Ali, A. I. M.; Ge, H., *Chem* **2020**, *6*, 2591–2657.

(67) Rogge, T.; Kaplaneris, N.; Chatani, N.; Kim, J.; Chang, S.; Punji, B.; Schafer, L. L.;

Musaev, D. G.; Wencel-Delord, J.; Roberts, C. A.; Sarpong, R.; Wilson, Z. E.; Brimble, M. A.; Johansson, M. J.; Ackermann, L., *Nat. Rev. Meth.* **2021**, *1*, 43.

(68) Dalton, T.; Faber, T.; Glorius, F. *ACS Cent. Sci.* **2021**, *7*, 245–261.

(69) Song, W.; Ackermann, L., *Angew. Chem. Int. Ed.* **2012**, *51*, 8251–8254.

(70) Kozhushkov, S. I.; Potukuchi, H. K.; Ackermann, L., *Catal. Sci. Technol.* **2013**, *3*, 562–571.

(71) Li, H.; Sun, C.; Yu, M.; Yu, D.; Li, B.; Shi, Z., *Chem. Eur. J.* **2011**, *17*, 3593–3597.

(72) Liu, W.; Cao, H.; Xin, J.; Jin, L.; Lei, A., *Chem. Eur. J.* **2011**, *17*, 3588–3592.

(73) Leitch, J. A.; Bhonoah, Y.; Frost, C. G., *ACS Catal.* **2017**, *7*, 5618–5627.

(74) Sandtorv, A. H., *Adv. Synth. Catal.* **2015**, *357*, 2403–2435.

(75) Hirano, K.; Miura, M., *Synlett* **2011**, *201*, 294–307.

(76) Albano, G.; Punzi, A.; Capozzi, M. A. M.; Farinola, G. M., *Green Chem.* **2022**, *24*, 1809–1894.

(77) Tiwari, V. K.; Kamal, N.; Kapur, M., *Org. Lett.* **2015**, *17*, 1766–1769.

(78) Ackermann, L.; Lygin, A. V. *Org. Lett.* **2011**, *13*, 3332–3335.

(79) Wu, H.; Liu, T.; Cui, M.; Li, Y.; Jian, J.; Wang, H.; Zeng, Z., *Org. Biomol. Chem.* **2017**, *15*, 536–540.

(80) Wang, X.; Lane, B. S.; Sames, D., *J. Am. Chem. Soc.* **2005**, *127*, 4996–4997.

(81) Ghobrial, M.; Mihovilovic, M. D.; Schnürch, M., *Beilstein J. Org. Chem.* **2014**, *10*, 2186–2199.

(82) Islam, S.; Larrosa, I., *Chem. Eur. J.* **2013**, *19*, 15093–15096.

(83) Punji, B.; Song, W.; Shevchenko, G. A.; Ackermann, L., *Chem. Eur. J.* **2013**, *19*, 10605–10610.

(84) Zhu, X.; Su, J. H.; Du, C.; Wang, Z. L.; Ren, C. J.; Niu, J. L.; Song, M. P., *Org. Lett.* **2017**, *19*, 596–599.

(85) Jacob, N.; Zaid, Y.; Oliveira, J. C. A.; Ackermann, L.; Wencel-Delord, J., *J. Am. Chem. Soc.* **2022**, *144*, 798–806.

(86) Phipps, R. J.; Grimster, N. P.; Gaunt, M. J., *J. Am. Chem. Soc.* **2008**, *130*, 8172–8174.

(87) Modha, S. G.; Greaney, M. F., *J. Am. Chem. Soc.* **2015**, *137*, 1416–1419.

(88) Jagtap, R. A.; Soni, V.; Punji, B., *ChemSusChem* **2017**, *10*, 2242–2248.

(89) Pandey, D. K.; Vijaykumar, M.; Punji, B., *J. Org. Chem.* **2019**, *84*, 12800–12808.

(90) Yang, Y.; Li, R.; Zhao, Y.; Zhao, D.; Shi, Z., *J. Am. Chem. Soc.* **2016**, *138*, 8734–8737.

(91) Yang, Y.; Gao, P.; Zhao, Y.; Shi, Z., *Angew. Chem. Int. Ed.* **2017**, *56*, 3966–3971.

(92) Niu, Y.; Yan, C.-X.; Yang, X.-X.; Bai, P.-B.; Zhou, P.-P.; Yang, S.-D., *Org. Chem. Front.* **2022**, *9*, 1023–1032.

Objective of the Present Study

In recent years, significant efforts have been devoted to the chemoselective hydrogenation of various unsaturated bonds, as well as C–H bond functionalization, using base metal catalysts. Particularly, the chemoselective hydrogenation of unsaturated moieties has mostly employed noble metal catalysts under transfer hydrogenation conditions. However, the harsh reaction conditions, higher hydrogen pressure, and generation of stoichiometric waste remain major limitations for industrial applications. Therefore, the objective of the present work was to develop environmentally benign and efficient protocols for the chemoselective hydrogenation of various unsaturated moieties by designing a mixed-donor pincer manganese complex. Specifically, achieving chemoselective C=C, C=O, and C=C bond hydrogenation of conjugated carbonyl compounds, conjugated imines, epoxy ketones, and ketoamides is crucial for late-stage functionalization, thereby minimizing unnecessary synthetic steps such as protection and deprotection. Additionally, we aimed to develop a direct arylation method using aryl iodides under copper catalysis.

The results obtained from the present studies are discussed in Chapters 2-5. In Chapter 2, the synthesis of mixed donor pincer manganese complexes and their application for chemoselective C=C, C=O, and C=N bond hydrogenation of α,β-unsaturated compounds is described. In this, we developed a chemoselective hydrogenation process at room temperature using moderate hydrogen pressure. Using this method various C=C bonds in α,β-unsaturated ketones could be selectively reduced to form saturated ketones with the tolerance of multiple reducing functional groups including terminal alkene, alkyne, epoxide, nitrile, nitro, and heteroarenes. In addition, C=O and C=N bonds in α,β-unsaturated aldehydes, and α,β-unsaturated imines could be selectively reduced to corresponding ally alcohol and allyl amine respectively. The experimental and DFT calculations suggest that the reaction proceeds via metal-ligand cooperation for H_2 activation and the manganese hydride adds in 1,4 fashion to the α,β-unsaturated ketone (Michael addition).

The chemoselective hydrogenation of the C=O bond in α,β-epoxy ketones using synthesized manganese(I) complexes was achieved under mild reaction conditions, and these results were discussed in Chapter 3. Various α,β-epoxy ketones were hydrogenated to provide the corresponding α-hydroxy epoxide with excellent selectivity. In addition to aromatic epoxy ketones, aliphatic epoxy ketones derived from different aromatic derivatives, as well as steroid molecules, were efficiently hydrogenated with good functional group tolerance, including reducible functional groups such as alkenes, alkynes, sulfones, alcohols, and acetyl groups. Furthermore, the synthetic utility of this protocol was demonstrated by protecting the alcoholic group with acetyl and silyl protecting groups. Additionally, the epoxide ring can be selectively reacted with sodium azide in good yield and selectivity. To extend the application of the developed catalyst, we examined the chemoselective hydrogenation of α-ketoamides to synthetically important α-hydroxy amides with excellent selectivity and these were described in Chapter 4. This method represents the first hydrogenation of α-ketoamides using a 3d-metal catalyst with molecular hydrogen. Similar to the earlier process, the current method also proceeds via metal-ligand cooperative H_2 activation.

In Chapter 5, we demonstrated a user-friendly, ligand-free, and solvent-free protocol for copper-catalyzed regioselective C(2)–H arylation of indoles directly using aryl iodides with a copper catalyst. Under optimized reaction conditions, various substituted aryl iodides and indoles were coupled in good to excellent yields. The wide scope was presented which shows the functional group tolerance of a range of functionalities, including halides, ethers, substituted amines, pyrrolyl, indolyl, and carbazolyl groups. This method further can be employed to the arylation of other related heteroarenes including, pyrrole and thiophenes. We showcased the applicability of this method by successfully removing directing groups to achieve *NH*-indoles, which were further functionalized to obtain tryptamine derivatives in good yields. Initial studies to elucidate the catalytic pathway suggested that this reaction proceeds via the formation of radicals as intermediates.

Chapter 2

Synthesis of Mn(I) Complexes for Chemoselective Hydrogenation of *α,β*-Unsaturated Compounds

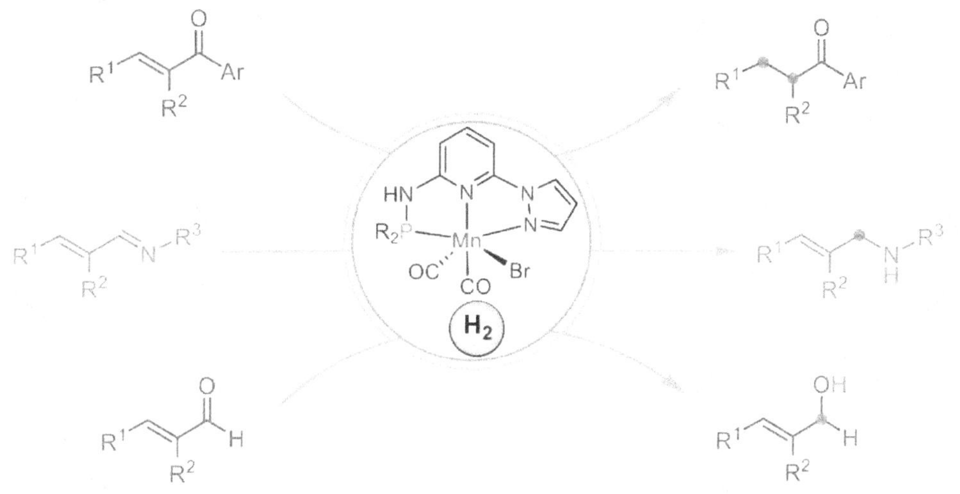

This chapter has been adapted from the publication "Room Temperature Chemoselective Hydrogenation of C=C, C=O and C=N Bonds by Using a Well-Defined Mixed Donor Mn(I) Pincer Catalyst" **Shabade, A. B.**; Sharma, D. M.; Bajpai, P.; Gonnade, R. G.; Vanka, K.; and Punji, B. *Chem. Sci.*, **2022**, *13*, 13764-13773.

2.1 INTRODUCTION

Chemoselective hydrogenation of unsaturated organic compounds is extremely important in academia and industry, as it is pivotal in preparing pharmaceutical intermediates, fragrances and various bulk products. Particularly, the catalytic reductions using molecular hydrogen under ambient conditions represent one of the economical, atom-efficient and environmentally benign transformations.[1,2] In this context, numerous heterogeneous catalysts are developed and demonstrated for the hydrogenation of various functional groups, which often require high reaction temperatures and/or pressures and result in poor selectivity.[3-5] Over the time, many well-defined, active and highly efficient homogeneous catalysts,[6] derived from the precious and non-precious transition metals, have been established for the hydrogenation.[7-12] In general, noble metal catalysis has dominated the field and accomplished very high catalytic turnovers and desired selectivity. However, the limited availability of noble metals in the lithosphere, expensiveness and underlying toxicity could limit their future wide applications in hydrogenation.

Sustainable and environmentally friendly development is one of the main goals of modern science. Accordingly, consideration has been made to the hydrogenations using catalysts based on non-precious metals that are affordable, readily available, and low-toxic.[13-16] Particularly, the hydrogenations employing iron,[17-20] cobalt[21-23] and nickel[19] are substantially explored and disclosed as active and highly efficient for reducing carbonyls, imines and nitriles.[24-27] Being the less toxic, cheap and third most abundant transition metal, the manganese-based catalyst has been implemented for the hydrogenation of aldehydes/ketones by Beller,[28] Kempe,[29] Sortais,[30-33] Kirchner[34-36] and others (Scheme 2.1a).[37-43] Similarly, the hydrogenation of imines (or C=N bond) using bidentate or pincer-ligated manganese catalysts was established.[44-48] Meanwhile, the independent development of chiral pincer-manganese catalysts by Clarke,[49] Beller,[50-51] Han/Ding[52-53] and others[54-57] led to the asymmetric hydrogenation of ketones. In a significant advancement, Liu has shown the asymmetric hydrogenation of C=N bond in heteroaromatics using chiral Mn-catalyst.[58-59] Despite all developments on the Mn-catalyzed hydrogenations of multiple polar bonds, the selective hydrogenation of alkene is highly challenging and extremely rare due to the associated high bond enthalpies. The group of Kirchner and Khusnutdinova independently showed the bidentate PP- and PN-ligated manganese (I) catalysts for the reduction of alkenes using 50 bar and 30 bar H_2 pressures, respectively, at elevated temperature (Scheme 2.1b).[60, 61]

Similarly, Topf has demonstrated alkene hydrogenation in conjugated carboxylic derivatives using 30-50 bar H_2 at 100-120 °C; however, this protocol failed to provide selective hydrogenation in conjugated ketones.[39] Though, manganese-based catalysts promoted the hydrogenations of many

C=O and C=N bonds and certain C=C bonds;[31,39,60,61] most of the reactions proceed at a high H_2 pressure and at elevated temperature, which is a significant drawback for practical applications. Additionally, most Mn-catalyzed hydrogenations require a large amount of strong base (KO^{t}Bu) as an additive. As yet, a chemoselective hydrogenation protocol for reducing one unsaturated functional moiety in the presence of the other using beneficial Mn-catalyst under ambient H_2 pressure and temperature is unknown.[62-65]

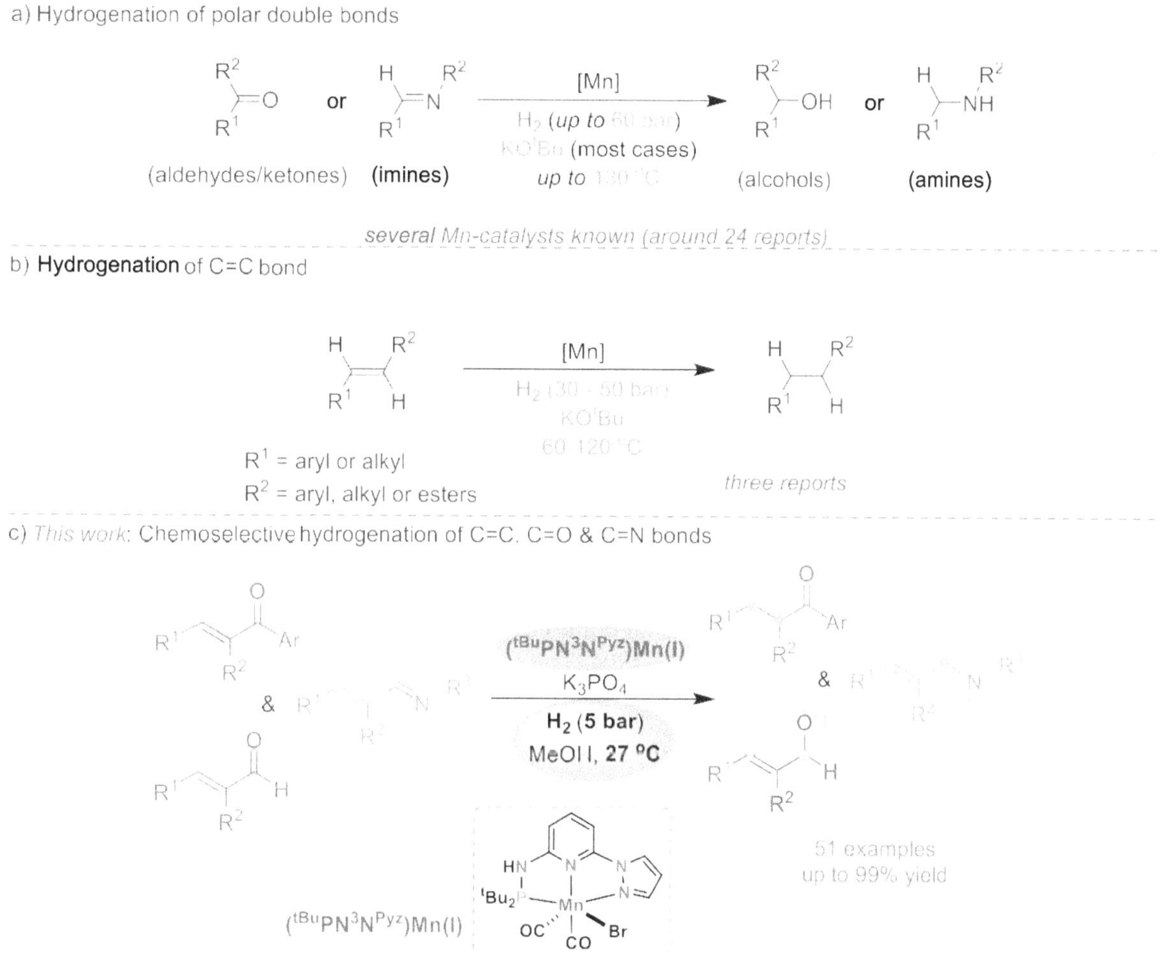

Scheme 2.1 Manganese-Catalyzed Hydrogenation of Unsaturated Bonds: a) Carbonyls and Imines, b) C=C Bonds, c) Chemoselective Hydrogenation of C=C, C=O, C=N Bonds.

To achieve more sustainability, efficiency and selectivity in Mn-catalysis; in this chapter, we developed the mixed-donor (PN^{3}N)Mn(I) complexes and disclosed the hydrogenation of C=C, C=O, and C=N bonds chemoselectively using 5 bar H_2 and a mild base K_3PO_4 at room temperature (Scheme 2.1c). The notable features of the present method are (i) using 3rd most plentiful transition metal as a catalyst, (ii) high chemoselectivity for reducing C=O, C=C, and C=N bonds, (iii) hydrogenation using 5 bar H_2 and at room temperature (27 °C), (iv) use of mild base and atom-efficient H_2 source, and (v) broad substrates scope with excellent tolerance of hydrogen-sensitive functionalities.

2.2 RESULTS AND DISCUSSION

2.2.1 Synthesis of Mn(I) Complexes

We were interested in synthesizing mixed-donor PN^3N ligands and pincer manganese complexes. It was envisioned that a mixed-donor ligand with different electronic features of donating center could suitably administer the electronic requirement at the metal center and can stabilize the active catalytic species. Furthermore, the inclusion of *NH* on the side arm may promote the ligand's non-innocent behavior through the aromatization/dearomatization process, which would result in the heterolytic activation of H_2 molecules. With this in mind, the isopropyl-tagged PN^3N ligand **L2** was synthesized in a few steps starting from 2,6-dibromopyridine, following the protocol similar to the synthesis of phenyl or *tert*-butyl PN^3N ligands **L1** and **L3**.[67,68] Similar to the synthesis of **Mn-3** complex,[110] treatment of pincer ligands **L1** and **L2** with $Mn(CO)_5Br$ in THF at room temperature afforded dicarbonyl Mn(I) complexes in good yields (Scheme 2.2). These complexes were extensively described by elemental analysis, FT-IR, ESI-MS, and multinuclear NMR spectroscopy. The $^{31}P\{^1H\}$-NMR spectra of complexes **Mn-1**, and **Mn-2** show peaks at 134.5, 136.9 and 158.4, 159.8, respectively. Two peaks for each complex at very similar chemical shift values indicate the formation of two geometrical isomers for each complex. Interestingly, all complexes displayed three IR peaks for carbonyls ranging between 1857-2053 cm^{-1}. Similarly, a closer look at the $^{13}C\{^1H\}$-NMR spectra of complexes indicates the three carbonyl signals. Moreover, the NMR spectra [both ^1H and $^{13}C\{^1H\}$] of all complexes display two sets of peaks. All these observations support the formation of two geometrical isomeric species in each complex **Mn-1**, and **Mn-2**. Considering the presence of three signals for carbonyls in each complex, we assume that one isomer of the Mn-complex would have two carbonyls that are *trans* to –Br and –N_{py} ligands accounting for two peaks (as indicated in X-ray structure), whereas the other isomer would have two carbonyls *trans* to each other and display a single carbonyl peak (Scheme 2.2). Even though the assumed isomers are highly convincing, the probability of the mixture of a neutral (with –Br coordination) and a cationic (with Br as an anion) manganese species cannot be completely ruled out.[11a,13b,16,17a] All the complexes are further characterized by ESI-MS that show two prominent isotopic masses for ^{79}Br and ^{81}Br containing complexes. **Mn-3** was prepared following the reported method.[110] The structure of **Mn-2** was verified by X-ray crystallography (Figure 2.1). As expected, the coordination geometry around the manganese is distorted octahedral with –Br and one –CO unit *trans* to each other, making the

C9–Mn1–Br angle almost linear (175.54(15)°). The Mn1–C10 bond length (1.797(4) Å) is marginally more than the bond length of Mn1–C9 bond length (1.773(4) Å), which might be due to the greater *trans* influence of $-N_{py}$ than the –Br. Although each synthesized manganese complexes contain two isomeric species, it is anticipated that the active catalyst generated in presence of base will consist of a single isomer (proved and discussed vide infra).

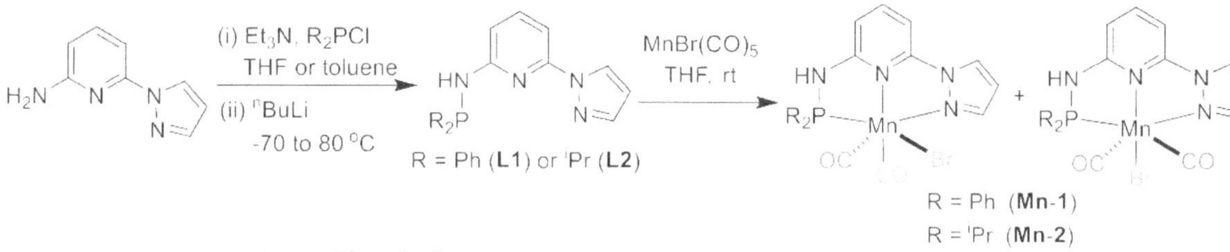

Scheme 2.2 Synthesis of ($^{R2}PN^3N^{Pyz}$) Manganese Pincer Complexes.

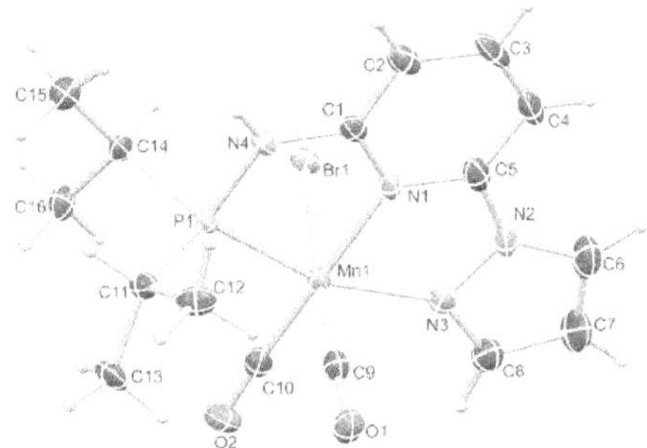

Figure 2.1 ORTEP of Compound Mn-2. showing the atom-numbering scheme. Displacement ellipsoids are drawn at the 50% probability level. Selected bond length (Å): Br1–Mn1, 2.5916(7); Mn1–C9, 1.773(4); Mn1–C10, 1.797(4); Mn1–N3, 2.011(3); Mn1–N1, 2.012(3); Mn1–P1, 2.2462(12). Selected bond angles (°): N3–Mn1–P1, 160.49(11); C10–Mn1–N1, 175.88(16); C9–Mn1–Br1, 175.54(14); C9–Mn1–C10, 87.40(19); C9–Mn1–N1, 96.23(16); N1–Mn1–Br1, 84.57(9).

2.2.2 Optimization for Catalytic Hydrogenation

After synthesizing the sterically and electronically distinct mixed-donor (PN^3N) manganese complexes, we have investigated their catalytic behavior for the chemoselective reduction of conjugated ketones using molecular hydrogen. We first screened the activity of (PN^3N)Mn(I) complexes **Mn-1** to **Mn-3** for C=C bond hydrogenation in (*E*)-Chalcone (**4a**) using a catalytic amount of KO^tBu and 30 bar hydrogen pressure in methanol at 50 °C (Table 2.1). The use of **Mn-1** as catalyst led to 75% conversion of **4a**, wherein chemoselective C=C hydrogenated product **4** was obtained in 65%, and complete hydrogenated compound **4'** in 10% (entry 1). The electronically rich complex **Mn-2** gave complete conversion of **4a**; however, the product **4'** was obtained in 80%, and the remaining was allylic alcohol with only C=O hydrogenation (entry 2). Interestingly, the bulky ^{t}Bu-substituted complex **Mn-3** gave a complete conversion of **4a** with more chemoselectivity for **4** (entry 3). The formation of allylic alcohol in the presence of catalyst **Mn-2** suggests the probability of a sequential C=O hydrogenation-allylic alcohol isomerization leading to the uncontrolled hydrogenated compound **4'**. The low steric in complex **Mn-2** compared to **Mn-3** might play a crucial role in chemoselective hydrogenation (1,2 hydrogenation of C=O *versus* 1,4 hydrogenation of C=C). Notably, the reaction at lower hydrogen pressure and room temperature (27 °C) significantly improved the chemoselective C=C bond hydrogenation without altering the overall conversion (entries 4, 5). The use of other bases, such as NaO^{t}Bu, LiO^{t}Bu, and KOAc, led to low conversion, whereas the presence of mild bases K_2CO_3 or K_3PO_4 provided complete conversion with good chemoselectivity for **4** using 10 bar H_2 and at room temperature (entries 7-11). The activity of manganese catalyst using a catalytic amount of mild base K_3PO_4 is notable, as most of the Mn-catalysis generally employs a strong base like KO^{t}Bu. The chemoselectivity was further increased with less reaction time (entries 12, 13). The chemoselective hydrogenation of **4a** also proceeded smoothly using 5 bar hydrogen pressure at room temperature (27 °C) and provided **4** in 96% isolated yield just in an hour (entry 14). Lower conversion of **4a** was observed when the hydrogen pressure was reduced from 5 bar to 2 bar (entry 15). In addition, lowering the catalyst and base loading resulted in inferior results (entries 16-18). Notably, the manganese catalysts **Mn-1** and **Mn-2** were less effective for the hydrogenation under the optimized conditions (5 bar H_2/27 °C/1 h), and provided < 19% of product **4** (entries 19, 20). Therefore, all chemoselective hydrogenations were conducted employing the **Mn-3** catalyst and 5 bar H_2 pressure at room temperature with the best-optimized reaction time. In-situ generated $Mn(CO)_5Br$/**L3** catalyst system is less effective, affording the hydrogenated product in 39%; highlighting the importance of a well-defined manganese catalyst (entry 21). However, a Mn(II) precursor with **L3** ligand ($MnCl_2$/**L3** or $MnBr_2$/**L3**) did not provide the hydrogenation (entries 22, 23). Similarly, the $Mn(CO)_5Br$ precursor and bidentate *N*-donor or *P*-donor ligand systems were ineffective (entries 24-27). The hydrogenation did not proceed in the absence of catalyst, base or H_2, which suggests the importance of these components for the reaction (entries 28-30).

Table 2.1 Optimization of Reaction Conditions. [a]

(4a) → [Mn] (5.0 mol%), base (10 mol%), H_2 (bar), MeOH, T (°C), t (h) → (4) + (4')

Entry	[Mn]	base	H_2 (bar)	T (°C)/ t (h)	Conv (%)[b]	4 (%)[b]	4' (%)[b]
1	**Mn-1**	KO^{t}Bu	30	50/20	75	65	10
2 [c]	**Mn-2**	KO^{t}Bu	30	50/20	100	--	80
3	**Mn-3**	KO^{t}Bu	30	50/20	100	63 (60)	37
4	**Mn-3**	KO^{t}Bu	30	27/20	100	81 (79)	19
5	**Mn-3**	KO^{t}Bu	20	27/20	100	86	14
6	**Mn-3**	KO^{t}Bu	10	27/20	100	91 (88)	9
7	**Mn-3**	NaO^{t}Bu	10	27/20	40	39	trace
8	**Mn-3**	LiO^{t}Bu	10	27/20	74	73	trace
9	**Mn-3**	K_2CO_3	10	27/20	100	74	26
10	**Mn-3**	KOAc	10	27/20	20	19	trace
11	**Mn-3**	K_3PO_4	10	27/20	100	88	12
12	**Mn-3**	K_3PO_4	10	27/8	100	95	5
13	**Mn-3**	K_3PO_4	10	27/1	99	95	4
14	**Mn-3**	**K_3PO_4**	**5**	**27/1**	**100**	**98 (96)**	**2**
15	**Mn-3**	K_3PO_4	2	27/1	53	53	--
16 [d]	**Mn-3**	K_3PO_4	5	27/1	50	49	1
17 [e]	**Mn-3**	K_3PO_4	5	27/1	68	68	trace
18 [f]	**Mn-3**	K_3PO_4	5	27/1	45	44	1
19	**Mn-1**	K_3PO_4	5	27/1	trace	trace	--
20	**Mn-2**	K_3PO_4	5	27/1	23	19	--
21	$Mn(CO)_5Br$/**L3**	K_3PO_4	5	27/1	39	39	--
22	$MnBr_2$/**L3**	K_3PO_4	5	27/1	--	--	--
23	$MnCl_2$/**L3**	K_3PO_4	5	27/1	--	--	--
24	$Mn(CO)_5Br$/bpy	K_3PO_4	5	27/1	--	--	--
25	$Mn(CO)_5Br$/phen	K_3PO_4	5	27/1	--	--	--
26	$Mn(CO)_5Br$/dppf	K_3PO_4	5	27/1	--	--	--
27	$Mn(CO)_5Br$/dppbz	K_3PO_4	5	27/1	--	--	--

28	--	K_3PO_4	5	27/1	--	--	--
29	**Mn-3**	--	5	27/1	--	--	--
30	**Mn-3**	K_3PO_4	--	27/1	--	--	--

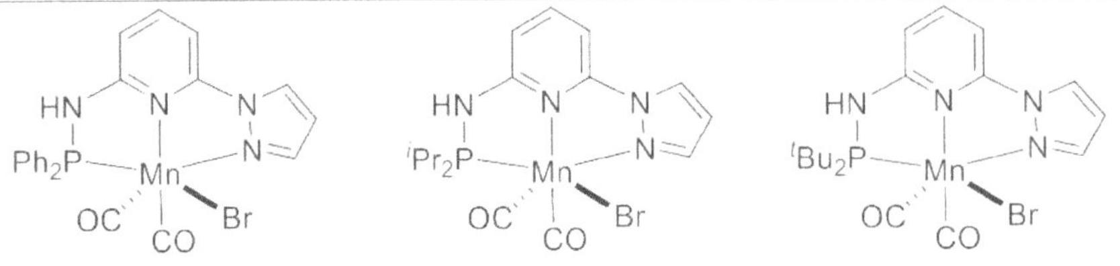

(Mn-1) (**Mn-2**) (**Mn-3**)

[a] Reaction Conditions: **4a** (0.042 g, 0.20 mmol), base (0.02 mmol), [Mn] catalyst (0.01 mmol, 5 mol %), solvent (1.0 mL). [b]GC conversion, isolated yield is given in parentheses. [c]20% allylic alcohol was observerd. [d] Using 6.0 mol% K_3PO_4. [e] Using 3 mol% of **Mn-3**. [f] Using 3 mol% of **Mn-3** and 6 mol% of K_3PO_4. bpy = 2,2'-bipyridine, phen = 1,10-phenanthroline, dppf = 1,1′-bis(diphenylphosphino)ferrocene, dppbz = 1,2-bis(diphenylphosphino)benzene. All three manganese complexes contain mixture of two geometrical isomers.

2.2.3 Substrate Scope of the Hydrogenation

After successfully optimizing the reaction parameters for chemoselective C=C bond hydrogenation in chalcone, we have explored the reaction scope using catalyst **Mn-3**, catalytic K_3PO_4 and 5 bar H_2 at room temperature (Scheme 2.3). Depending upon the substrates, the hydrogenations were performed for different time intervals, and the best yields were reported. First, we checked the scope of conjugated ketones with various substitutions on the benzoyl ring. Thus, the chalcones containing electron-donating alkyl and alkoxy substitutions at the *para* position of benzoyl reacted smoothly to give a good to an excellent yield of saturated ketones, **5-7**. The halogen substitutions, –Cl, –Br, –I, –CF_3, were well tolerated at *para* of the benzoyl moiety affording desired saturated ketones (**8-11**) in good yields. The tolerance of such functionalities is noteworthy as they can be employed for late-stage diversification. In addition to the *para*-substituted chalcones, the electron-rich and electron-deficient *ortho*-substituted compounds smoothly participated in the hydrogenation (**12** and **13**). Chalcone having a phenolic *–OH* at the *meta* position of benzoyl, reacted slowly and afforded the product **14** in 33% yield.

Interestingly, unsaturated ketones with sensitive and reducible functionalities, such as terminal alkene, alkyne, and epoxide, reacted chemoselectively to give compounds **15-17**. The unsaturated ketones with the naphthyl moiety also reacted with good yields (**18**, **19**). A higher-scale hydrogenation of compound **4a** (0.5 g, 2.4 mmol) provided the product **4** in 87% isolated yield

(Scheme 2.3, in parenthesis), highlighting the potential practical application.

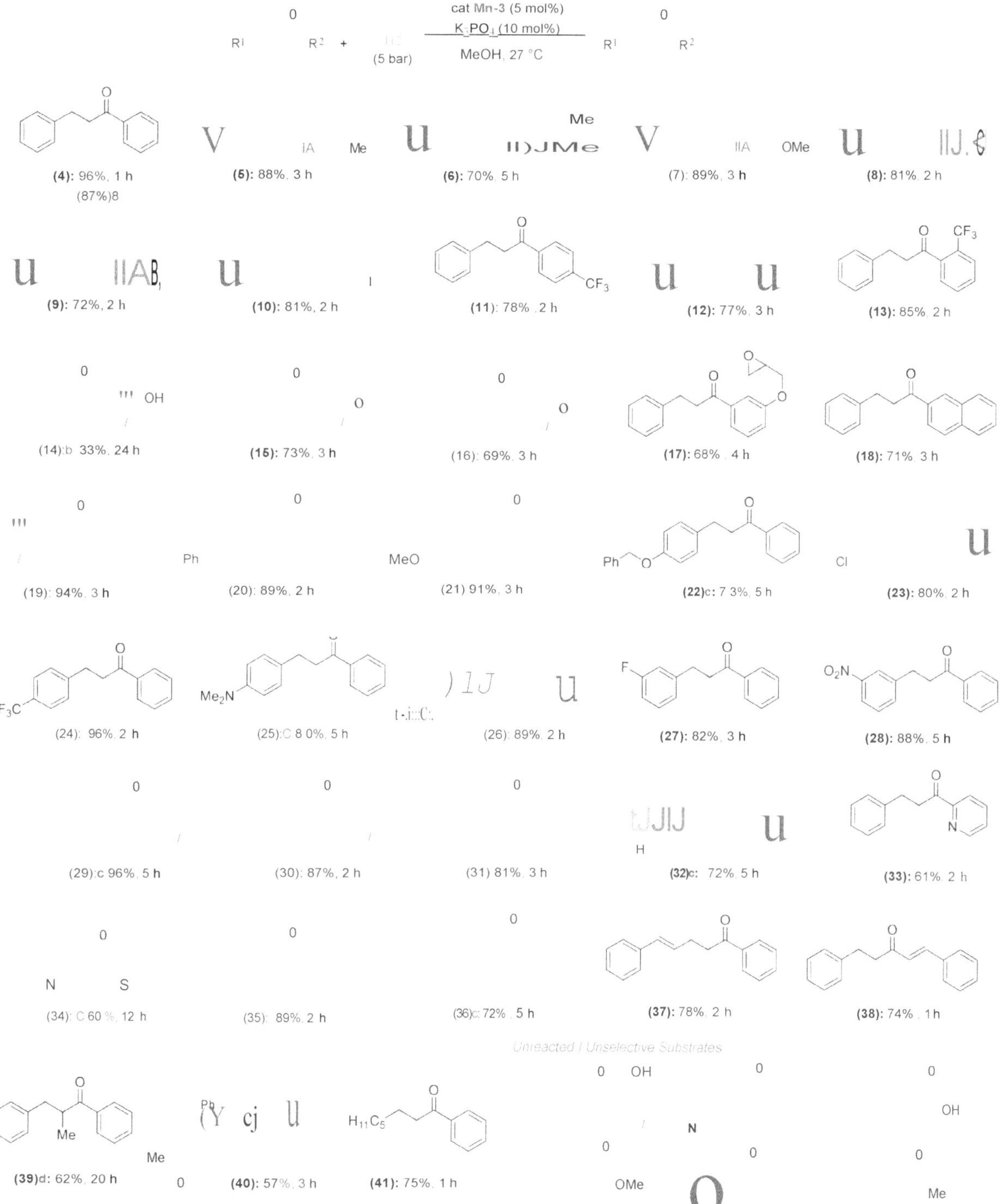

Scheme 2.3 Scope of Mn-Catalyzed C=C Bond Hydrogenation of *α,β*–Unsaturated Ketones. Reaction conditions: Substrate, *α,β*-unsaturated ketone (0.20 mmol), **Mn-3** (0.005 g, 0.01 mmol, 5 mol%), K_3PO_4 (0.0043 g, 0.02 mmol), MeOH (1.0 mL), H_2 (5 bar). Yields are of isolated

compounds. [a] Reaction on 0.5 g scale. [b] Reaction at 50 °C. [c] Mixture of MeOH:DCM (4:1) used. [d] Reaction at 10 bar H_2 and 50 °C.

After this, we moved to check the effect and tolerability of different substitutions on the alkenyl-arenes towards the chemoselective hydrogenation. Substrates with electron-rich substituents such as phenyl, methoxy and benzyloxy reacted efficiently, producing an excellent yield of products, **20-22**. The survival of benzyl protection of phenolic *–OH* is notable, as such substrates are prone to hydrogenolysis under hydrogenation conditions. Similarly, the –Cl and $–CF_3$ groups remained unaffected and delivered the saturated halo-ketones **23** and **24** in 80% and 96% yields, respectively. An amine functionality that could poison the catalysis by binding to the metal is also sustained under optimized conditions (**25**). To our surprise, the highly desirous and hydrogenation-sensitive functionalities, –CN and $–NO_2$ groups, could be tolerated to afford the products **26** and **28** in around 88% yields. A range of conjugated ketones derived from heteroarenes, including furanyl, thiophenyl, indolyl, pyrrolyl and pyridinyl, were successfully hydrogenated to afford the saturated heteroaryl ketones (**30-35**). The chemoselective hydrogenation of these heteroaryl-containing compounds is remarkable, as the heteroaryl rings often interfere with the reaction due to their coordinating ability to metal. The hydrogenation of ketones having unprotected *NH* indolyl and pyrrolyl opens up a new avenue as they can further be diversified. An unsaturated ketone containing a ferrocene backbone provided selective hydrogenation to **36** in 72% yield. Interestingly, the hydrogenation of a ketone containing extended conjugation provided selectively semi-hydrogenated product **37** in 78% yield. Similarly, the substrate where the carbonyl group is in conjugation with two alkenes, selectively one C=C bond was hydrogenated and provided a good yield of product, **38**. This hydrogenation protocol is also suitable for α,β-unsaturated ketones having trisubstituted alkene to provide saturated ketones, albeit in moderate yields (**39**, **40**). A β-alkyl-α,β-unsaturated ketone, (*E*)-1-phenyloct-2-en-1-one could be hydrogenated to give compound **41** in 75% yield. Interestingly, in all these cases, an excellent chemoselective C=C bond hydrogenation was observed in the presence of other H_2-sensitive functionalities. Such chemoselective hydrogenation employing Mn-catalyst is extremely rare.[31,36,39] Unfortunately, the conjugated ketones with free -OH at the *ortho*, carboxylate, ester and amide derivatives failed to participate in the reaction under the optimized conditions. However, the substrate (*E*)-4-phenylbut-3-en-2-one gave a mixture of highly unselective hydrogenated products.

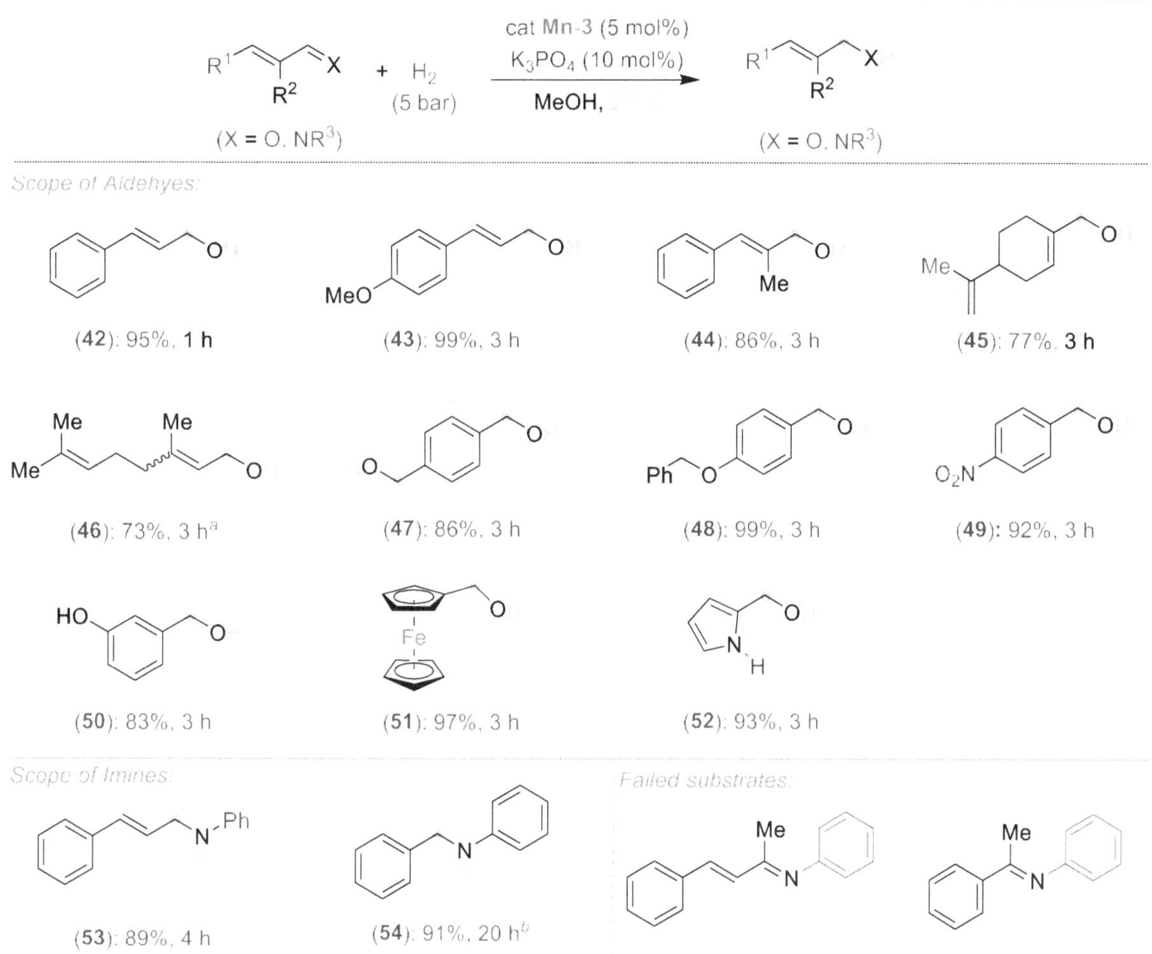

Scheme 2.4 Scope of Mn-Catalyzed Hydrogenation of Aldehydes and Imines. Reaction conditions: Substrate (0.20 mmol), **Mn-3** (0.005 g, 0.01 mmol, 5 mol%), K_3PO_4 (0.0043 g, 0.02 mmol), H_2 (5 bar), MeOH (1.0 mL). Yields are of isolated compounds. [a] Obtained as a mixture of *E* and *Z* isomers as the starting compound was also mixture of both. [b] Reaction performed at 50 °C.

After exploring the scopes and limitations of selective hydrogenation of C=C bond in unsaturated ketones, we were eager to know the reactivity of the synthesized manganese complex on the hydrogenation of enals and conjugated imines (Scheme 2.4). Surprisingly, under the standard reaction conditions, selective C=O bond hydrogenation of cinnamaldehyde derivatives was observed, leading to the 3-phenylprop-2-en-1-ols (**42-44**) in excellent yields. Aliphatic and acyclic conjugated aldehydes also participated in the selective hydrogenation to unsaturated alcohols without harming the alkenyl groups, thus leading to the products **45** and **46** in 77% and 73% yields, respectively. The aromatic aldehydes containing benzyloxy, nitro, –OH, and pyrrolyl groups were also smoothly hydrogenated to the corresponding alcohols at room temperature (**48-52**). The tolerance of free *–OH* and *–NH* groups is highly impressive. Similarly, an α,β-unsaturated imine was chemoselectively hydrogenated to unsaturated amine (**53**) in good yield. Even a simple unconjugated *N*-aryl imine could be hydrogenated to amine in high yield (**54**). However, the attempted hydrogenations of ketone-derived imine analogues failed under the optimized conditions and the unreacted starting compounds were quantitatively recovered (Scheme 2.4). We assume that the steric around the keto- derived imine inhibited its approach towards the manganese center, leading to an unsuccessful reaction. The chemoselective hydrogenations of C=C bond over the ketone carbonyl and that of C=O bond of aldehyde and C=N bond of imine over the C=C by the newly developed catalyst are notable. Particularly, mild reaction conditions and the use of catalytic K_3PO_4 base are significant. This catalyst can further be applied to novel catalytic approaches considering its advantage over other similar catalysts. Though, an excellent chemoselectivity was observed in the reduction of *α,β*-unsaturated ketones, a trace formation of both C=C and C=O reductions was unavoidable in some cases.

2.2.4 Mechanistic Aspect

2.2.4.1 Controlled Mechanistic Experiments

We have performed a few controlled experiments to understand the operating mode of manganese catalyst. First, the *N*-methyl substituted ligand (**L3-Me**) and corresponding manganese complex, (^{tBu}PN(Me)N^{Pyz})Mn(CO)$_3$Br (**Mn-3^{Me}**) were synthesized and thoroughly described using a range of analytical methods (Scheme 2.5a). An attempted hydrogenation of **4a** employing complex **Mn-3^{Me}** as catalyst under the standard hydrogenation conditions did not provide hydrogenated product (Scheme 2.5b), and the starting compound (**4a**) was quantitatively recovered. This finding supports the important role of the *N–H* proton in the complex (^{tBu}PN^{3}(H)N^{Pyz})Mn(CO)$_2$Br (**Mn-3**) during hydrogenation. Furthermore, the treatment of **Mn-3** with stoichiometric KO^{t}Bu produced dearomatized active intermediate **A** (Figure 2.2). The ^{31}P{^{1}H} NMR spectrum of **A** showed a single peak (against two isomeric peaks for **Mn-3**), and two CO signals were observed in IR as well as ^{13}C{^{1}H} NMR spectra of **A** (against three peaks for CO in **Mn-3**). Interestingly, the employment of species **A** as a catalyst in the hydrogenation without the use of additional base provided quantitative yield of **4**. All these results suggest that the species **A** acts as active catalyst and the catalytic reaction proceeds through metal-ligand cooperation (dearomatization/aromatization) pathway.

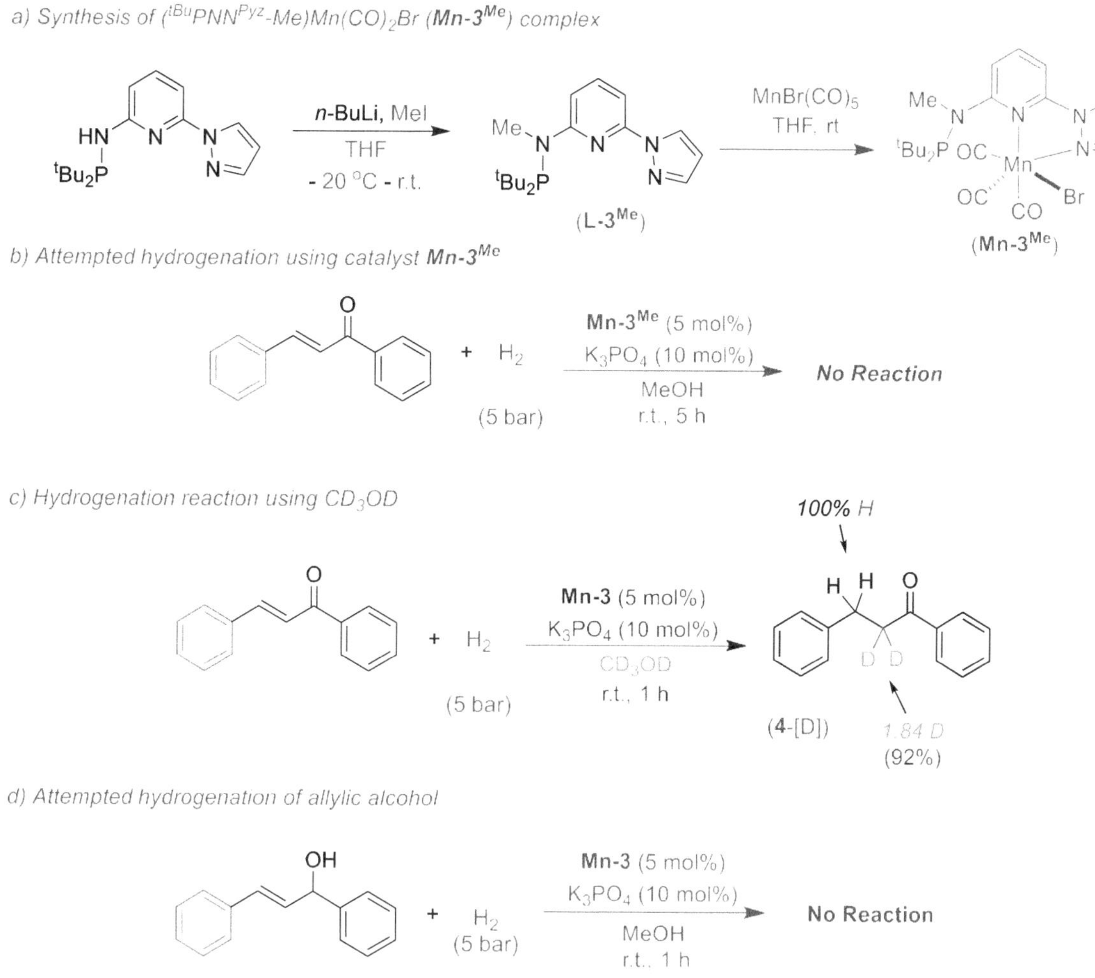

a) Synthesis of (^{tBu}PNN^{Pyz}-Me)Mn(CO)$_2$Br (**Mn-3^{Me}**) complex

n-BuLi, MeI, THF, - 20 °C - r.t.; (L-3^{Me}); MnBr(CO)$_5$, THF, rt; (Mn-3^{Me})

b) Attempted hydrogenation using catalyst **Mn-3^{Me}**

+ H_2 (5 bar); Mn-3^{Me} (5 mol%), K_3PO_4 (10 mol%), MeOH, r.t., 5 h; No Reaction

c) Hydrogenation reaction using CD_3OD

+ H_2 (5 bar); Mn-3 (5 mol%), K_3PO_4 (10 mol%), CD_3OD, r.t., 1 h; (4-[D]); 100% H; 1.84 D (92%)

d) Attempted hydrogenation of allylic alcohol

+ H_2 (5 bar); Mn-3 (5 mol%), K_3PO_4 (10 mol%), MeOH, r.t., 1 h; No Reaction

Scheme 2.5 Synthesis of **Mn-3^{Me}** and Controlled Mechanistic Experiments.

A hydrogenation reaction was performed using CD_3OD as a solvent to thoroughly understand the hydrogenation process (Scheme 2.5c). The isolated hydrogenated product **4**-[D] shows 92% deuterium incorporation at the *alpha*-methylene position (see ^{1}H, ^{13}C and deuterium NMR spectra in the figure 2.5). Moreover, the reaction did not proceed in an aprotic solvent (THF, dioxane or CH_2Cl_2). These observations indicate the necessity of a protic solvent as a proton source and tentatively supports a Mn-enolate intermediate. All these findings are consistent with the low energy barrier observed for the protonation step (discussed vide *infra*). An attempted hydrogenation of allylic alcohol under the standard reaction conditions failed to give hydrogenated product (Scheme 2.5d). These finding highlights that the hydrogenation of (*E*)-chalcone (**4a**) does not proceed *via* the 1,2-hydrogen addition, instead a 1,4-addition of hydrogen occurs. The observed chemoselectivity and controlled studies allowed us to propose the followings: (i) the **Mn-3** catalyst prefers 1,2 hydrogen addition to C=O or C=N when steric on carbonyl's/imine's carbon is low (C=O of aldehyde and C=N of imine preferred over the C=C), whereas (ii) the **Mn-3** catalyst allowed 1,4 hydrogen addition when carbonyl's carbon is doubly substituted (i.e. in *α,β*–unsaturated ketones) due to more steric constraint. The Mn–H might fail to approach the carbonyl's carbon in *α,β*–unsaturated ketones due to steric, instead it can access the *β*-carbon *via* 1,4-hydrogen addition. Notably, the **Mn-2** catalyst, which is less bulky than the **Mn-3**, allowed 1,2 hydrogen addition even in *α,β*–unsaturated ketone to form allylic alcohol. These findings are noteworthy in the consideration of catalyst developments for chemoselective functionalization.

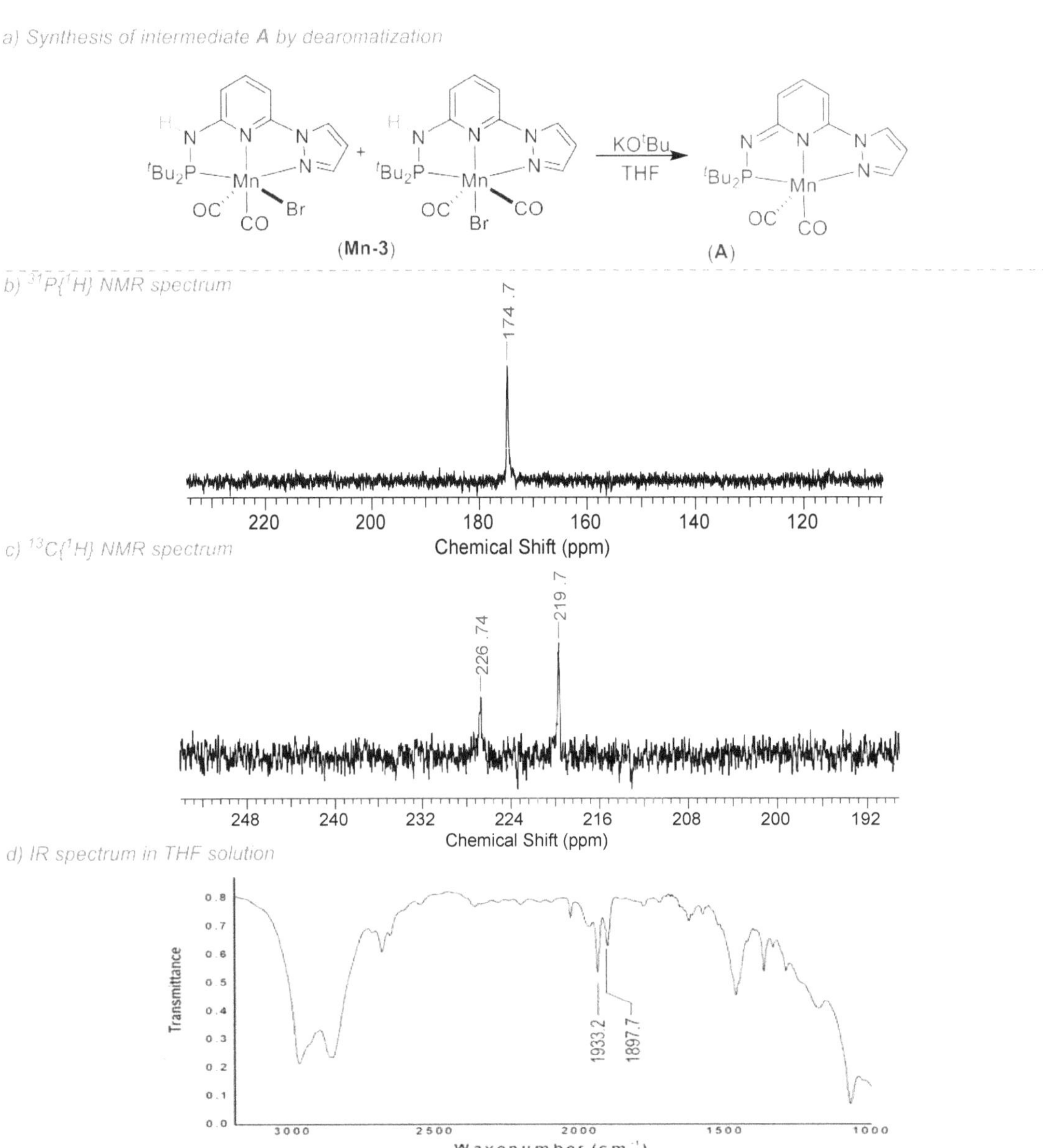

Figure 2.2 Synthesis of Active Intermediate **A** and Characterization Spectra.

2.2.4.2 DFT Calculations

We have investigated the reaction mechanism of **Mn-3** catalyzed hydrogenation of *α,β*-unsaturated carbonyl compound **4a** using density functional theory (DFT) calculations (Figure 2.3). Initially, the generation of the active catalyst **A** would occur when the pre-catalyst **Mn-3** reacts with K_3PO_4 (Figure 2.3(I)). In this step, the Mn–Br and N–H bonds break, forming the catalyst **A**, K_2HPO_4 and KBr. This step is thermodynamically favorable (ΔG = -19.9 kcal/mol). Next, the H_2 molecule adds to catalyst **A**, forming the intermediate **B** *via* a limit of 14.7 kcal/mol (**TS-1**). Then, the reaction proceeds through **TS-2** with a limit of 21.7 kcal/mol, in which the H–H bond breaks, and the Mn–H and N–H bonds form, generating the intermediate **C**. The overall barrier for the H_2 activation (31.3 kcal/mol) seems reasonable, considering that a minimum of 5 bar H_2 pressure is essential for the reaction. In the next step, the reaction crosses the barrier of 9.1 kcal/mol (**TS-3**), wherein **4a** reacts with **C** to form the intermediate **D** by hydride (H^-) migration from Mn-H to **4a**. In the next step, the intermediate **D** rearranges and forms a new intermediate **E**. This step is found to be thermodynamically favorable (ΔG = -5.2 kcal/mol). Starting with the intermediate **E**, two approaches are considered for protonation (Figure 2.3(II)). In the first possibility, the methanol protonates the substrate step-wise, as shown in Figure 2.3(II). Thus, MeOH can provide a proton to the substrate and coordinates with Mn concurrently. In this process, the Mn–O bond breaks and a different Mn–O bond forms via the transition state **TS-4** with a barrier of 10.7 kcal/mol, leads to form intermediate **F**. The reaction crosses a limit of 1.1 kcal/mol (**TS-5**) in the following step, leading to product **4** and intermediate **G**. Then, the reaction could proceed in two different ways: with the assistance of solvent (MeOH) **TS-6** and without the assistance of solvent **TS-6'**, which has barriers of 6.8 kcal/mol and 14.6 kcal/mol, respectively. The energy values show that, with the help of a solvent, the process will pass through the transition state **TS-6** (MeOH), result in to the formation of intermediate **H**. In next step, **H** converts into **A** after releasing the methanol. This step is thermodynamically favourable (ΔG = -4.4 kcal/mol). The overall low limit for protonation step tentatively supports the experimental observation, wherein a reversible protonation was assumed.

In the second possibility of protonation, the solvent MeOH directly shuttles the protons between Mn–O and N–H through a transition state **TS-7** with a barrier of 11.0 kcal/mol, result to form intermediate **I** (Figure 2.3(II)). In the next step, intermediate **I** convert into the active species **A** and product **4**. This step is thermodynamically favourable (ΔG = -18.5 kcal/mol). A perusal of the two pathways based on turnover frequency (TOF) analysis indicates that the TOF would be the same for both approaches (Figure 2.3(II)), because the main intermediate (TDI) (**A**) along the pathways and the main transition state (TDTS) (**TS-2**) are present in the early part of the cycles, which are common to both the proposed protonation pathways (for more information on the TOF analysis, see the Supporting Information). The calculated energy barrier values indicated a maximum barrier for H_2 activation, which would be feasible at room temperature considering the reaction's 5.0 bar H_2 pressure requirement. Consequently, it is possible to suggest that the H_2 activation is a plausible turnover-limiting step. Notably, the protonation of the final compound by MeOH is very facile, which corroborates the experimental findings.

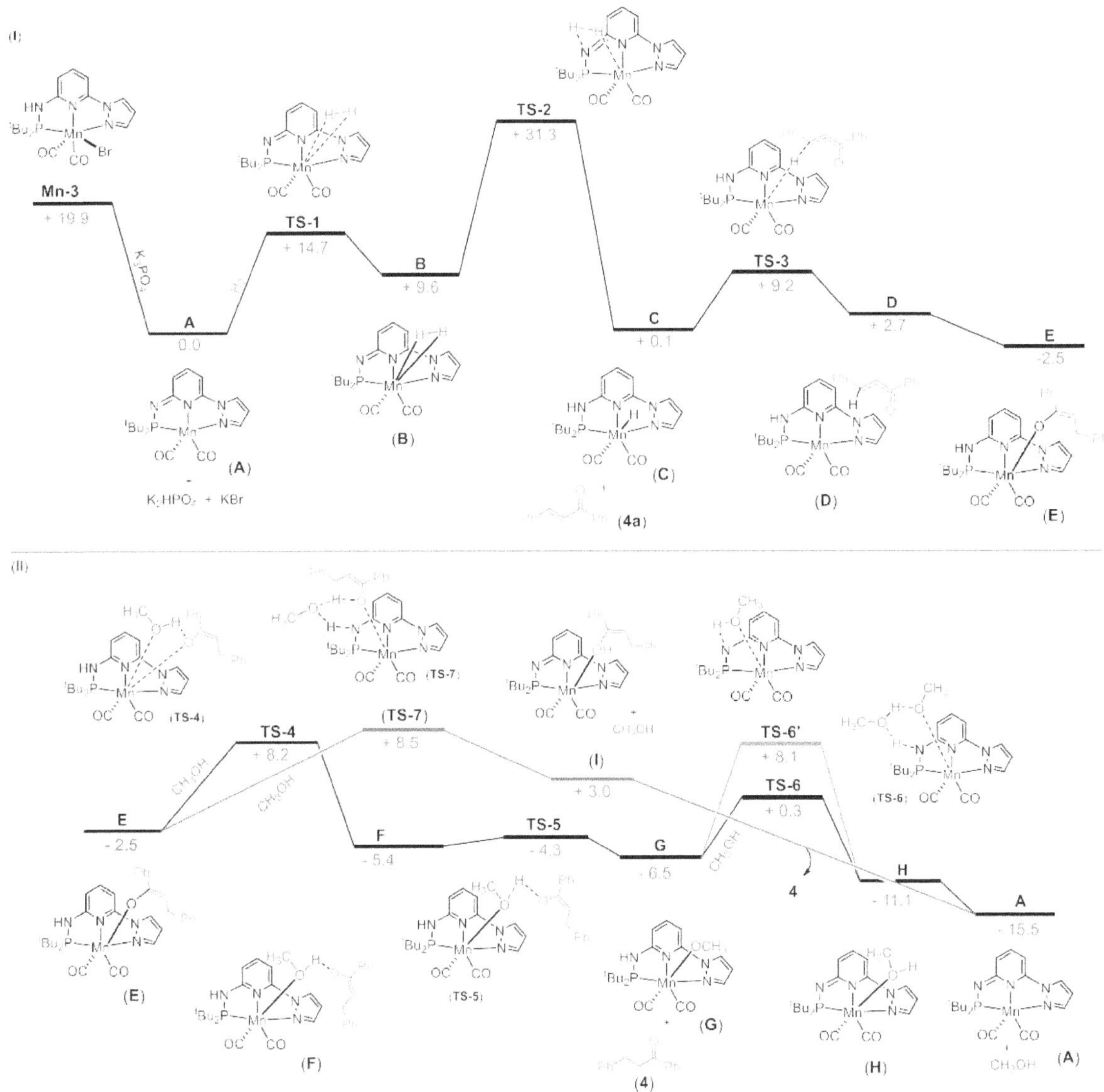

Figure 2.3 Free Energy Profile for the ($^{tBu}PNN^{Pyz}$)Mn(I)-Catalyzed Hydrogenation: (I) **Mn-3** to intermediate **E**, (II) Intermediate **E** to regeneration of active catalyst **A**. The free energy values are given in kcal/mol.

2.2.4.3 Plausible Catalytic Cycle

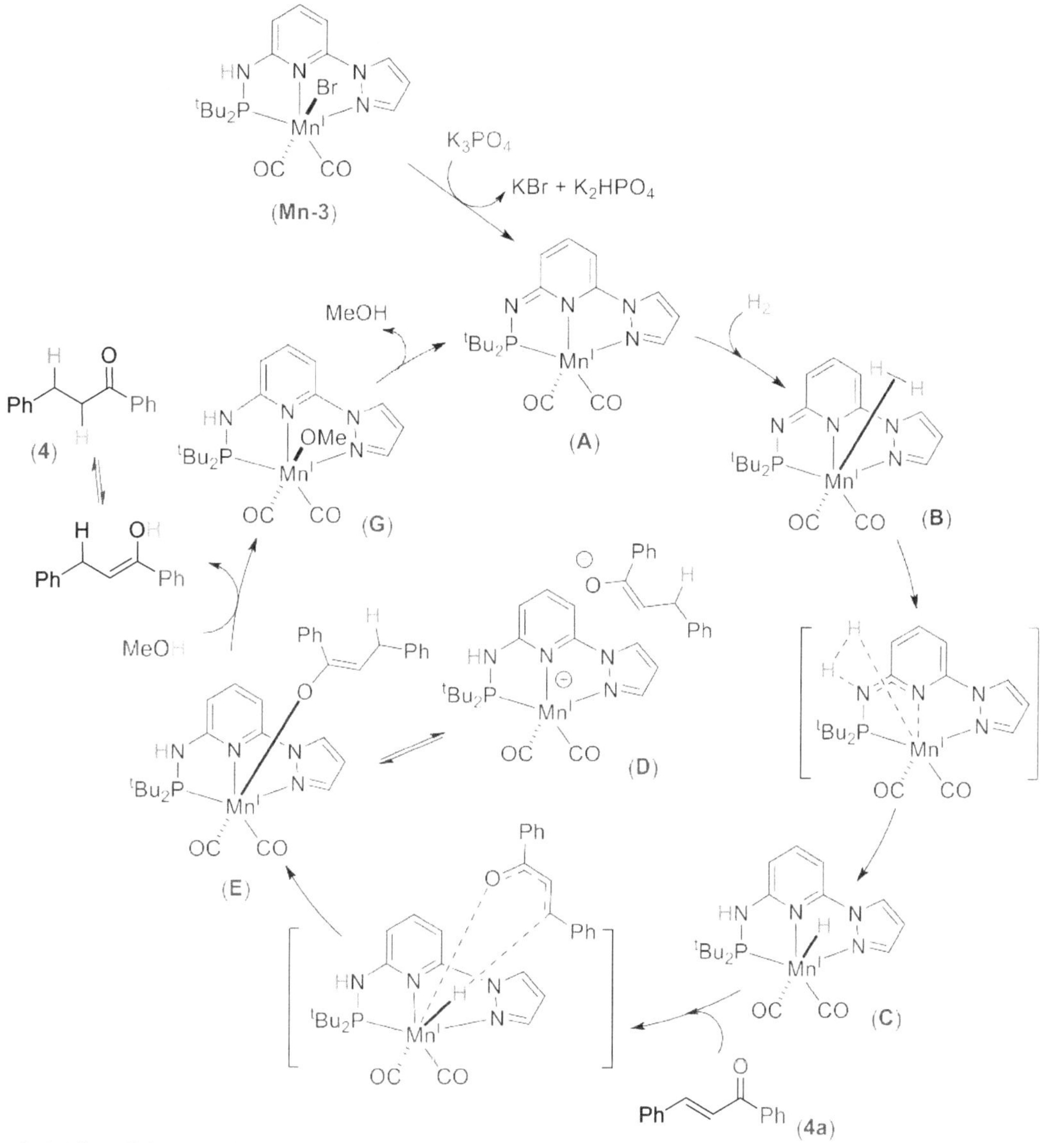

Figure 2.4 Plausible Catalytic Cycle for Mn(I) Catalyzed Chemoselective C=C Bond Hydrogenation of *α,β*-Unsaturated Ketones.

For the Mn-catalyzed chemoselective hydrogenation of *α,β*-unsaturated ketones, we suggested a tentative catalytic cycle based on experimental results, DFT calculations, and prior literatur.[34,46,61] we proposed a tentative catalytic cycle for the Mn-catalyzed chemoselective hydrogenation of *α,β*-unsaturated ketones (Figure 2.4). Initially, the complex **Mn-3** would transform into the active catalyst **A** by dearomatization in the presence of K_3PO_4. We have identified the species **A** by NMR and IR analyses. The similar conversion of metal complex into dearomatized species is well documented.[34,61,69,70] Moreover, the species **A** as an active catalyst was verified. The H_2 molecule will coordinate to Mn(I) species, followed by metal-ligand (M-L) cooperative activation of H_2 leading to species **C** through transition state **TS-2**. Next, the reaction of substrate **4a** with **C** leads to the 1,4-hydride migration resulting in intermediate **D** (or **E**). The protonation of the semi-hydrogenated species in intermediate **D** (or **E**) by methanol provides the hydrogenated product **4**. Finally, the resulted intermediate **G** would lead to the regeneration of active catalyst **A**. The DFT energy calculations supported all these elementary catalytic steps. The H_2 activation by the Mn(I) species has a high barrier and is a pivotal step in the hydrogenation process. Therefore, the H_2 activation by Mn(I)-ligand cooperation can be assumed as the turnover-limiting step.

2.3 CONCLUSIONS

In this chapter, we have shown how the well-defined pincer-ligated Mn(I) complex catalyzes the chemoselective hydrogenation of C=C, C=O, and C=N bonds in *α,β*-unsaturated ketones, aldehydes, and imines. The employment of a mild base, moderate H_2 pressure and room temperature (27 °C) in the Mn-catalyzed hydrogenation are highly advantageous to the commonly employed KO^tBu base and extreme reaction conditions. The mixed donor Mn(I) complexes were developed and carefully characterized using a range of methods. Though both ^{i}Pr and ^{t}Bu-substituted Mn(I) complexes are efficient for the reduction of conjugated ketones, the $(^{tBu2}PNN^{Pyz})Mn(CO)_2Br$ complex as a catalyst provided exceptional chemoselectivity for the hydrogenation of alkene (1,4-hydrogen addition). Thus, using the beneficial molecular hydrogen and a mild K_3PO_4 base, the $(^{tBu2}PNN^{Pyz})$Mn(I) catalyst could hydrogenate diverse unsaturated ketones to saturated ketones at room temperature with the compatibility of sensitive groups, including halides, alkoxy, alcohol, oxirane, acetyl, cyano, nitro, isolated alkene and alkyne. The C=O bond in aldehydes and C=N bond in imines were preferentially hydrogenated (1,2-hydrogen addition) over the C=C bond using the **Mn-3** catalyst. A comprehensive mechanistic investigation by controlled studies endorsed the metal-ligand cooperation of ligand in the Mn catalyst. The DFT energy calculations highlighted a probable turnover-limiting H_2 activation with the facile progress of other elementary steps. Particularly, the H_2 provided a hydride source, and solvent MeOH acts as a proton-source for the hydrogenation. The DFT energy calculations unanimously supported the proposed mechanistic cycle.

2.4 EXPERIMENTAL SECTION

All the manipulations were conducted under an argon atmosphere either in a glove box or using standard Schlenk techniques in pre-dried glassware. The catalytic reactions were performed in flame-dried reaction vessels with a Teflon screw cap. Solvents were dried over Na/benzophenone or CaH_2 and distilled prior to use. Liquid reagents were flushed with argon prior to use. All other chemicals were obtained from commercial sources and were used without further purification. High-resolution mass spectrometry (HRMS) mass spectra were recorded on a Thermo Scientific Q-Exactive, Accela 1250 pump. NMR: (^{1}H and ^{13}C) spectra were recorded at 400 or 500 MHz (^{1}H), 100 or 125 MHz {^{13}C, DEPT (distortionless enhancement by polarization transfer)}, 377MHz (^{19}F), respectively in $CDCl_3$ solutions, if not otherwise specified; chemical shifts (δ) are given in ppm. The ^{1}H and ^{13}C NMR spectra are referenced to residual solvent signals ($CDCl_3$:δ H = 7.26 ppm, δ C = 77.2 ppm).

GC Method. Gas Chromatography analyses were performed using a Shimadzu GC-2010 gas chromatograph equipped with a Shimadzu AOC-20s auto sampler and a Restek RTX-5 capillary column (30 m x 0.25 mm x 0.25 μm). The instrument was set to an injection volumeof 1 μL, an inlet split ratio of 10:1, and inlet and detector temperatures of 250 and 320 °C, respectively. UHP-grade argon was used as carrier gas with a flow rate of 30 mL/min. The temperature program used for all the analyses is as follows: 80 °C, 1 min; 30 °C/min to 200 °C, 2 min; 30 °C/min to 260 °C, 3 min; 30 °C/min to 300 °C, 15 min.

2.4.1 Synthesis of Ligand and Manganese Complexes

Synthesis of *N*-(di-*iso*-propylphosphaneyl)-6-(1*H*-pyrazol-1-yl)pyridin-2-amine (L2):

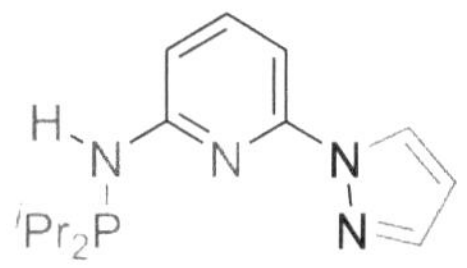

In a 50 mL oven dried Schlenk flask, 6-(1*H*-pyrazol-1-yl)pyridin-2-amine (0.20 g, 1.25 mmol) was dissolved in THF (10 mL) followed by the addition of freshly distilled Et_3N (0.21 mL, 1.5 mmol) under inert atmosphere. The reaction mixture was cooled to 0 °C and chlorodiisopropylphosphine (0.21 g, 1.37 mmol) was added dropwise. The reaction mixture was allowed to room temperature and stirred for 1 h. Further the reaction mixture was cooled to -78 °C and *n*-BuLi (0.94 mL, 1.5 mmol; 1.6 M in THF) was added dropwise resulting in a colorless solution. Then the reaction was allowed to room temperature and stirred for 1 h, followed by heating at 60 °C for 16 h. The volatiles were evaporated under vacuum and 30 mL of toluene was added into it. The filtration and evaporation of toluene extract gave *N*-(di-*iso*-propylphosphaneyl)-6-(1*H*-pyrazol-1-yl)pyridin-2-amine (**L2**; 0.27 g, 78%) as colorless liquid. ^{1}H-NMR (500 MHz, C_6D_6): δ = 8.65 (d, J = 2.1 Hz, 1H, Ar–H), 7.68-7.66 (m, 2H, Ar–H), 7.13 (t, J = 8.1 Hz, 1H, Ar–H), 6.95 (dd, J = 7.9, 2.2 Hz, 1H, Ar–H), 6.14 (dd, J = 2.4, 1.5 Hz, 1H), 4.55 (br s, 1H, *NH*), 1.41 (d sept, J = 6.9, 1.7 Hz, 2H, CH), 0.90 (dd, J = 16.2, 7.3 Hz, 6H, CH_3), 0.86 (dd, J = 11.0, 7.0 Hz, 6H, CH_3). ^{13}C{^{1}H}-NMR (125 MHz, C_6D_6): δ = 160.4 (d, J

= 20.0 Hz, C_q), 151.4 (C_q), 142.2 (CH), 140.6 (CH), 127.0 (CH), 107.7 (CH), 106.4 (d, J = 17.2 Hz, CH), 103.0 (CH), 26.8 (d, J = 12.4 Hz, 2C, CH), 19.2 (CH_3), 19.0 (CH_3), 17.5 (CH_3), 17.4 (CH_3). $^{31}P\{^1H\}$-NMR (162 MHz, C_6D_6): 48.6 (s).

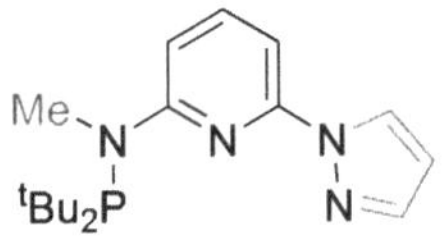

N-(Di-tert-butylphosphaneyl)-N-methyl-6-(1H-pyrazol-1-yl)pyridin-2-amine ($L3^{Me}$): In a 50 mL round bottom flask, N-(di-*tert*-butylphosphaneyl)-6-(1H-pyrazol-1-yl)pyridin-2-amine (0.10 g, 0.33 mmol) in THF (5 mL) was cooled to -20 °C and n-BuLi (0.25 mL, 0.40 mmol; 1.6 M in hexane) was added followed by the addition of MeI (0.055 g, 0.39 mmol). The reaction mixture was allowed to the room temperature and stirred for overnight. The volatiles were evaporated under vacuo and 20 mL of water was added. The organic layer was extracted in EtOAc and the crude product was subjected to the column chromatography on silica gel (petroleum ether/EtOAc:10/1) to yield **$L3^{Me}$** (0.071 g, 68%) as white solid. ^1H-NMR (500 MHz, $CDCl_3$): δ = 8.30 (d, J = 2.4 Hz, 1H, Ar–H), 7.66 (s, 1H, Ar–H), 7.36 (dt, J = 7.9, 2.1 Hz, 1H, Ar–H), 7.04 (d, J = 7.5 Hz, 1H, Ar–H), 6.51 (d, J = 8.1 Hz, 1H), 6.37 (t, J = 2.1 Hz, 1H, Ar–H), 1.83 (d, J = 11.1 Hz, 3H, CH_3), 1.34 (d, J = 14.1 Hz, 18H, CH_3). $^{13}C\{^1H\}$-NMR (100 MHz, $CDCl_3$): δ = 164.4 (d, J = 9.2 Hz, C_q), 149.8 (C_q), 140.9 (CH), 138.6 (d, J = 3.8 Hz, CH), 126.3 (CH), 116.1 (d, J = 22.1 Hz, CH), 106.5 (CH), 98.5 (CH), 36.4 (d, J = 65.6 Hz, 2C, C_q), 27.2 (6C, CH_3), 5.6 (d, J = 42.0 Hz, CH_3). $^{31}P\{^1H\}$-NMR (162 MHz, $CDCl_3$): 36.9 (s).

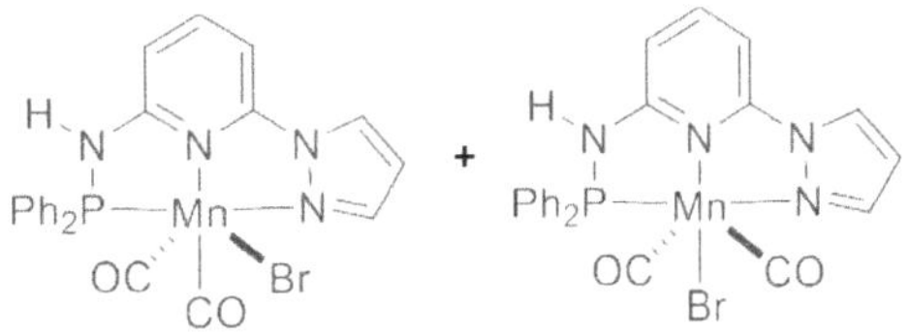

Synthesis and Characterization of Mn-1: In a 25 mL round bottom flask N-(diphenylphosphaneyl)-6-(1H-pyrazol-1-yl)pyridin-2-amine (**L1**; 0.084 g, 0.244 mmol) and $Mn(CO)_5Br$ (0.067 g, 0.244 mmol) were added inside the glove box. The reaction flask was taken out and THF (3 mL) was added. The reaction mixture was stirred under inert atmosphere at room temperature (27 °C) for 20 h, wherein the desired complex was precipitated out. The solid compound was separated from mother liquor and was washed with n-hexane (10 mL x 3). The resulted complex was dried under vacuum to give yellow powder of **Mn-1** (0.090 g, 69%). **FT-IR** (ν_{CO}, cm^{-1}): 1935, 1865, 1857. **^1H-NMR** (500 MHz, DMSO-d_6): (*one isomer*): δ = 10.34 (br s, 1H), 9.15 (br s, 1H),

8.57 (br s, 1H), 8.10 (br s, 1H), 7.77-7.42 (m, 11H), 7.28 (br s, 1H), 6.92 (br s, 1H); (*other isomer*): δ = 10.02 (br s, 1H), 8.99 (br s, 1H), 8.22 (br s, 1H), 7.96 (br s, 1H), 7.77-7.42 (m, 11H), 7.16 (br s, 1H), 6.77 (br s, 1H). **^{13}C{^{1}H}-NMR** (125 MHz, DMSO-d_6): (*for both isomers*) δ = 229.3, 225.3, 223.8, 160.0, 147.4, 146.1, 144.9, 141.5, 139.9, 136.2, 130.9, 129.8, 129.5, 128.9, 128.1, 127.7, 127.0, 125.7, 110.9, 106.7, 99.3. **^{31}P{^{1}H}-NMR** (162 MHz, DMSO-d_6, ppm): 136. 9 (s), 134.5 (s). **ESI-MS** (–ve mode): *m/z* [M–H]$^-$ Calcd for [$C_{22}H_{17}{}^{79}BrMnN_4O_2P$–H] 532.9569, Found 532.9575; *m/z* [M–H]$^-$ Calcd for ($C_{22}H_{17}{}^{81}BrMnN_4O_2P$–H) 534.9549, Found 534.9554.

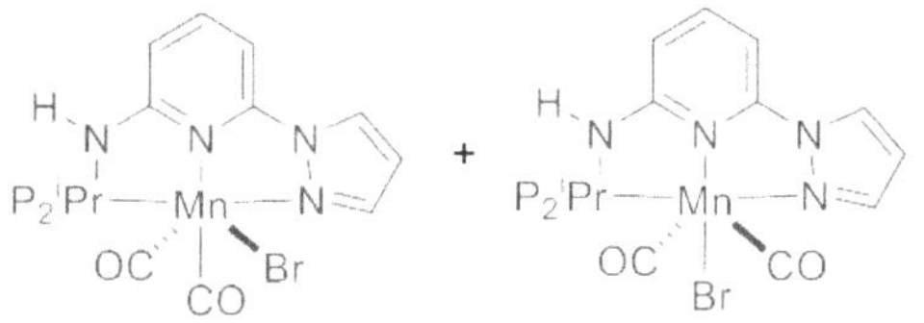

Synthesis and Characterization of Mn-2: This complex was synthesized following a procedure similar to the synthesis of **Mn-1**, using *N*-(di-*iso*-propylphosphaneyl)-6-(1*H*-pyrazol-1-yl)pyridin-2-amine (**L2**; 0.060 g, 0.217 mmol) and $Mn(CO)_5Br$ (0.060 g, 0.218 mmol). The complex **Mn-2** was obtained as an orange powder. Yield: 0.090 g (89%). **FT-IR** (ν_{CO}, cm^{-1}): 1940, 1872, 1857. **^{1}H-NMR** (400 MHz, DMSO-d_6): (*one isomer*): δ = 9.15 (br s, 1H), 8.92 (br s, 1H), 8.49 (br s, 1H), 7.94 (br s, 1H), 7.53 (br s, 1H), 7.03 (br s, 1H), 6.91 (br s, 1H), 1.30-1.18 (m, 14H); (*other isomer*): δ = 9.13 (br s, 1H), 8.80 (br s, 1H), 8.10 (br s, 1H), 7.79 (br s, 1H), 7.40 (br s, 1H), 6.91 (br s, 1H), 6.72 (br s, 1H), 1.18-1.09 (m, 14H). **^{13}C{^{1}H}-NMR** (100 MHz, DMSO-d_6): (*one isomer*): δ = 230.2 (d, J = 15.3 Hz, CO), 226.6 (d, J = 22.9 Hz, CO), 161.0 (d, J = 9.5 Hz, C_q), 147.4 (C_q), 146.1 (CH), 141.1 (CH), 130.9 (CH), 111.4 (CH), 106.5 (CH), 98.8 (CH), 29.7, 26.4, 18.1, 17.4; (*other isomer*): δ = 227.0 (d, J = 19.1 Hz, CO), 225.8 (d, J = 22.9 Hz, CO), 160.6 (d, J = 9.5 Hz, C_q), 147.4 (C_q), 144.3 (CH), 139.5 (CH), 129.3 (CH), 110.6 (CH), 105.2 (CH), 98.1 (CH), 26.5, 25.8, 17.3, 16.5. **^{31}P{^{1}H}-NMR** (162 MHz, DMSO-d_6): 159.8 (s), 158.4 (s). **ESI-MS** (–ve mode): *m/z* [M–H]$^-$ Calcd for [$C_{16}H_{21}{}^{79}BrMnN_4O_2P$–H] 464.9882, Found 464.9887; *m/z* [M–H]$^-$ Calcd for [$C_{16}H_{21}{}^{81}BrMnN_4O_2P$–H] 466.9862, Found 466.9866.

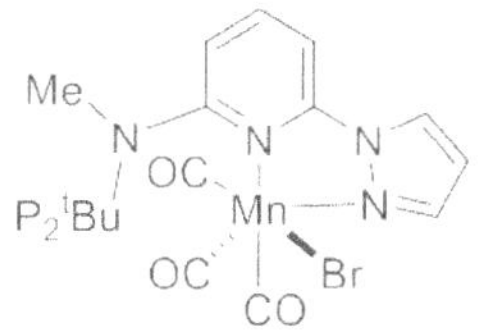

Synthesis and characterization of Mn-3^{Me}: This complex was synthesized following the procedure similar to the synthesis of **Mn-1**, using *N*-(di-tert-butylphosphaneyl)-*N*-methyl-6-(1*H*-pyrazol-1-

yl)pyridin-2-amine (0.042 g, 0.132 mmol) and $Mn(CO)_5Br$ (0.036 g, 0.131 mmol), and the reaction mixture was stirred at room temperature (27 °C) for 16 h. The complex **Mn-3^{Me}** was obtained as an orange powder. Yield: 0.060 g (85%). **FT-IR** (ν_{CO}, cm^{-1}): 2018, 1924, 1901. **^{1}H-NMR** (500 MHz, acetone-d_6): δ = 8.67 (br s, 1H, Ar–H), 8.21 (br s, 1H, Ar–H), 7.45 (br s, 1H, Ar–H), 7.02 (br s, 1H, Ar–H), 6.73 (br s, 1H, Ar–H), 6.63 (br s, 1H, Ar–H), 1.93 (s, 3H, CH_3), 1.43 (d, J = 12.8 Hz, 18H, CH_3). **^{13}C{^{1}H}-NMR** (125 MHz, DMSO-d_6): δ = 224.2 (CO), 223.9 (CO), 223.1 (CO), 164.9 (C_q), 148.3 (C_q), 144.0 (CH), 138.1 (CH), 129.7 (CH), 112.0 (CH), 110.5 (CH), 96.4 (CH), 36.2 (d, J = 63.6 Hz, 2C C_q), 25.8 (d, J = 30.5 Hz, 6C, CH_3), 5.1 (d, J = 42.0 Hz, CH_3). **^{31}P{^{1}H}-NMR** (162 MHz, acetone-d_6): 41.5 (s).

2.4.2 Representative Procedure for Hydrogenation and Characterization Data

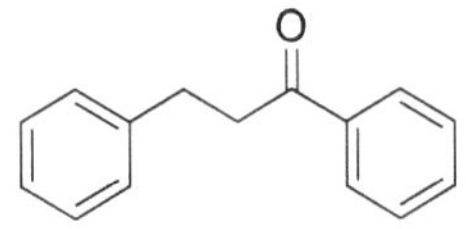

Synthesis of 1,3-diphenylpropan-1-one (4): To a dry vial with magnetic bar was introduced **Mn-3** (0.005 g, 0.01 mmol), K_3PO_4 (0.0043 g, 0.02 mmol), and (E)-chalcone (**4a**; 0.042 g, 0.202 mmol) inside the glove box. The reaction vial was transferred to an autoclave under argon atmosphere. Then MeOH (1.0 mL) was added and the autoclave was pressurized with H_2 (5 bar) and vented for five times. Finally, the autoclave was pressurized with 5 bar H_2 and stirred (700 rpm) at room temperature (27 °C) for 1 h. After reaction time, the reaction mixture was concentrated and subjected to column chromatography on silica gel (petroleum ether/EtOAc: 70/1) to yield **4** (0.041 g, 97%) as white solid. ^{1}H-NMR (500 MHz, $CDCl_3$): δ = 7.95 (d, J = 7.4 Hz, 2H, Ar–H), 7.54 (t, J = 7.4 Hz, 1H, Ar–H), 7.44 (t, J = 7.6 Hz, 2H, Ar–H), 7.31-7.18 (m, 5H, Ar–H), 3.29 (t, J = 7.7 Hz, 2H, CH_2), 3.06 (t, J = 7.7 Hz, 2H, CH_2). ^{13}C{^{1}H}-NMR (125 MHz, $CDCl_3$): δ = 199.4 (CO), 144.4 (C_q), 137.0 (C_q), 133.2 (CH), 128.8 (2C, CH), 128.7 (2C, CH), 128.6 (2C, CH), 128.2 (2C, CH), 126.3 (CH), 40.6 (CH_2), 30.3 (CH_2). HRMS (ESI): m/z Calcd for $C_{15}H_{14}O + H^+$ $[M + H]^+$ 211.1117; Found 211.1115. The ^{1}H and ^{13}C spectra are consistent with those reported in the literature.[71]

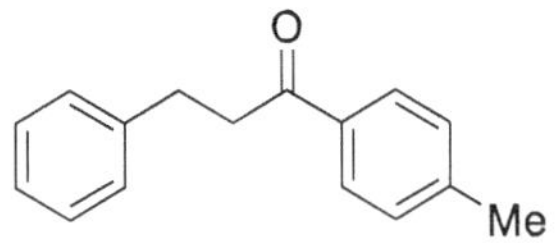

3-Phenyl-1-(p-tolyl)propan-1-one (5): The representative procedure was followed, using substrate **5a** (0.045 g, 0.202 mmol) and the reaction mixture was stirred at room temperature (27 °C) for 3 h. Purification by column chromatography on silica gel (petroleum ether/EtOAc: 50/1) yielded **5** (0.040

g, 88%) as brown solid. ^{1}H-NMR (400 MHz, $CDCl_3$): δ = 7.85 (d, J = 8.3 Hz, 2H, Ar–H), 7.30-7.27 (m, 2H, Ar–H), 7.25-7.17 (m, 5H, Ar–H), 3.26 (t, J = 7.8 Hz, 2H, CH_2), 3.05 (t, J = 7.7 Hz, 2H, CH_2), 2.39 (s, 3H, CH_3). $^{13}C\{^1H\}$-NMR (100 MHz, $CDCl_3$): δ = 199.0 (CO), 143.9 (C_q), 141.5 (C_q), 134.5 (C_q), 129.4 (2C, CH), 128.6 (2C, CH), 128.6 (2C, CH), 128.3 (2C, CH), 126.2 (CH), 40.5 (CH_2), 30.4 (CH_2), 21.8 (CH_3). HRMS (ESI): *m/z* Calcd for $C_{16}H_{16}O + H^+$ $[M + H]^+$ 225.1274; Found 225.1272. The ^{1}H and $^{13}C\{^1H\}$ spectra are consistent with those reported in the literature.[72]

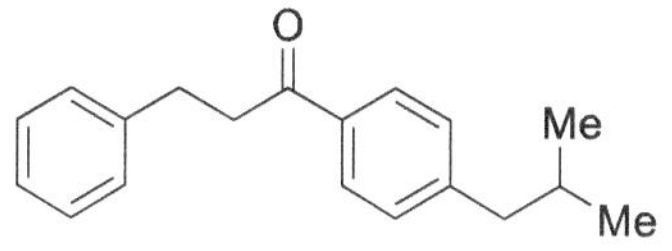

Me

Me

1-(4-*Iso*butylphenyl)-3-phenylpropan-1-one (6): The representative procedure was followed, using substrate **6a** (0.053 g, 0.20 mmol) and the reaction mixture was stirred at room temperature (27 °C) for 5 h. Purification by column chromatography on silica gel (petroleum ether/EtOAc: 50/1) yielded **6** (0.037 g, 70%) as yellow oil. ^{1}H-NMR (500 MHz, $CDCl_3$): δ = 7.88 (d, J = 8.1 Hz, 2H, Ar–H), 7.31-7.28 (m, 2H, Ar–H), 7.26-7.18 (m, 5H, Ar–H), 3.28 (t, J = 7.7 Hz, 2H, CH_2), 3.06 (t, J = 7.8 Hz, 2H, CH_2), 2.52 (d, J = 7.1 Hz, 2H, CH_2), 1.89 (sept, J = 6.8 Hz, 1H, CH), 0.90 (d, J = 6.6 Hz, 6H, CH_3). $^{13}C\{^1H\}$-NMR (125 MHz, $CDCl_3$): δ = 199.1 (CO), 147.7 (C_q), 141.6 (C_q), 134.8 (C_q), 129.5 (2C, CH), 128.7 (2C, CH), 128.6 (2C, CH), 128.2 (2C, CH), 126.3 (CH), 45.6 (CH_2), 40.5 (CH_2), 30.4 (CH_2), 30.3 (CH), 22.5 (2C, CH_3). HRMS (ESI): *m/z* Calcd for $C_{19}H_{22}O + H^+$ $[M + H]^+$ 267.1743; Found 276.1740. The ^{1}H and $^{13}C\{^1H\}$ spectra are consistent with those reported in the literature.[73]

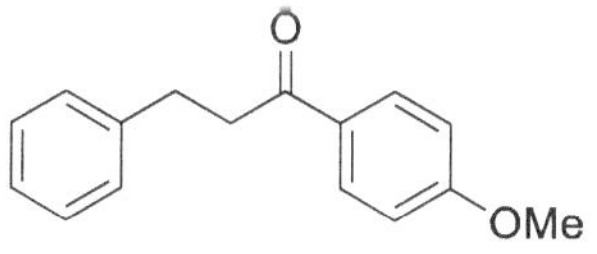

OMe

1-(4-Methoxyphenyl)-3-phenylpropan-1-one (7): The representative procedure was followed, using substrate **7a** (0.048 g, 0.201 mmol) and the reaction mixture was stirred at room temperature (27 °C) for 3 h. Purification by column chromatography on silica gel (petroleum ether/EtOAc: 30/1) yielded **7** (0.043 g, 89%) as white solid. ^{1}H-NMR (400 MHz, $CDCl_3$): δ = 7.94 (d, J = 9.0 Hz, 2H, Ar–H), 7.32-7.24 (m, 4H, Ar–H), 7.20 (t, J = 7.0 Hz, 1H, Ar–H), 6.92 (d, J = 9.0 Hz, 2H, Ar–H), 3.85 (s, 3H, CH_3), 3.25 (t, J = 7.8 Hz, 2H, CH_2), 3.06 (t, J = 7.8 Hz, 2H, CH_2). $^{13}C\{^1H\}$-NMR (100 MHz, $CDCl_3$): δ = 198.0 (CO), 163.6 (C_q), 141.6 (C_q), 130.4 (2C, CH), 130.1 (C_q), 128.6 (2C, CH), 128.6 (2C, CH), 126.2 (CH), 113.9 (2C, CH), 55.6 (CH_3), 40.2 (CH_2), 30.5 (CH_2). HRMS (ESI): *m/z* Calcd for $C_{16}H_{16}O_2 + H^+$ $[M + H]^+$ 241.1223; Found 241.1220. The ^{1}H and $^{13}C\{^1H\}$ spectra are consistent with those reported in the literature.[74]

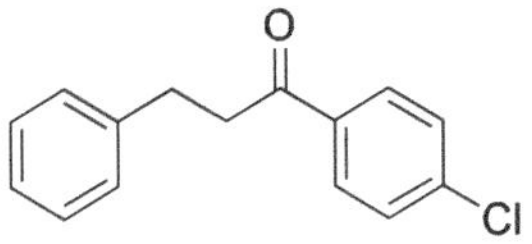

1-(4-Chlorophenyl)-3-phenylpropan-1-one (8): The representative procedure was followed, using substrate **8a** (0.049 g, 0.202 mmol) and the reaction mixture was stirred at room temperature (27 °C) for 2 h. Purification by column chromatography on silica gel (petroleum ether/EtOAc: 50/1) yielded **8** (0.040 g, 81%) as white solid. ^{1}H-NMR (400 MHz, $CDCl_3$): δ = 7.88 (d, J = 8.6 Hz, 2H, Ar–H), 7.41 (d, J = 8.6 Hz, 2H, Ar–H), 7.31-7.27 (m, 2H, Ar–H), 7.24-7.18 (m, 3H, Ar–H), 3.26 (t, J = 7.7 Hz, 2H, CH_2), 3.05 (t, J = 7.7 Hz, 2H, CH_2). ^{13}C{^{1}H}-NMR (100 MHz, $CDCl_3$): δ = 198.1 (CO), 141.2 (C_q), 139.6 (C_q), 135.3 (C_q), 129.6 (2C, CH), 129.1 (2C, CH), 128.7 (2C, CH), 128.6 (2C, CH), 126.4 (CH), 40.6 (CH_2), 30.2 (CH_2). HRMS (ESI): *m/z* Calcd for $C_{15}H_{13}ClO + H^+$ $[M + H]^+$ 245.0728; Found 245.0726. The ^{1}H and ^{13}C{^{1}H} spectra are consistent with those reported in the literature.[75]

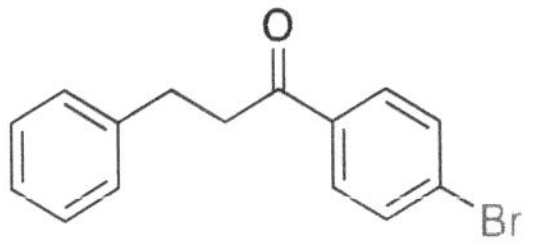

1-(4-Bromophenyl)-3-phenylpropan-1-one (9): The representative procedure was followed, using substrate **9a** (0.058 g, 0.202 mmol) and the reaction mixture was stirred at room temperature (27 °C) for 2 h. Purification by column chromatography on silica gel (petroleum ether/EtOAc: 100/1) yielded **9** (0.042 g, 72%) as white solid. ^{1}H-NMR (400 MHz, $CDCl_3$): δ = 7.80 (d, J = 8.6 Hz, 2H, Ar–H), 7.58 (d, J = 8.6 Hz, 2H, Ar–H), 7.31-7.18 (m, 5H, Ar–H), 3.25 (t, J = 7.6 Hz, 2H, CH_2), 3.05 (t, J = 7.6 Hz, 2H, CH_2). ^{13}C{^{1}H}-NMR (100 MHz, $CDCl_3$): δ = 198.3 (CO), 141.2 (C_q), 135.7 (C_q), 132.1 (2C, CH), 129.7 (2C, CH), 128.7 (2C, CH), 128.6 (2C, CH), 128.4 (C_q), 126.4 (CH), 40.6 (CH_2), 30.2 (CH_2). HRMS (ESI): *m/z* Calcd for $C_{15}H_{13}BrO + H^+$ $[M + H]^+$ 289.0223, 291.0208; Found 289.0216, 291.0197. The ^{1}H and ^{13}C{^{1}H} spectra are consistent with those reported in the literature.[76]

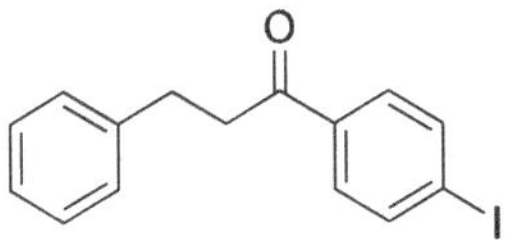

1-(4-Iodophenyl)-3-phenylpropan-1-one (10): The representative procedure was followed, using substrate **10a** (0.067 g, 0.201 mmol) and the reaction mixture was stirred at room temperature (27 °C) for 2 h. Purification by column chromatography on silica gel (petroleum ether/EtOAc: 50/1) yielded **10** (0.055 g, 81%) as white solid. ^{1}H-NMR (400 MHz, $CDCl_3$): δ = 7.80 (d, J = 8.4 Hz, 2H, Ar–H), 7.64 (d, J = 8.4 Hz, 2H, Ar–H), 7.29 (t, J = 7.4 Hz, 2H, Ar–H), 7.24-7.18 (m, 3H, Ar–H), 3.24 (t, J = 7.6 Hz, 2H, CH_2), 2.93 (t, J = 7.6 Hz, 2H, CH_2). ^{13}C{^{1}H}-NMR (100 MHz, $CDCl_3$): δ = 198.6 (CO), 141.2 (C_q), 138.1 (2C, CH), 136.2 (C_q), 129.6 (2C, CH), 128.7 (2C, CH), 128.6 (2C, CH), 126.4 (CH), 101.2 (C_q), 40.5 (CH_2), 30.2 (CH_2). HRMS (ESI): m/z Calcd for $C_{15}H_{13}IO + H^+$ [M + H]$^+$ 337.0084; Found 337.0080. The ^{1}H and ^{13}C{^{1}H} spectra are consistent with those reported in the literature.[77]

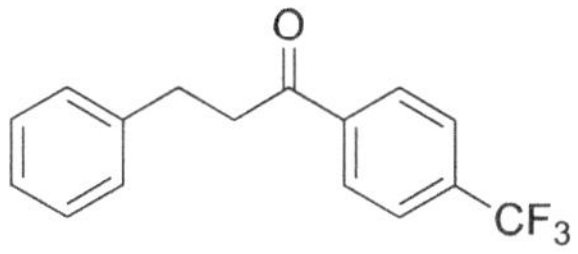

3-Phenyl-1-(4-(trifluoromethyl)phenyl)propan-1-one (11): The representative procedure was followed, using substrate **11a** (0.056 g, 0.203 mmol) and the reaction mixture was stirred at room temperature (27 °C) for 2 h. Purification by column chromatography on silica gel (petroleum ether/EtOAc: 100/1) yielded **11** (0.044 g, 78%) as yellow oil. ^{1}H-NMR (500 MHz, $CDCl_3$): δ = 8.04 (d, J = 8.2 Hz, 2H, Ar–H), 7.71 (d, J = 8.2 Hz, 2H, Ar–H), 7.33-7.28 (m, 2H, Ar–H), 7.26-7.20 (m, 3H, Ar–H), 3.33 (t, J = 7.6 Hz, 2H, CH_2), 3.08 (t, J = 7.6 Hz, 2H, CH_2). ^{13}C{^{1}H}-NMR (125 MHz, $CDCl_3$): δ = 198.4 (CO), 141.0 (C_q), 139.7 (C_q), 134.5 (q, $^{2}J_{C-F}$ = 32.8 Hz, C_q), 128.8 (2C, CH), 128.6 (2C, CH), 128.5 (2C, CH), 126.5 (CH), 125.9 (q, $^{3}J_{C-F}$ = 3.8 Hz, 2C, CH), 123.8 (q, $^{1}J_{C-F}$ = 272.6 Hz, C_q), 40.9 (CH_2), 30.1 (CH_2). ^{19}F-NMR (377 MHz, $CDCl_3$): δ = −63.1 (s). HRMS (ESI): m/z Calcd for $C_{16}H_{13}F_3O + H^+$ [M + H]$^+$ 279.0991; Found 279.0988. The ^{1}H and ^{13}C{^{1}H} spectra are consistent with those reported in the literature.[75]

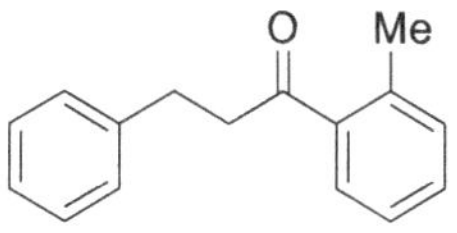

3-Phenyl-1-(o-tolyl)propan-1-one (12): The representative procedure was followed, using substrate **12a** (0.045 g, 0.202 mmol) and the reaction mixture was stirred at room temperature (27 °C) for 3 h. Purification by column chromatography on silica gel (petroleum ether/EtOAc: 50/1) yielded **12** (0.035 g, 77%) as yellow oil. ^{1}H-NMR (400 MHz, $CDCl_3$): δ = 7.59 (d, J = 7.9 Hz, 1H, Ar–H), 7.35 (t, J = 7.4 Hz, 1H, Ar–H), 7.28 (t, J = 7.4 Hz, 2H, Ar–H), 7.24-7.17 (m, 5H, Ar–H), 3.22 (t, J = 7.7 Hz, 2H, Ar–H), 3.04 (t, J = 7.6 Hz, 2H, CH_2), 2.46 (s, 3H, CH_3). ^{13}C{^{1}H}-NMR (100 MHz, $CDCl_3$):

δ = 203.6 (C_q), 141.4 (C_q), 138.3 (C_q), 138.1 (C_q), 132.1 (CH), 131.4 (CH), 128.7 (2C, CH), 128.6 (2C, CH), 128.5 (CH), 126.3 (CH), 125.8 (CH), 43.4 (CH_2), 30.5 (CH_2), 21.4 (CH_3). HRMS (ESI): *m/z* Calcd for $C_{16}H_{16}O + H^+$ $[M + H]^+$ 225.1274; Found 225.1271. The ^{1}H and $^{13}C\{^1H\}$ spectra are consistent with those reported in the literature.[78]

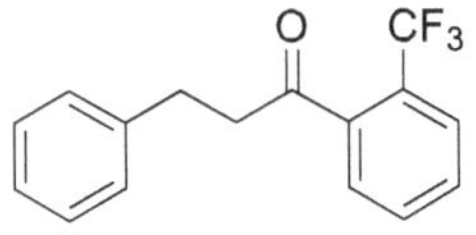

3-Phenyl-1-(2-(trifluoromethyl)phenyl)propan-1-one (13): The representative procedure was followed, using substrate **13a** (0.056 g, 0.203 mmol) and the reaction mixture was stirred at room temperature (27 °C) for 2 h. Purification by column chromatography on silica gel (petroleum ether/EtOAc: 50/1) yielded **13** (0.048 g, 85%) as yellow oil. ^{1}H-NMR (400 MHz, $CDCl_3$): δ = 7.69 (d, J = 7.8 Hz, 1H, Ar–H), 7.57-7.50 (m, 2H, Ar–H), 7.31-7.27 (m, 3H, Ar–H), 7.24-7.18 (m, 3H, Ar–H), 3.16 (t, J = 7.6 Hz, 2H, CH_2), 3.05 (t, J = 7.4 Hz, 2H, CH_2). $^{13}C\{^1H\}$-NMR (100 MHz, $CDCl_3$): δ = 203.6 (CO), 140.8 (C_q), 140.5 (q, $^3J_{C-F}$ = 1.9 Hz, C_q), 132.0 (CH), 130.2 (CH), 128.7 (2C, CH), 128.5 (2C, CH), 127.1 (q, $^2J_{C-F}$ = 32.1 Hz, C_q), 127.0 (CH), 126.9 (q, $^3J_{C-F}$ = 5.0 Hz, CH), 126.4 (CH), 122.4 (q, $^1J_{C-F}$ = 273.9 Hz, C_q), 45.0 (q, $^5J_{C-F}$ = 1.5 Hz, CH_2), 30.0 (CH_2). ^{19}F-NMR (377 MHz, $CDCl_3$): δ = −58.1 (s). HRMS (ESI): *m/z* Calcd for $C_{16}H_{13}F_3O + H^+$ $[M + H]^+$ 279.0991; Found 279.0986. The ^{1}H and $^{13}C\{^1H\}$ spectra are consistent with those reported in the literature.[78]

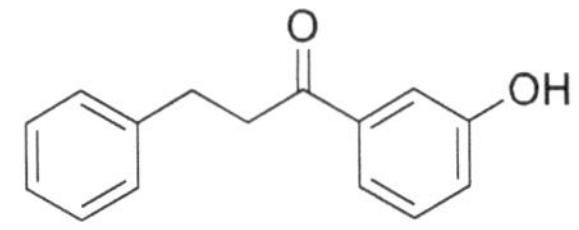

1-(3-Hydroxyphenyl)-3-phenylpropan-1-one (14): The representative procedure was followed, using substrate **14a** (0.045 g, 0.201 mmol) and the reaction mixture was stirred at 50 °C for 24 h. Purification by column chromatography on silica gel (petroleum ether/EtOAc: 10/1) yielded **14** (0.015 g, 33%) as a white solid. ^{1}H-NMR (400 MHz, $CDCl_3$): δ = 7.53 (s, 1H, Ar–H), 7.50 (d, J = 7.9 Hz, 1H, Ar–H), 7.34-7.27 (m, 3H, Ar–H), 7.24-7.18 (m, 3H, Ar–H), 7.07 (d, J = 7.9 Hz, 1H, Ar–H), 6.21-6.09 (br s, 1H, OH), 3.28 (t, J = 7.3 Hz, 2H, CH_2), 3.05 (t, J = 7.3 Hz, 2H, CH_2). $^{13}C\{^1H\}$-NMR (100 MHz, $CDCl_3$): δ = 200.0 (CO), 156.4 (C_q), 141.3 (C_q), 138.4 (C_q), 130.1 (CH), 128.7 (2C, CH), 128.6 (2C, CH), 126.4 (CH), 120.9 (CH), 120.8 (CH), 114.7 (CH), 40.8 (CH_2), 30.3 (CH_2). HRMS (ESI): *m/z* Calcd for $C_{15}H_{14}O_2 + H^+$ $[M + H]^+$ 227.1067; Found 227.1065.[79]

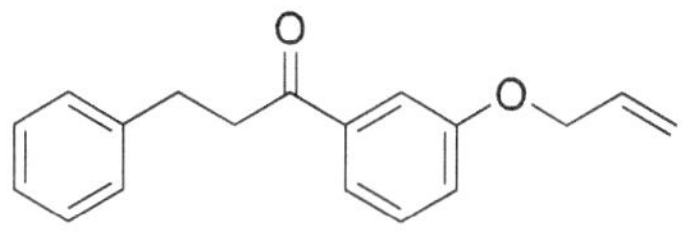

1-(3-(Allyloxy)phenyl)-3-phenylpropan-1-one (15): The representative procedure was followed, using substrate **15a** (0.053 g, 0.201 mmol) and the reaction mixture was stirred at room temperature (27 °C) for 3 h. Purification by column chromatography on silica gel (petroleum ether/EtOAc: 50/1) yielded **15** (0.039 g, 73%) as yellow oil. ^{1}H-NMR (400 MHz, $CDCl_3$): δ = 7.54 (d, J = 7.8 Hz, 1H, Ar–H), 7.51-7.50 (m, 1H, Ar–H), 7.35 (t, J = 7.9 Hz, 1H, Ar–H), 7.30 (t, J = 7.3 Hz, 2H, Ar–H), 7.26-7.19 (m, 3H, Ar–H), 7.12 (dd, J = 8.3, 2.0 Hz, 1H, Ar–H), 6.10-6.01 (m, 1H, CH), 5.43 (dd, J = 17.3, 1.5 Hz, 1H, CH), 5.31 (dd, J = 10.6, 1.4 Hz, 1H, CH), 4.57 (d, J = 5.3 Hz, 2H, CH_2), 3.28 (t, J = 7.7 Hz, 2H, CH_2), 3.06 (t, J = 7.6 Hz, 2H, CH_2). $^{13}C\{^{1}H\}$-NMR (100 MHz, $CDCl_3$): δ = 199.4 (CO), 159.3 (C_q), 141.7 (C_q), 138.7 (C_q), 133.3 (CH), 130.1 (CH), 129.0 (2C, CH), 129.0 (2C, CH), 126.6 (CH), 121.3 (CH), 120.6 (CH), 118.4 (CH_2), 113.8 (CH), 69.4 (CH_2), 41.0 (CH_2), 30.6 (CH_2). HRMS (ESI): m/z Calcd for $C_{18}H_{18}O_2 + H^+$ $[M + H]^+$ 267.1380; Found 267.1373.

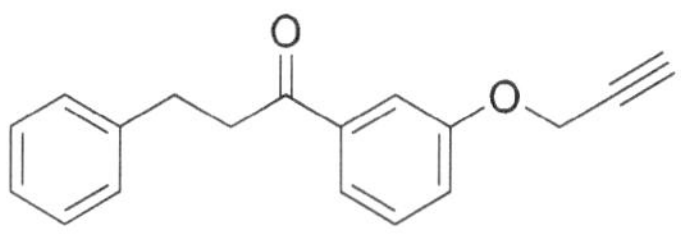

3-Phenyl-1-(3-(prop-2-yn-1-yloxy)phenyl)propan-1-one (16): The representative procedure was followed, using substrate **16a** (0.053 g, 0.202 mmol) and the reaction mixture was stirred at room temperature (27 °C) for 3 h. Purification by column chromatography on silica gel (petroleum ether/EtOAc: 50/1) yielded **16** (0.037 g, 69%) as yellow oil. ^{1}H-NMR (400 MHz, $CDCl_3$): δ = 7.57 (d, J = 8.1 Hz, 2H, Ar–H), 7.38 (t, J = 7.9 Hz, 1H, Ar–H), 7.30 (t, J = 7.4 Hz, 2H, Ar–H), 7.26-7.16 (m, 4H, Ar–H), 4.73 (d, J = 2.3 Hz, 2H, CH_2), 3.28 (t, J = 7.7 Hz, 2H, CH_2), 3.06 (t, J = 7.6 Hz, 2H, CH_2), 2.53 (t, J = 2.2 Hz, 1H, CH). $^{13}C\{^{1}H\}$-NMR (100 MHz, $CDCl_3$): δ = 198.9 (CO), 157.9 (C_q), 141.4 (C_q), 138.4 (C_q), 129.8 (CH), 128.7 (2C, CH), 128.6 (2C, CH), 126.3 (CH), 121.6 (CH), 120.4 (CH), 113.7 (CH), 78.2 (C_q), 76.1 (CH), 56.1 (CH_2), 40.7 (CH_2), 30.3 (CH_2). HRMS (ESI): m/z Calcd for $C_{18}H_{16}O_2 + H^+$ $[M + H]^+$ 265.1223; Found 265.1217.

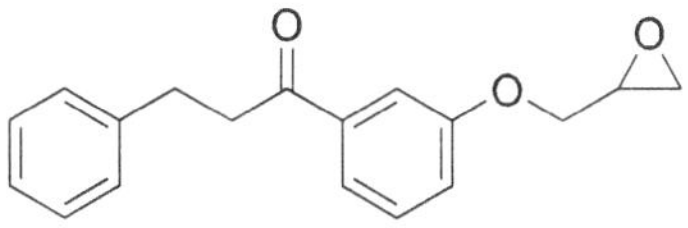

1-(3-(Oxiran-2-ylmethoxy)phenyl)-3-phenylpropan-1-one (17): The representative procedure was followed, using substrate **17a** (0.057 g, 0.203 mmol) and the reaction mixture was stirred at room temperature (27 °C) for 4 h. Purification by column chromatography on silica gel (petroleum ether/EtOAc: 30/1) yielded **17** (0.039 g, 68%) as yellow oil. ^{1}H-NMR (400 MHz, $CDCl_3$): δ = 7.56 (d, J = 7.8 Hz, 1H, Ar–H), 7.50 (t, J = 1.9 Hz, 1H, Ar–H), 7.36 (t, J = 7.9 Hz, 1H, Ar–H), 7.30 (t, J = 7.3 Hz, 2H, Ar–H), 7.26-7.19 (m, 3H, Ar–H), 7.14 (dd, J = 2.5, 0.6 Hz, 1H, Ar–H), 4.30 (dd, J = 11.0, 2.9 Hz, 1H, CH), 3.97 (dd, J = 11.0, 5.9 Hz, 1H, CH), 3.39-3.35 (m, 1H, CH), 3.29 (t, J = 7.6 Hz, 2H, CH_2), 3.06 (t, J = 7.6 Hz, 2H, CH_2), 2.92 (t, J = 4.5 Hz, 1H, CH_2), 2.67 (dd, J = 4.9 Hz, 2.63 Hz, 1H, CH_2). $^{13}C\{^{1}H\}$-NMR (100 MHz, $CDCl_3$): δ = 199.1 (CO), 158.9 (C_q), 141.4 (C_q), 138.4 (C_q), 129.9 (CH), 128.7 (2C, CH), 128.6 (2C, CH), 126.3 (CH), 121.4 (CH), 120.3 (CH), 113.2 (CH), 69.1 (CH_2), 50.2 (CH), 44.7 (CH_2), 40.7 (CH_2), 30.3 (CH_2). HRMS (ESI): m/z Calcd for $C_{18}H_{18}O_3 + H^+$ $[M + H]^+$ 283.1329; Found 283.1223.

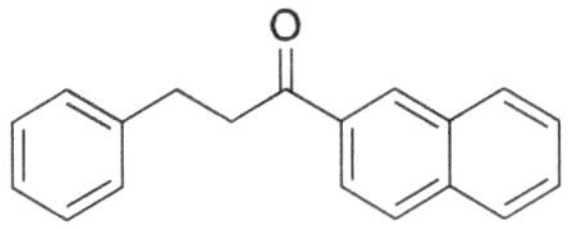

1-(Naphthalen-2-yl)-3-phenylpropan-1-one (18): The representative procedure was followed, using substrate **18a** (0.052 g, 0.201 mmol) and the reaction mixture was stirred at room temperature (27 °C) for 3 h. Purification by column chromatography on silica gel (petroleum ether/EtOAc: 50/1) yielded **18** (0.037 g, 71%) as brown solid. ^{1}H-NMR (500 MHz, $CDCl_3$): δ = 8.44 (s, 1H, Ar–H), 8.02 (dd, J = 8.6, 1.6 Hz, 1H, Ar–H), 7.92-7.84 (m, 3H, Ar–H), 7.59-7.50 (m, 2H, Ar–H), 7.33-7.19 (m, 5H, Ar–H), 3.42 (t, J = 7.8 Hz, 2H, CH_2), 3.12 (t, J = 7.7 Hz, 2H, CH_2). $^{13}C\{^{1}H\}$-NMR (125 MHz, $CDCl_3$): δ = 199.3 (CO), 141.5 (C_q), 135.7 (C_q), 134.3 (C_q), 132.7 (C_q), 129.8 (CH), 129.7 (CH), 128.7 (2C, CH), 128.6 (2C, CH), 128.6 (CH), 128.6 (CH), 127.9 (CH), 126.9 (CH), 126.3 (CH), 124.0 (CH), 40.7 (CH_2), 30.4 (CH_2). HRMS (ESI): m/z Calcd for $C_{19}H_{16}O + H^+$ $[M + H]^+$ 261.1274; Found 261.1270. The ^{1}H and $^{13}C\{^{1}H\}$ spectra are consistent with those reported in the literature.[80]

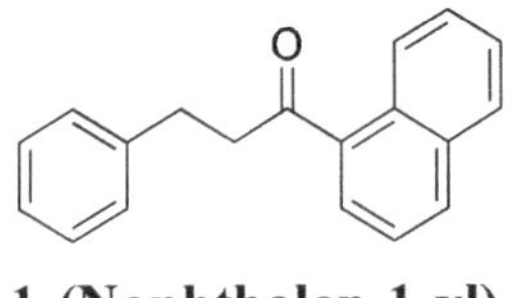

1-(Naphthalen-1-yl)-3-phenylpropan-1-one (19): The representative procedure was followed, using substrate **19a** (0.052 g, 0.201 mmol) and the reaction mixture was stirred at room temperature (27 °C) for 3 h. Purification by column chromatography on silica gel (petroleum ether/EtOAc: 50/1) yielded **19** (0.049 g, 94%) as white solid. ^{1}H-NMR (400 MHz, $CDCl_3$): δ = 8.54 (d, J = 8.5 Hz, 1H, Ar–H), 7.94 (d, J = 8.3 Hz, 1H, Ar–H), 7.85 (d, J = 8.3 Hz, 1H, Ar–H), 7.79 (dd, J = 7.1, 0.9 Hz, 1H, Ar–H), 7.58-7.49 (m, 2H, Ar–H), 7.44 (t, J = 7.7 Hz, 1H, Ar–H), 7.30-7.17 (m, 5H, Ar–H), 3.36 (t, J = 7.7 Hz, 2H, CH_2), 3.12 (t, J = 7.6 Hz, 2H, CH_2). $^{13}C\{^{1}H\}$-NMR (100 MHz, $CDCl_3$): δ = 203.7

(CO), 141.3 (C_q), 136.1 (C_q), 134.1 (C_q), 132.7 (CH), 130.3 (C_q), 128.7 (2C, CH), 128.6 (2C, CH), 128.6 (CH), 128.0 (CH), 127.6 (CH), 126.6 (CH), 126.3 (CH), 125.9 (CH), 124.5 (CH), 44.0 (CH_2), 30.7 (CH_2). HRMS (ESI): *m/z* Calcd for $C_{19}H_{16}O + H^+$ $[M + H]^+$ 261.1274; Found 261.1270. The ^1H and $^{13}C\{^1H\}$ spectra are consistent with those reported in the literature.[81]

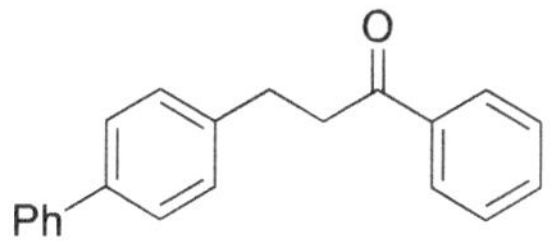

3-([1,1'-Biphenyl]-4-yl)-1-phenylpropan-1-one (20): The representative procedure was followed, using substrate **20a** (0.057 g, 0.20 mmol) and the reaction mixture was stirred at room temperature (27 °C) for 2 h. Purification by column chromatography on silica gel (petroleum ether/EtOAc: 50/1) yielded **20** (0.051 g, 89%) as yellow solid. ^1H-NMR (500 MHz, $CDCl_3$): $\delta = 8.01$ (d, $J = 7.1$ Hz, 2H, Ar–H), 7.63-7.56 (m, 5H, Ar–H), 7.50-7.44 (m, 4H, Ar–H), 7.38-7.34 (m, 3H, Ar–H), 3.37 (t, $J = 7.7$ Hz, 2H, CH_2), 3.15 (t, $J = 7.6$ Hz, 2H, CH_2). $^{13}C\{^1H\}$-NMR (125 MHz, $CDCl_3$): $\delta = 199.3$ (CO), 141.1 (C_q), 140.6 (C_q), 139.3 (C_q), 137.0 (C_q), 133.2 (CH), 129.0 (2C, CH), 128.9 (2C, CH), 128.8 (2C, CH), 128.2 (2C, CH), 127.4 (2C, CH), 127.3 (CH), 127.1 (2C, CH), 40.5 (CH_2), 29.9 (CH_2). HRMS (ESI): *m/z* Calcd for $C_{21}H_{18}O + H^+$ $[M + H]^+$ 287.1430; Found 287.1425. The ^1H and $^{13}C\{^1H\}$ spectra are consistent with those reported in the literature.[74]

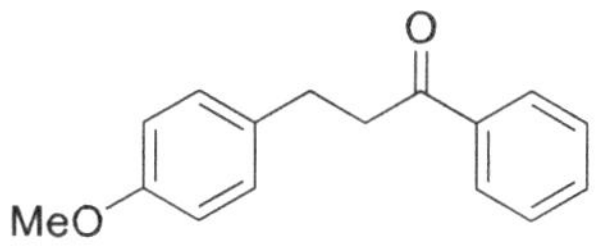

3-(4-Methoxyphenyl)-1-phenylpropan-1-one (21): The representative procedure was followed, using substrate **21a** (0.048 g, 0.201 mmol) and the reaction mixture was stirred at room temperature (27 °C) for 3 h. Purification by column chromatography on silica gel (petroleum ether/EtOAc: 30/1) yielded **21** (0.044 g, 91%) as yellow solid. ^1H-NMR (400 MHz, $CDCl_3$): $\delta = 7.98$-7.95 (m, 2H, Ar–H), 7.56 (vt, $J = 7.4$ Hz, 1H, Ar–H), 7.48-7.43 (m, 2H, Ar–H), 7.18 (d, $J = 8.8$ Hz, 2H, Ar–H), 6.85 (d, $J = 8.8$ Hz, 2H, Ar–H), 3.79 (s, 3H, CH_3), 3.28 (t, $J = 7.7$ Hz, 2H, CH_2), 3.02 (t, $J = 7.6$ Hz, 2H, CH_2). $^{13}C\{^1H\}$-NMR (100 MHz, $CDCl_3$): $\delta = 199.5$ (CO), 158.1 (C_q), 137.0 (C_q), 133.5 (C_q), 133.2 (CH), 129.5 (2C, CH), 128.7 (2C, CH), 128.2 (2C, CH), 114.1 (2C, CH), 55.4 (CH_3), 40.9 (CH_2), 29.4 (CH_2). HRMS (ESI): *m/z* Calcd for $C_{16}H_{16}O_2 + H^+$ $[M + H]^+$ 241.1223; Found 241.1218. The ^1H and $^{13}C\{^1H\}$ spectra are consistent with those reported in the literature.[74]

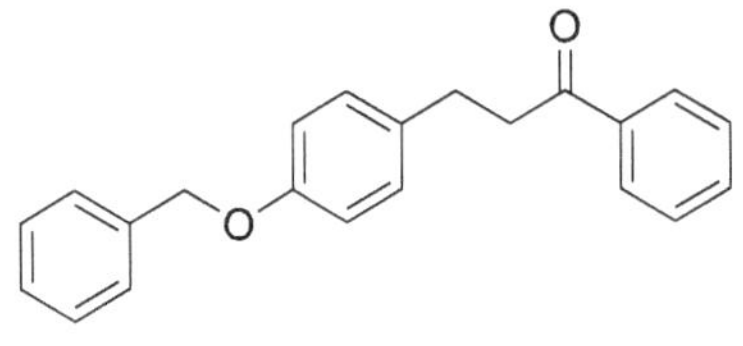

3-(4-(Benzyloxy)phenyl)-1-phenylpropan-1-one (22): The representative procedure was followed, using substrate **22a** (0.063 g, 0.20 mmol), 1.0 mL of MeOH:DCM (4:1) and the reaction mixture was stirred at room temperature (27 °C) for 5 h. Purification by column chromatography on silica gel (petroleum ether/EtOAc: 30/1) yielded **22** (0.046 g, 73%) as yellow oil. ^{1}H-NMR (400 MHz, $CDCl_3$): δ = 7.96 (d, J = 7.7 Hz, 2H, Ar–H), 7.56 (vt, J = 7.4 Hz, 1H, Ar–H), 7.48-7.31 (m, 7H, Ar–H), 7.18 (d, J = 8.8 Hz, 2H, Ar–H), 6.92 (d, J = 8.6 Hz, 2H, Ar–H), 5.05 (s, 2H, CH_2), 3.28 (t, J = 7.6 Hz, 2H, CH_2), 3.02 (t, J = 7.7 Hz, 2H, CH_2). $^{13}C\{^1H\}$-NMR (100 MHz, $CDCl_3$): δ = 199.6 (CO), 157.4 (C_q), 137.3 (C_q), 137.1 (C_q), 133.8 (C_q), 133.2 (CH), 129.6 (2C, CH), 128.8 (2C, CH), 128.7 (2C, CH), 128.2 (2C, CH), 128.1 (CH), 127.6 (2C, CH), 115.1 (2C, CH), 70.2 (CH_2), 40.9 (CH_2), 29.5 (CH_2). HRMS (ESI): *m/z* Calcd for $C_{22}H_{20}O_2 + H^+$ $[M + H]^+$ 317.1536; Found 317.1531. The ^{1}H and $^{13}C\{^1H\}$ spectra are consistent with those reported in the literature.[82]

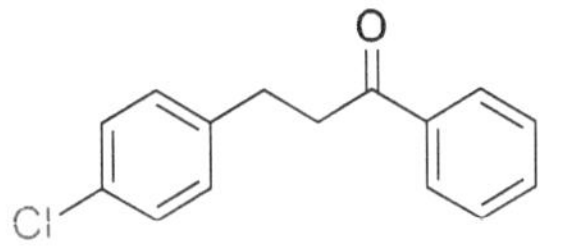

3-(4-Chlorophenyl)-1-phenylpropan-1-one (23): The representative procedure was followed, using substrate **23a** (0.049 g, 0.20 mmol) and the reaction mixture was stirred at room temperature (27 °C) for 2 h. Purification by column chromatography on silica gel (petroleum ether/EtOAc: 50/1) yielded **23** (0.039 g, 80%) as yellow solid. ^{1}H-NMR (400 MHz, $CDCl_3$): δ = 7.94 (d, J = 7.4 Hz, 2H, Ar–H), 7.55 (t, J = 7.4 Hz, 1H, Ar–H), 7.45 (t, J = 7.6 Hz, 2H, Ar–H), 7.25 (d, J = 8.4 Hz, 2H, Ar–H), 7.17 (d, J = 8.4 Hz, 2H, Ar–H), 3.27 (t, J = 7.6 Hz, 2H, CH_2), 3.03 (t, J = 7.6 Hz, 2H, CH_2). $^{13}C\{^1H\}$-NMR (100 MHz, $CDCl_3$): δ = 199.0 (CO), 139.9 (C_q), 136.9 (C_q), 133.3 (CH), 132.0 (C_q), 130.0 (2C, CH), 128.8 (2C, CH), 128.7 (2C, CH), 128.2 (2C, CH), 40.3 (CH_2), 29.5 (CH_2). HRMS (ESI): *m/z* Calcd for $C_{15}H_{13}ClO - H^+$ $[M - H]^+$ 243.0571; Found 243.0570. The ^{1}H and $^{13}C\{^1H\}$ spectra are consistent with those reported in the literature.[78]

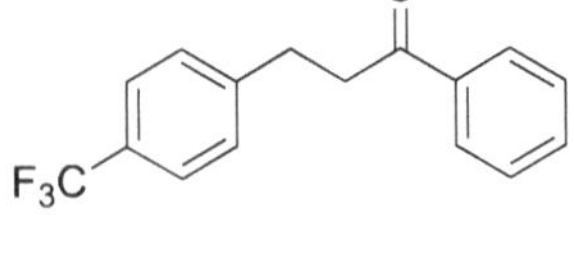

1-Phenyl-3-(4-(trifluoromethyl)phenyl)propan-1-one (24): The representative procedure was followed, using substrate **24a** (0.056 g, 0.203 mmol) and the reaction mixture was stirred at room temperature (27 °C) for 2 h. Purification by column chromatography on silica gel (petroleum ether/EtOAc: 50/1) yielded **24** (0.054 g, 96%) as a white solid. ^{1}H-NMR (500 MHz, $CDCl_3$): δ = 7.96 (d, J = 7.8 Hz, 2H, Ar–H), 7.59-7.54 (m, 3H, Ar–H), 7.46 (t, J = 7.6 Hz, 2H, Ar–H), 7.38 (d, J = 8.0 Hz, 2H, Ar–H), 3.33 (t, J = 7.6 Hz, 2H, CH_2), 3.14 (t, J = 7.5 Hz, 2H, CH_2). $^{13}C\{^{1}H\}$-NMR (125 MHz, $CDCl_3$): δ = 198.7 (CO), 145.6 (q, $^{5}J_{C-F}$ = 1.5 Hz, C_q), 136.8 (C_q), 133.4 (CH), 129.0 (2C, CH), 128.8 (2C, CH), 128.7 (q, $^{2}J_{C-F}$ = 32.1 Hz, C_q), 128.2 (2C, CH), 125.6 (q, $^{3}J_{C-F}$ = 3.8 Hz, 2C, CH), 124.5 (q, $^{1}J_{C-F}$ = 271.1 Hz, CF_3), 40.0 (CH_2), 29.9 (CH_2). ^{19}F-NMR (377 MHz, $CDCl_3$): δ = −62.4 (s). HRMS (ESI): *m/z* Calcd for $C_{16}H_{13}F_3O + H^+$ $[M + H]^+$ 279.0991; Found 279.0985. The ^{1}H and $^{13}C\{^{1}H\}$ spectra are consistent with those reported in the literature.[74]

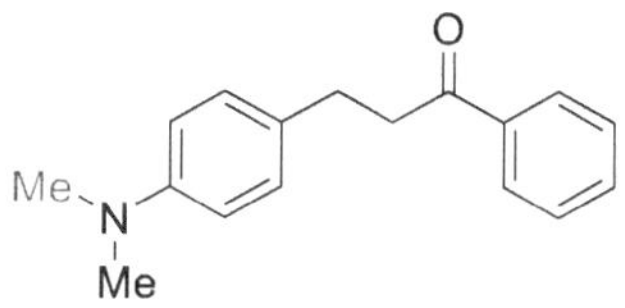

3-(4-(Dimethylamino)phenyl)-1-phenylpropan-1-one (25): The representative procedure was followed, using substrate **25a** (0.051 g, 0.203 mmol), 1.0 mL of MeOH:DCM (4:1) and the reaction mixture was stirred at room temperature (27 °C) for 5 h. Purification by column chromatography on silica gel (petroleum ether/EtOAc: 30/1) yielded **25** (0.041 g, 80%) as yellow oil. ^{1}H-NMR (400 MHz, $CDCl_3$): δ = 7.97 (d, J = 7.8 Hz, 2H, Ar–H), 7.56 (t, J = 7.3 Hz, 1H, Ar–H), 7.46 (t, J = 7.6 Hz, 2H, Ar–H), 7.15 (d, J = 8.63 Hz, 2H, Ar–H), 6.72 (d, J = 8.8 Hz, 2H, Ar–H), 3.27 (t, J = 7.8 Hz, 2H, CH_2), 2.99 (t, J = 7.8 Hz, 2H, CH_2), 2.93 (s, 6H, CH_3). $^{13}C\{^{1}H\}$-NMR (100 MHz, $CDCl_3$): δ = 199.9 (CO), 149.4 (C_q), 137.1 (C_q), 133.1 (CH), 129.4 (C_q), 129.2 (2C, CH), 128.7 (2C, CH), 128.2 (2C, CH), 113.2 (2C, CH), 41.1 (CH_2), 41.0 (CH_3), 29.4 (CH_2). HRMS (ESI): *m/z* Calcd for $C_{17}H_{19}NO + H^+$ $[M + H]^+$ 254.1539; Found 254.1536. The ^{1}H and $^{13}C\{^{1}H\}$ spectra are consistent with those reported in the literature.[76]

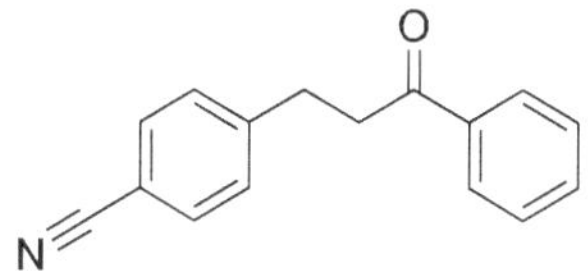

4-(3-Oxo-3-phenylpropyl)benzonitrile (26): The representative procedure was followed, using substrate **26a** (0.047 g, 0.201 mmol) and the reaction mixture was stirred at room temperature (27 °C) for 2 h. Purification by column chromatography on silica gel (petroleum ether/EtOAc: 30/1) yielded **26** (0.042 g, 89%) as white solid. ^{1}H-NMR (500 MHz, $CDCl_3$): δ = 7.94 (d, J = 7.2 Hz, 2H,

Ar–H), 7.54-7.58 (m, 3H, Ar–H), 7.54 (t, J = 7.7 Hz, 2H, Ar–H), 7.36 (d, J = 8.3 Hz, 2H, Ar–H), 3.32 (t, J = 7.4 Hz, 2H, CH_2), 3.13 (t, J = 7.4 Hz, 2H, CH_2). $^{13}C\{^1H\}$-NMR (125 MHz, $CDCl_3$): δ = 198.4 (CO), 147.1 (C_q), 136.7 (C_q), 133.5 (CH), 132.4 (2C, CH), 129.5 (2C, CH), 128.8 (2C, CH), 128.1 (2C, CH), 119.1 (C_q), 110.2 (C_q), 39.5 (CH_2), 30.1 (CH_2). HRMS (ESI): m/z Calcd for $C_{16}H_{13}NO + H^+$ $[M + H]^+$ 236.1070; Found 236.1067. The ^1H and $^{13}C\{^1H\}$ spectra are consistent with those reported in the literature.[83]

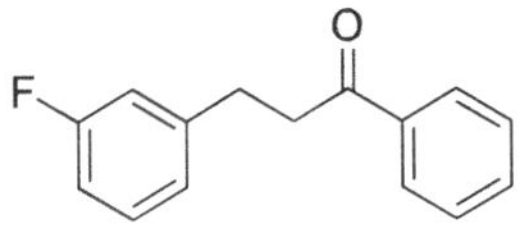

3-(3-Fluorophenyl)-1-phenylpropan-1-one (27): The representative procedure was followed, using substrate **27a** (0.046 g, 0.203 mmol) and the reaction mixture was stirred at room temperature (27 °C) for 3 h. Purification by column chromatography on silica gel (petroleum ether/EtOAc: 50/1) yielded **27** (0.038 g, 82%) as yellow solid. ^1H-NMR (400 MHz, $CDCl_3$): δ = 7.96 (d, J = 7.8 Hz, 2H, Ar–H), 7.57 (tt, J = 7.4, 1.9 Hz, 1H, Ar–H), 7.46 (t, J = 7.6 Hz, 2H, Ar–H), 7.28-7.23 (m, 1H, Ar–H), 7.03 (d, J = 7.6 Hz, 1H, Ar–H), 6.98-6.95 (m, 1H, Ar–H), 6.90 (td, J = 8.5, 2.4 Hz, 1H, Ar–H), 3.31 (t, J = 7.6 Hz, 2H, CH_2), 3.08 (t, J = 7.6 Hz, 2H, CH_2). $^{13}C\{^1H\}$-NMR (100 MHz, $CDCl_3$): δ = 198.9 (CO), 163.1 (d, $^1J_{C-F}$ = 245.4 Hz, C_q), 144.0 (d, $^3J_{C-F}$ = 6.9 Hz, C_q), 136.9 (C_q), 133.3 (CH), 130.1 (d, $^3J_{C-F}$ = 8.4 Hz, CH), 128.8 (2C, CH), 128.2 (2C, CH), 124.2 (d, $^4J_{C-F}$ = 3.1 Hz, CH), 115.5 (d, $^2J_{C-F}$ = 20.6 Hz, CH), 113.2 (d, $^2J_{C-F}$ = 21.4 Hz, CH), 40.1 (CH_2), 29.9 (d, $^4J_{C-F}$ = 1.5 Hz, CH_2). ^{19}F-NMR (377 MHz, $CDCl_3$): δ = −113.5 (s). HRMS (ESI): m/z Calcd for $C_{15}H_{13}FO + H^+$ $[M + H]^+$ 229.1023; Found 229.1021. The ^1H and $^{13}C\{^1H\}$ spectra are consistent with those reported in the literature.[84]

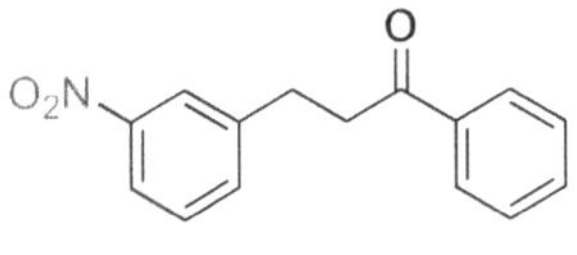

3-(3-Nitrophenyl)-1-phenylpropan-1-one (28): The representative procedure was followed, using substrate **28a** (0.051 g, 0.201 mmol) and the reaction mixture was stirred at room temperature (27 °C) for 5 h. Purification by column chromatography on silica gel (petroleum ether/EtOAc: 50/1) yielded **28** (0.045 g, 88%) as a white solid. ^1H-NMR (400 MHz, $CDCl_3$): δ = 8.12 (s, 1H, Ar–H), 8.05 (d, J = 8.1 Hz, 1H, Ar–H), 7.95 (d, J = 7.8 Hz, 2H, Ar–H), 7.61 (d, J = 7.8 Hz, 1H, Ar–H), 7.57 (t, J = 7.4 Hz, 1H, Ar–H), 7.48-7.43 (m, 3H, Ar–H), 3.37 (t, 7.4 Hz, 2H, CH_2), 3.19 (t, J = 7.3 Hz,

2H, CH_2). $^{13}C\{^1H\}$-NMR (100 MHz, $CDCl_3$): δ = 198.4 (CO), 148.5 (C_q), 143.5 (C_q), 136.7 (C_q), 135.1 (CH), 133.5 (CH), 129.5 (CH), 128.9 (2C, CH), 128.1 (2C, CH), 123.4 (CH), 121.5 (CH), 39.7 (CH_2), 29.6 (CH_2). HRMS (ESI): *m/z* Calcd for $C_{15}H_{13}NO_3 + H^+$ $[M + H]^+$ 256.0973; Found 256.0961. The ^1H and $^{13}C\{^1H\}$ spectra are consistent with those reported in the literature.[84]

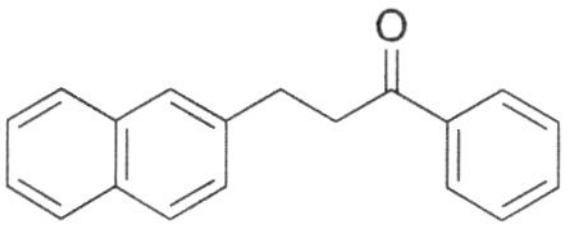

3-(Naphthalen-2-yl)-1-phenylpropan-1-one (29): The representative procedure was followed, using substrate **29a** (0.052 g, 0.201 mmol), 1.0 mL of MeOH:DCM (4:1)] and the reaction mixture was stirred at room temperature (27 °C) for 5 h. Purification by column chromatography on silica gel (petroleum ether/EtOAc: 50/1) yielded **29** (0.050 g, 96%) as a white solid. ^1H-NMR (400 MHz, $CDCl_3$): δ = 7.99 (d, J = 8.3 Hz, 2H, Ar–H), 7.81 (t, J = 7.8 Hz, 3H, Ar–H), 7.70 (s, 1H, Ar–H), 7.56 (t, J = 7.4 Hz, 1H, Ar–H), 7.49-7.39 (m, 5H, Ar–H), 3.40 (t, J = 7.7 Hz, 2H, CH_2), 3.25 (t, J = 7.7 Hz, 2H, CH_2) $^{13}C\{^1H\}$-NMR (100 MHz, $CDCl_3$): δ = 199.4 (CO), 139.0 (C_q), 137.0 (C_q), 133.8 (C_q), 133.3 (CH), 132.3 (C_q), 128.8 (2C, CH), 128.3 (CH), 128.2 (2C, CH), 127.8 (CH), 127.6 (CH), 127.4 (CH), 126.7 (CH), 126.2 (CH), 125.5 (CH), 40.5 (CH_2), 30.4 (CH_2). HRMS (ESI): *m/z* Calcd for $C_{19}H_{16}O + H^+$ $[M + H]^+$ 261.1274; Found 261.1270. The ^1H and $^{13}C\{^1H\}$ spectra are consistent with those reported in the literature.[81]

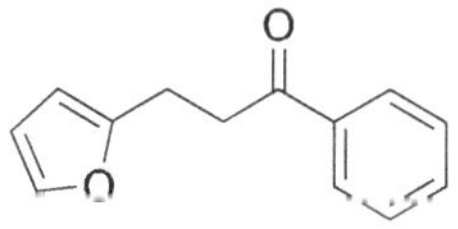

3-(Furan-2-yl)-1-phenylpropan-1-one (30): The representative procedure was followed, using substrate **30a** (0.040 g, 0.202 mmol) and the reaction mixture was stirred at room temperature (27 °C) for 2 h. Purification by column chromatography on silica gel (petroleum ether/EtOAc: 30/1) yielded **30** (0.035 g, 87%) as yellow oil. ^1H-NMR (400 MHz, $CDCl_3$): δ = 7.98 (d, J = 7.8 Hz, 2H, Ar–H), 7.56 (tt, J = 7.4, 1.9 Hz, 1H, Ar–H), 7.46 (t, J = 7.6 Hz, 2H, Ar–H), 7.31 (d, J = 1.8 Hz, 1H, Ar–H), 6.29 (dd, J = 3.1, 1.9 Hz, 1H, Ar–H), 6.05 (dd, J = 3.1, 0.8 Hz, 1H, Ar–H), 3.34 (t, J = 7.4 Hz, 2H, CH_2), 3.10 (t, J = 7.5 Hz, 2H, CH_2). $^{13}C\{^1H\}$-NMR (100 MHz, $CDCl_3$): δ = 198.8 (CO), 154.9 (C_q), 141.2 (CH), 136.9 (C_q), 133.3 (CH), 128.8 (2C, CH), 128.2 (2C, CH), 110.4 (CH), 105.5 (CH), 37.1 (CH_2), 22.6 (CH_2). HRMS (ESI): *m/z* Calcd for $C_{13}H_{12}O_2 + H^+$ $[M + H]^+$ 201.0910; Found 201.0908. The ^1H and $^{13}C\{^1H\}$ spectra are consistent with those reported in the literature.[85]

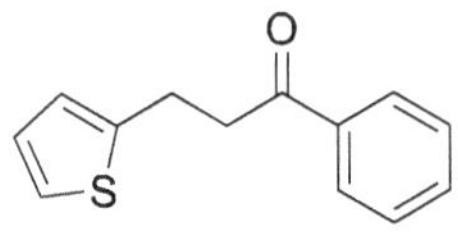

1-Phenyl-3-(thiophen-2-yl)propan-1-one (31): The representative procedure was followed, using substrate **31a** (0.043 g, 0.201 mmol) and the reaction mixture was stirred at room temperature (27 °C) for 3 h. Purification by column chromatography on silica gel (petroleum ether/EtOAc: 30/1) yielded **31** (0.035 g, 81%) as green oil. ^{1}H-NMR (400 MHz, $CDCl_3$): δ = 7.96 (d, J = 7.4 Hz, 2H, Ar–H), 7.56 (t, J = 7.4 Hz, 1H, Ar–H), 7.45 (t, J = 7.7 Hz, 2H, Ar–H), 7.11 (d, J = 5.1 Hz, 1H, Ar–H), 6.91 (dd, J = 5.0, 3.5 Hz, 1H, Ar–H), 6.86 (d, J = 3.3 Hz, 1H, Ar–H), 3.38-3.34 (m, 2H, CH_2), 3.31-3.27 (m, 2H, CH_2). $^{13}C\{^{1}H\}$-NMR (100 MHz, $CDCl_3$): δ = 198.7 (CO) 144.0 (C_q), 136.9 (C_q), 133.3 (CH), 128.8 (2C, CH), 128.2 (2C, CH), 127.0 (CH), 124.8 (CH), 123.5 (CH), 40.7 (CH_2), 24.4 (CH_2). The ^{1}H and $^{13}C\{^{1}H\}$ spectra are consistent with those reported in the literature.[74]

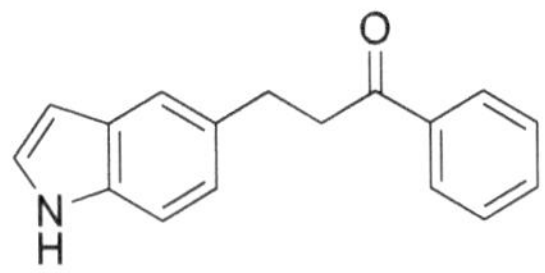

3-(1*H*-Indol-5-yl)-1-phenylpropan-1-one (32): The representative procedure was followed, using substrate **32a** (0.050 g, 0.202 mmol), 1.0 mL of MeOH:DCM (4:1) and the reaction mixture was stirred at room temperature (27 °C) for 5 h. Purification by column chromatography on silica gel (petroleum ether/EtOAc: 20/1) yielded **32** (0.036 g, 72%) as a yellow solid. ^{1}H-NMR (400 MHz, $CDCl_3$): δ = 8.19 (br s, 1H, *N*H), 7.99 (d, J = 7.4 Hz, 2H, Ar–H), 7.58-7.53 (m, 2H, Ar–H), 7.46 (t, J = 7.6 Hz, 2H, Ar–H), 7.33 (d, J = 8.4 Hz, 1H, Ar–H), 7.19 (t, J = 2.8 Hz, 1H, Ar–H), 7.11 (dd, J = 8.4, 1.6 Hz, 1H, Ar–H), 6.51 (t, J = 8.4 Hz, 1H, Ar–H), 3.36 (d, J = 7.8 Hz, 2H, CH_2), 3.18 (t, J = 7.7 Hz, 2H, CH_2). $^{13}C\{^{1}H\}$-NMR (100 MHz, $CDCl_3$): δ = 200.0 (CO), 137.1 (C_q), 134.7 (C_q), 133.1 (CH), 132.7 (C_q), 128.7 (2C, CH), 128.3 (C_q), 128.2 (2C, CH), 124.7 (CH), 123.0 (CH), 120.0 (CH), 111.2 (CH), 102.4 (CH), 41.7 (CH_2), 30.5 (CH_2). HRMS (ESI): m/z Calcd for $C_{17}H_{15}NO + H^+$ $[M + H]^+$ 250.1226; Found 250.1225.

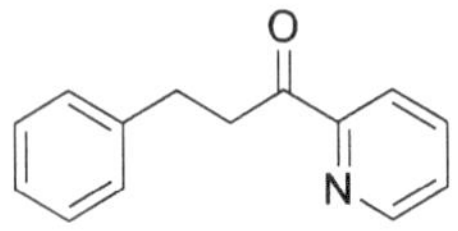

3-Phenyl-1-(pyridin-2-yl)propan-1-one (33): The representative procedure was followed, using substrate **33a** (0.042 g, 0.201 mmol) and the reaction mixture was stirred at room temperature (27 °C) for 2 h. Purification by column chromatography on silica gel (petroleum ether/EtOAc: 30/1) yielded **33** (0.026 g, 61%) as a green oil. ^{1}H-NMR (400 MHz, $CDCl_3$): δ = 8.66 (d, J = 4.6 Hz, 1H,

Ar–H), 8.05 (d, J = 7.8 Hz, 1H, Ar–H), 7.83 (td, J = 7.8, 1.7 Hz, 1H, Ar–H), 7.47-7.44 (m, 1H, Ar–H), 7.28 (d, J = 4.5 Hz, 4H, Ar–H), 7.21-7.17 (m, 1H, Ar–H), 3.58 (t, J = 7.7 Hz, 2H, CH_2), 3.08 (t, J = 7.7 Hz, 2H, CH_2). $^{13}C\{^1H\}$-NMR (100 MHz, $CDCl_3$): δ = 201.2 (CO), 153.5 (C_q), 149.1 (CH), 141.6 (C_q), 137.0 (CH), 128.6 (2C, CH), 128.6 (2C, CH), 127.3 (CH), 126.1 (CH), 122.0 (CH), 39.6 (CH_2), 30.0 (CH_2). HRMS (ESI): m/z Calcd for $C_{14}H_{13}NO + H^+$ $[M + H]^+$ 212.1070; Found 212.1068. The ^1H and $^{13}C\{^1H\}$ spectra are consistent with those reported in the literature.[86]

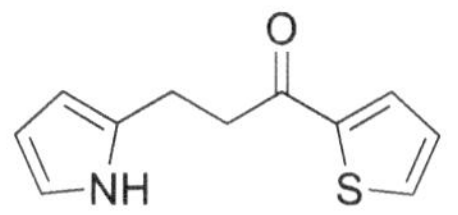

3-(1*H*-Pyrrol-2-yl)-1-(thiophen-2-yl)propan-1-one (34): The representative procedure was followed, using substrate **34a** (0.041 g, 0.202 mmol), 1.0 mL of MeOH:DCM (4:1) and the reaction mixture was stirred at room temperature (27 °C) for 12 h. Purification by column chromatography on silica gel (petroleum ether/EtOAc: 20/1) yielded **34** (0.025 g, 60%) as brown solid. ^1H-NMR (400 MHz, $CDCl_3$): δ = 8.65 (br s, 1H, *N*H), 7.72 (dd, J = 3.8, 0.8 Hz, 1H, Ar–H), 7.65 (dd, J = 4.9, 0.8 Hz, 1H, Ar–H), 7.13 (dd, J = 4.8, 4.0 Hz, 1H, Ar–H), 6.68-6.67 (m, 1H, Ar–H), 6.12-6.09 (m, 1H, Ar–H), 5.96 (s, 1H, Ar–H), 3.29 (t, J = 6.4 Hz, 2H, CH_2), 3.05 (t, J = 6.4 Hz, 2H, CH_2). $^{13}C\{^1H\}$-NMR (100 MHz, $CDCl_3$): δ = 193.6 (CO), 144.0 (C_q), 134.0 (CH), 132.3 (CH), 131.5 (C_q), 128.4 (CH), 117.0 (CH), 108.0 (CH), 105.6 (CH), 40.1 (CH_2), 21.8 (CH_2). HRMS (ESI): m/z Calcd for $C_{11}H_{11}NOS + H^+$ $[M + H]^+$ 206.0634; Found 206.0630.

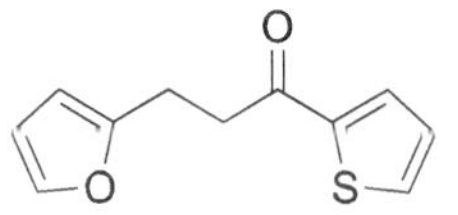

3-(Furan-2-yl)-1-(thiophen-2-yl)propan-1-one (35): The representative procedure was followed, using substrate **35a** (0.041 g, 0.201 mmol) and the reaction mixture was stirred at room temperature (27 °C) for 2 h. Purification by column chromatography on silica gel (petroleum ether/EtOAc: 30/1) yielded **35** (0.037 g, 89%) as brown liquid. ^1H-NMR (400 MHz, $CDCl_3$): δ = 7.71 (d, J = 3.9 Hz, 1H, Ar–H), 7.63 (d, J = 5.0 Hz, 1H, Ar–H), 7.30 (d, J = 1.1 Hz, 1H, Ar–H), 7.12 (dd, J = 4.9, 3.9 Hz, 1H, Ar–H), 6.27 (dd, J = 2.9, 2.0 Hz, 1H, Ar–H), 6.05 (d, J = 3.0 Hz, 1H, Ar–H), 3.26 (t, J = 7.5 Hz, 2H, CH_2), 3.08 (t, J = 7.5 Hz, CH_2). $^{13}C\{^1H\}$-NMR (100 MHz, $CDCl_3$): δ = 191.7 (CO), 154.6 (C_q), 144.1 (C_q), 141.3 (CH), 133.8 (CH), 132.0 (CH), 128.3 (CH), 110.4 (CH), 105.6 (CH), 37.7 (CH_2), 22.8 (CH_2). HRMS (ESI): m/z Calcd for $C_{11}H_{10}O_2S + H^+$ $[M + H]^+$ 207.0474; Found 207.0470.

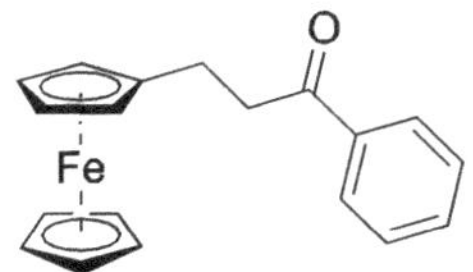

3-Ferrocenyl-1-phenylpropan-1-one (36): The representative procedure was followed, using substrate **36a** (0.064 g, 0.202 mmol), 1.0 mL of MeOH:DCM (4:1) and the reaction mixture was stirred at room temperature (27 °C) for 5 h. Purification by column chromatography on silica gel (petroleum ether/EtOAc: 30/1) yielded **36** (0.046 g, 72%) as a yellow solid. ^{1}H-NMR (400 MHz, $CDCl_3$): δ = 7.97 (d, J = 7.8 Hz, 2H, Ar–H), 7.57 (t, J = 7.4 Hz, 1H, Ar–H), 7.47 (t, J = 7.6 Hz, 2H, Ar–H), 4.14 (s, 5H, Ar–H), 4.13 (s, 2H, Ar–H), 4.08 (s, 2H, Ar–H), 3.20 (t, J = 7.7 Hz, 2H, CH_2) 2.79 (t, J = 7.7 Hz, 2H, CH_2). $^{13}C\{^1H\}$-NMR (100 MHz, $CDCl_3$): δ = 199.7 (CO), 137.1 (C_q), 133.2 (CH), 128.8 (2C, CH), 128.2 (2C, CH), 88.2 (C_q), 68.8 (5C, CH), 68.3 (2C, CH), 67.6 (2C, CH), 40.5 (CH_2), 24.3 (CH_2). HRMS (ESI): m/z Calcd for $C_{19}H_{18}FeO + H^+$ $[M + H]^+$ 318.0702; Found 318.0699. The ^{1}H and $^{13}C\{^1H\}$ spectra are consistent with those reported in the literature.[76]

(*E*)-1,5-Diphenylpent-4-en-1-one (37): The representative procedure was followed, using substrate **37a** (0.047 g, 0.201 mmol) and the reaction mixture was stirred at room temperature (27 °C) for 2 h. Purification by column chromatography on silica gel (petroleum ether/EtOAc: 50/1) yielded **37** (0.037 g, 78%) as brown solid. ^{1}H-NMR (400 MHz, $CDCl_3$): δ = 8.01 (d, J = 7.8 Hz, 2H, Ar–H), 7.58 (t, J= 7.3 Hz, 1H, Ar–H), 7.48 (t, J= 7.6 Hz, 2H, Ar–H), 7.36 (d, J= 7.3 Hz, 2H, Ar–H), 7.30 (t, J= 7.6 Hz, 2H, Ar–H), 7.21 (t, J= 7.1 Hz, 1H, Ar–H), 6.48 (d, J= 15.8 Hz, 1H, CH), 6.31 (dt, J= 15.9, 6.8 Hz, 1H, CH) 3.17 (t, J = 7.4 Hz, 2H, CH_2), 2.68 (q, J = 6.9 Hz, 2H, CH_2). $^{13}C\{^1H\}$-NMR (100 MHz, $CDCl_3$): δ = 199.5 (CO), 137.6 (C_q), 137.1 (C_q), 133.2 (CH), 130.9 (CH), 129.3 (CH), 128.8 (2C, CH), 128.7 (2C, CH), 128.2 (2C, CH), 127.2 (CH), 126.2 (2C, CH), 38.4 (CH_2), 27.7 (CH_2). HRMS (ESI): m/z Calcd for $C_{17}H_{16}O + H^+$ $[M + H]^+$ 237.1274; Found 237.1271. The ^{1}H and $^{13}C\{^1H\}$ spectra are consistent with those reported in the literature.[87]

(*E*)-1,5-Diphenylpent-1-en-3-one (38): The representative procedure was followed, using substrate **38a** (0.047 g, 0.201 mmol) and the reaction mixture was stirred at room temperature (27 °C) for 1 h.

Purification by column chromatography on silica gel (petroleum ether/EtOAc: 50/1) yielded **38** (0.035 g, 74%) as yellow solid. ^{1}H-NMR (400 MHz, $CDCl_3$): δ = 7.56-7.51 (m, 3H, Ar–H), 7.39-7.37 (m, 3H, Ar–H), 7.31-7.28 (m, 2H, Ar–H), 7.24-7.18 (m, 3H, Ar–H), 7.73 (d, J = 16.3 Hz, 1H, Ar–H), 3.00 (br s, 4H, CH_2). $^{13}C\{^{1}H\}$-NMR (100 MHz, $CDCl_3$): δ = 199.5 (CO), 142.9 (CH), 131.4 (C_q), 134.6 (C_q), 130.6 (CH), 129.1 (2C, CH), 128.7 (2C, CH), 128.6 (2C, CH), 128.4 (2C, CH), 126.3 (CH), 126.3 (CH), 42.6 (CH_2), 30.3 (CH_2). HRMS (ESI): m/z Calcd for $C_{17}H_{16}O + H^+$ $[M + H]^+$ 237.1274; Found 237.1273. The ^{1}H and $^{13}C\{^{1}H\}$ spectra are consistent with those reported in the literature.[88]

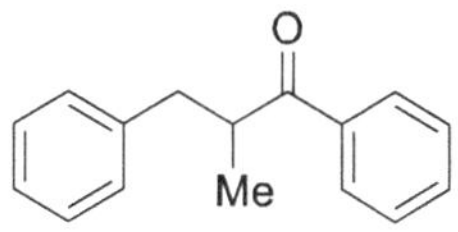

2-Methyl-1,3-diphenylpropan-1-one (39): The representative procedure was followed, using substrate **39a** (0.045 g, 0.202 mmol), 10 bar H_2 and the reaction mixture was stirred at 50 °C for 20 h. Purification by column chromatography on silica gel (petroleum ether/EtOAc: 50/1) yielded **39** (0.028 g, 62%) as colorless oil. ^{1}H-NMR (400 MHz, $CDCl_3$): δ = 7.94-7.91 (m, 2H, Ar–H), 7.55-7.51 (m, 1H, Ar–H), 7.43 (vt, J = 7.5 Hz, 2H, Ar–H), 7.28-7.24 (m, 2H, Ar–H), 7.20-7.15 (m, 3H, Ar–H), 3.79-3.70 (m, 1H, CH), 3.17 (dd, J = 13.7, 6.3 Hz, 1H, CH_2), 2.69 (dd, J = 13.7, 7.8 Hz, 1H, CH_2), 1.20 (d, J = 6.9 Hz, 3H, CH_3). $^{13}C\{^{1}H\}$-NMR (100 MHz, $CDCl_3$): δ = 203.9 (CO), 140.1 (C_q), 136.6 (C_q), 133.1 (CH), 129.3 (2C, CH), 128.8 (2C, CH), 128.5 (2C, CH), 128.5 (2C, CH), 126.4 (CH), 42.9 (CH), 39.5 (CH_2), 17.6 (CH_3). HRMS (ESI): m/z Calcd for $C_{16}H_{16}O + H^+$ $[M + H]^+$ 225.1274; Found 225.1268. The ^{1}H and $^{13}C\{^{1}H\}$ spectra are consistent with those reported in the literature.[89]

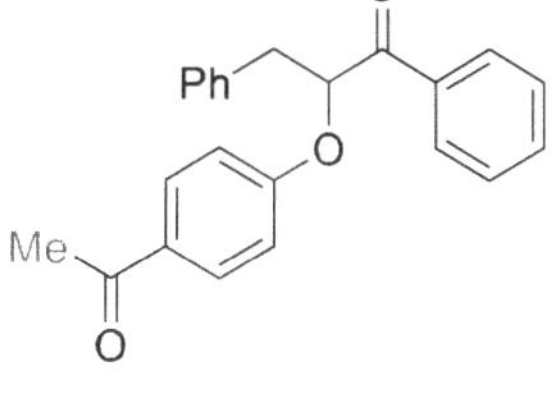

2-(4-Acetylphenoxy)-1,3-diphenylpropan-1-one (40): The representative procedure was followed, using substrate **40a** (0.069 g, 0.202 mmol) and the reaction mixture was stirred at room temperature (27 °C) for 3 h. Purification by column chromatography on silica gel (petroleum ether/EtOAc: 30/1) yielded **40** (0.040 g, 57%) as a yellow solid. ^{1}H-NMR (500 MHz, $CDCl_3$): δ = 8.04 (d, J = 7.9 Hz, 2H, Ar–H), 7.82 (d, J = 8.8 Hz, 2H, Ar–H), 7.61 (t, J = 7.4 Hz, 1H, Ar–H), 7.48 (t, J = 7.8 Hz, 2H, Ar–H), 7.32-7.23 (m, 5H, Ar–H), 6.84 (d, J = 8.9 Hz, 2H, Ar–H), 5.61 (dd, J = 7.4, 5.3 Hz, 1H, CH),

3.36-3.34 (m, 2H, CH_2), 2.48 (s, 3H, CH_3). $^{13}C\{^1H\}$-NMR (125 MHz, $CDCl_3$): δ = 197.4 (CO), 196.8 (CO), 161.5 (C_q), 136.6 (C_q), 134.4 (C_q), 134.2 (CH), 131.2 (C_q), 130.8 (2C, CH), 129.5 (2C, CH), 129.1 (2C, CH), 128.9 (2C, CH), 128.8 (2C, CH), 127.3 (CH), 115.1 (2C, CH), 81.8 (CH), 39.4 (CH_2), 26.5 (CH_3). HRMS (ESI): *m/z* Calcd for $C_{23}H_{20}O_3 + H^+$ $[M + H]^+$ 345.1485; Found 345.1478.

OH

(*E*)-3-Phenylprop-2-en-1-ol (41): The representative procedure was followed, using substrate **41a** (0.027 g, 0.204 mmol) and the reaction mixture was stirred at room temperature (27 °C) for 1 h. Purification by column chromatography on silica gel (petroleum ether/EtOAc: 10/1) yielded **41** (0.026 g, 95%) as yellow oil. ^1H-NMR (500 MHz, $CDCl_3$): δ = 7.38 (d, J = 7.3 Hz, 2H, Ar–H), 7.32 (t, J = 7.4 Hz, 2H, Ar–H), 7.24 (t, J = 7.2 Hz, 1H, Ar–H), 6.61 (d, J = 16.0 Hz, 1H, CH), 6.35 (dt, J = 15.8, 5.8 Hz, 1H, CH), 4.31 (dd, J = 5.8, 1.5 Hz, 2H, CH_2), 2.01 (br s, 1H, OH). $^{13}C\{^1H\}$-NMR (125 MHz, $CDCl_3$): δ = 136.8 (C_q), 131.2 (CH), 128.7 (2C, CH), 128.7 (CH), 127.8 (CH), 126.6 (2C, CH), 63.8 (CH_2). The ^1H and $^{13}C\{^1H\}$ spectra are consistent with those reported in the literature.[21]

OH
MeO

(*E*)-3-(4-Methoxyphenyl)prop-2-en-1-ol (42): The representative procedure was followed, using substrate **42a** (0.033 g, 0.203 mmol) and the reaction mixture was stirred at room temperature (27 °C) for 3 h. Purification by column chromatography on silica gel (petroleum ether/EtOAc: 10/1) yielded **42** (0.033 g, 99%) as a white solid. ^1H-NMR (400 MHz, $CDCl_3$): δ = 7.31 (d, J = 8.6 Hz, 2H, Ar–H), 6.85 (d, J = 8.8 Hz, 2H, Ar–H), 6.54 (d, J = 15.9 Hz, 1H, CH), 6.22 (dt, J = 15.9, 5.9 Hz, 1H, CH), 4.28 (dd, J = 4.9, 1.4 Hz, 2H, CH_2), 3.80 (s, 3H, CH_3), 1.88 (br s, 1H, OH). $^{13}C\{^1H\}$-NMR (100 MHz, $CDCl_3$): δ = 159.4 (C_q), 131.0 (CH), 129.6 (C_q), 127.8 (2C, CH), 126.4 (CH), 114.1 (2C, CH), 64.0 (CH_2), 55.4 (CH_3). HRMS (ESI): *m/z* Calcd for $C_{10}H_{12}O_2 + H^+$ $[M + H]^+$ 165.0910; Found 165.0908. The ^1H and $^{13}C\{^1H\}$ spectra are consistent with those reported in the literature.[90]

OH
Me

(*E*)-2-Methyl-3-phenylprop-2-en-1-ol (43): The representative procedure was followed, using substrate **43a** (0.030 g, 0.205 mmol) and the reaction mixture was stirred at room temperature (27 °C) for 3 h. Purification by column chromatography on silica gel (petroleum ether/EtOAc: 10/1) yielded **43** (0.026 g, 86%) as colorless liquid. ^1H-NMR (400 MHz, $CDCl_3$): δ = 7.33 (t, J = 7.4 Hz,

2H, Ar–H), 7.27 (d, $J = 7.1$ Hz, 2H, Ar–H), 7.21 (t, $J = 7.13$ Hz, 1H, Ar–H), 6.52 (s, 1H, CH), 4.18 (s, 2H, CH_2), 1.90 (s, 3H, CH_3), 1.82 (br s, 1H, OH). $^{13}C\{^1H\}$-NMR (100 MHz, $CDCl_3$): $\delta = 137.8$ (C_q), 137.7 (C_q), 129.0 (2C, CH), 128.3 (2C, CH), 126.6 (CH), 125.2 (CH), 69.1 (CH_2), 15.4 (CH_3). The ^1H and $^{13}C\{^1H\}$ spectra are consistent with those reported in the literature.[22]

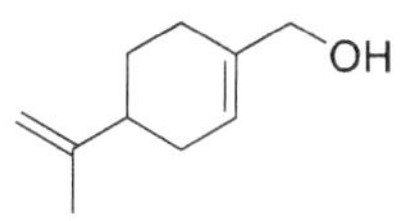

(4-(Prop-1-en-2-yl)cyclohex-1-en-1-yl)methanol (44): The representative procedure was followed, using substrate **44a** (0.031 g, 0.206 mmol) and the reaction mixture was stirred at room temperature (27 °C) for 3 h. Purification by column chromatography on silica gel (petroleum ether/EtOAc: 20/1) yielded **44** (0.024 g, 77%) as yellow oil. ^1H-NMR (400 MHz, $CDCl_3$): $\delta = 5.71$-5.67 (m, 1H, CH), 4.73-4.66 (m, 2H, CH_2), 4.01-3.97 (m, 2H, CH_2), 2.19-1.79 (m, 7H, CH), 1.74 (s, 3H, CH_3), 1.37 (br s, 1H, OH). $^{13}C\{^1H\}$-NMR (100 MHz, $CDCl_3$): $\delta = 150.0$ (C_q), 137.4 (C_q), 122.7 (CH), 108.8 (CH_2), 67.5 (CH_2), 41.3 (CH), 30.6 (CH_2), 27.7 (CH_2), 26.3 (CH_2), 21.0 (CH_3). HRMS (ESI): m/z Calcd for $C_{10}H_{16}O + H^+$ $[M + H]^+$ 153.1274; Found 153.11273. The ^1H and $^{13}C\{^1H\}$ spectra are consistent with those reported in the literature.[91]

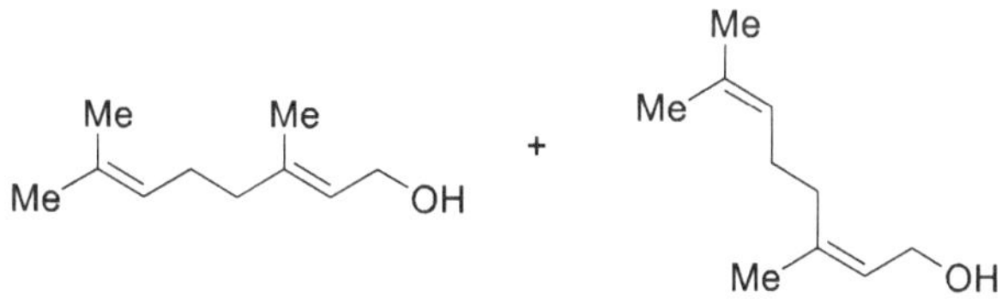

(*E/Z*)-3-(4-Methoxyphenyl)prop-2-en-1-ol (45): The representative procedure was followed, using isomeric mixture of **45a** (0.031 g, 0.204 mmol) and the reaction mixture was stirred at room temperature (27 °C) for 3 h. Purification by column chromatography on silica gel (petroleum ether/EtOAc: 10/1) yielded isomeric mixture (3:2) of **45** (0.023 g, 73%) as colorless oil. ^1H-NMR (400 MHz, $CDCl_3$): $\delta = 5.45$-5.41 (m, 1H, CH), 5.11-5.06 (m, 1H, CH), 4.15-4.07 (m, 2H, CH_2), 2.12-2.00 (m, 4H, CH_2), 1.75-1.67 (m, 6H, CH_3), 1.60 (s, 3H, CH_3). $^{13}C\{^1H\}$-NMR for one isomer of **46** (100 MHz, $CDCl_3$): $\delta = 139.9$ (C_q), 131.9 (C_q), 124.0 (CH), 123.5 (CH), 59.6 (CH_2), 39.7 (CH_2), 26.6 (CH_2), 25.8 (CH_3), 17.8 (CH_3), 16.4 (CH_3). For other isomer of **46**: (100 MHz, $CDCl_3$): $\delta =$ 140.1 (C_q), 132.6 (C_q), 124.6 (CH), 124.0 (CH), 59.1 (CH_2), 32.1 (CH_2), 26.7 (CH_2), 25.8 (CH_3), 23.6 (CH_3), 17.8 (CH_3). HRMS (ESI): m/z Calcd for $C_{10}H_{18}O - H^+$ $[M - H]^+$ 153.1274; Found 153.1273. The ^1H and $^{13}C\{^1H\}$ spectra are consistent with those reported in the literature.[92]

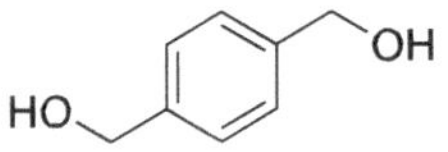

1,4-Phenylenedimethanol (46): The representative procedure was followed, using substrate **46a** (0.027 g, 0.201 mmol) and the reaction mixture was stirred at room temperature (27 °C) for 3 h. Purification by column chromatography on silica gel (petroleum ether/EtOAc: 10/1) yielded **46** (0.024 g, 86%) as white solid. ^{1}H-NMR (500 MHz, DMSO-d_6): δ = 7.26 (s, 4H, Ar–H), 5.12 (t, J = 5.7 Hz, 2H, OH), 4.48 (d, J = 5.6 Hz, 4H, CH_2). ^{13}C{^{1}H}-NMR (125 MHz, DMSO-d_6): δ = 140.9 (2C, C_q), 126.2 (4C, CH), 62.8 (2C, CH_2). The ^{1}H and ^{13}C{^{1}H} spectra are consistent with those reported in the literature.[93]

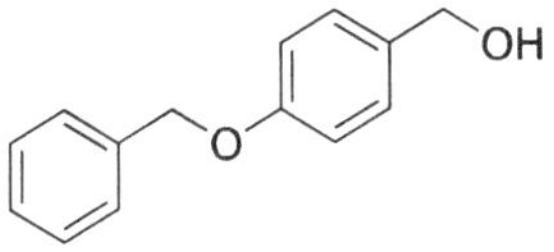

(4-(Benzyloxy)phenyl)methanol (47): The representative procedure was followed, using substrate **47a** (0.043 g, 0.203 mmol) and the reaction mixture was stirred at room temperature (27 °C) for 3 h. Purification by column chromatography on silica gel (petroleum ether/EtOAc: 10/1) yielded **47** (0.043 g, 99%) as a white solid. ^{1}H-NMR (400 MHz, $CDCl_3$): δ = 7.41 (d, J = 7.3 Hz, 2H, Ar–H), 7.36 (t, J = 7.3 Hz, 2H, Ar–H), 7.30 (t, J = 7.1 Hz, 1H, Ar–H), 7.25 (d, J = 8.6 Hz, 2H, Ar–H), 6.94 (d, J = 8.6 Hz, 2H, Ar–H), 5.04 (s, 2H, CH_2), 4.56 (s, 2H, CH_2), 1.91 (s, 1H, OH). ^{13}C{^{1}H}-NMR (100 MHz, $CDCl_3$): δ = 158.5 (C_q), 137.1 (C_q), 133.6 (C_q), 128.8 (2C, CH), 128.7 (2C, CH), 128.1 (CH), 127.6 (2C, CH), 115.1 (2C, CH), 70.2 (CH_2), 65.1 (CH_2). HRMS (ESI): m/z Calcd for $C_{14}H_{14}O_2 - H^+$ $[M-H]^+$ 213.0910; Found 213.0907. The ^{1}H and ^{13}C{^{1}H} spectra are consistent with those reported in the literature.[92]

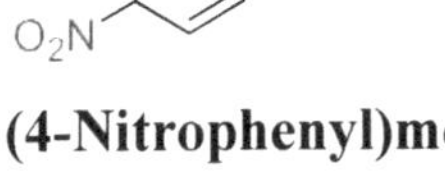

(4-Nitrophenyl)methanol (48): The representative procedure was followed, using substrate **48a** (0.031 g, 0.205 mmol) and the reaction mixture was stirred at room temperature (27 °C) for 3 h. Purification by column chromatography on silica gel (petroleum ether/EtOAc: 10/1) yielded **48** (0.029 g, 92%) as yellow solid. ^{1}H-NMR (400 MHz, $CDCl_3$): δ = 8.18 (d, J = 8.8 Hz, 2H, Ar–H), 7.51 (d, J = 8.9 Hz, 2H, Ar–H), 4.82 (s, 2H, CH_2), 2.28 (s, 1H, OH). ^{13}C{^{1}H}-NMR (100 MHz, $CDCl_3$): δ = 148.4 (C_q), 147.4 (C_q), 127.2 (2C, CH), 123.9 (2C, CH), 64.1 (CH_2). The ^{1}H and ^{13}C{^{1}H} spectra are consistent with those reported in the literature.[94]

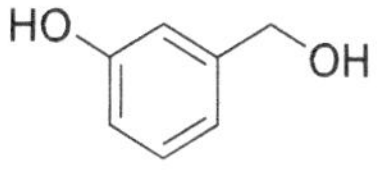

3-(Hydroxymethyl)phenol (49): The representative procedure was followed, using substrate **49a** (0.025 g, 0.205 mmol) and the reaction mixture was stirred at room temperature (27 °C) for 3 h. Purification by column chromatography on silica gel (petroleum ether/EtOAc: 10/1) yielded **49** (0.021 g, 83%) as white solid. ^{1}H-NMR (400 MHz, DMSO-*d*$_6$): δ = 9.30 (s, 1H, OH), 7.13 (t, *J* = 7.8 Hz, 1H, Ar–H), 6.78 (s, 1H, Ar–H), 6.75 (d, *J* = 7.5 Hz, 1H, Ar–H), 6.65 (dd, *J* = 8.0, 2.0 Hz, 1H, Ar–H), 5.12 (t, *J* = 5.8 Hz, 1H, OH), 4.45 (d, *J* = 5.8 Hz, 2H, CH_2). $^{13}C\{^1H\}$-NMR (100 MHz, DMSO-*d*$_6$): δ = 158.2 (C_q), 145.0 (C_q), 129.9 (CH), 117.8 (CH), 114.4 (CH), 114.2 (CH), 63.8 (CH_2). The ^{1}H and $^{13}C\{^1H\}$ spectra are consistent with those reported in the literature.[95]

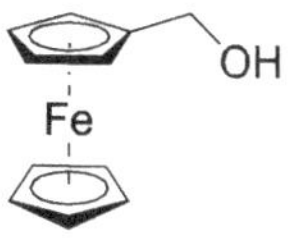

Ferrocenylmethanol (50): The representative procedure was followed, using substrate **51a** (0.043 g, 0.201 mmol) and the reaction mixture was stirred at room temperature (27 °C) for 3 h. Purification by column chromatography on silica gel (petroleum ether/EtOAc: 10/1) yielded **51** (0.042 g, 97%) as a yellow solid. ^{1}H-NMR (400 MHz, $CDCl_3$): δ = 4.32 (s, 2H, CH_2), 4.24 (t, *J* = 1.6 Hz, 2H, Ar–H), 4.19-4.17 (m, 7H, Ar–H), 1.71 (s, 1H, OH). $^{13}C\{^1H\}$-NMR (100 MHz, $CDCl_3$): δ = 88.5 (C_q), 68.5 (7C, CH), 68.1 (2C, CH), 60.9 (CH_2). HRMS (ESI): *m/z* Calcd for $C_{11}H_{12}FeO + H^+$ $[M + H]^+$ 217.0312; Found 217.0311. The ^{1}H and $^{13}C\{^1H\}$ spectra are consistent with those reported in the literature.[92]

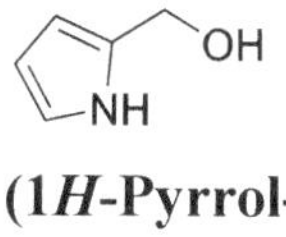

(1*H*-Pyrrol-2-yl)methanol (51): The representative procedure was followed, using substrate **51a** (0.019 g, 0.20 mmol) and the reaction mixture was stirred at room temperature (27 °C) for 3 h. Purification by column chromatography on neutral alumina (petroleum ether/EtOAc: 20/1) yielded **51** (0.018 g, 93%) as a colorless liquid. After purification the isolated product was stored in in refrigerator to avoid polymerization at room temperature. ^{1}H-NMR (400 MHz, $CDCl_3$): δ = 8.62 (br s, 1H, *NH*), 6.73 (s, 1H, Ar–H), 6.13 (d, *J* = 15.3 Hz, 2H, Ar–H), 4.54 (s, 2H, CH_2), 2.66 (br s, 1H, OH). $^{13}C\{^1H\}$-NMR (100 MHz, $CDCl_3$): δ = 131.1 (C_q), 118.8 (CH), 108.4 (CH), 107.3 (CH), 58.0 (CH_2).

***N*-Cinnamylaniline (52)**: The representative procedure was followed, using substrate **52a** (0.042 g, 0.203 mmol) and the reaction mixture was stirred at room temperature (27 °C) for 4 h. Purification by column chromatography on silica gel (petroleum ether/EtOAc: 50/1) yielded mixture of **52** and *N*-(3-phenylpropyl)aniline (0.038 g, 89%) as a yellow oil. For compound **52**: ^{1}H-NMR (400 MHz, $CDCl_3$): δ = 7.35 (d, J = 7.1 Hz, 2H, Ar–H), 7.29 (t, J = 7.3 Hz, 2H, Ar–H), 7.23-7.16 (m, 3H, Ar–H), 6.71 (t, J = 7.3 Hz, 1H, Ar–H), 6.65 (d, J = 7.9 Hz, 2H, Ar–H), 6.58-6.55 (m, 1H, CH), 6.31 (dt, J = 15.9, 5.7 Hz, 1H, CH), 3.92 (d, J = 5.4 Hz, 2H, CH_2), 3.68 (br s, 1H, *NH*). $^{13}C\{^{1}H\}$-NMR (100 MHz, $CDCl_3$): δ = 148.2 (C_q), 137.0 (C_q), 131.7 (CH), 129.4 (2C, CH), 128.7 (2C, CH), 127.7 (CH), 127.2 (CH), 126.5 (2C, CH), 117.8 (CH), 113.2 (2C, CH), 46.4 (CH_2). HRMS (ESI): *m/z* Calcd for $C_{15}H_{15}N - H^+$ $[M - H]^+$ 208.1121; Found 208.1120. The ^{1}H and $^{13}C\{^{1}H\}$ spectra are consistent with those reported in the literature.[90]

***N*-benzylaniline (53)**: The representative procedure was followed, using substrate **53a** (0.037 g, 0.204 mmol) and the reaction mixture was stirred at 50 °C for 20 h. Purification by column chromatography on silica gel (petroleum ether/EtOAc: 30/1) yielded **53** (0.034 g, 91%) as yellow oil. ^{1}H-NMR (400 MHz, $CDCl_3$): δ = 7.38-7.31 (m, 4H, Ar–H), 7.28-7.24 (m, 1H, Ar–H), 7.16 (dd, J = 8.6, 7.4 Hz, 2H, Ar–H), 6.71 (t, J = 7.3 Hz, 1H, Ar–H), 6.62 (dd, J = 8.6, 0.9 Hz, 2H, Ar–H), 4.31 (s, 2H, CH_2), 4.00 (br s, 1H, *NH*). $^{13}C\{^{1}H\}$-NMR (100 MHz, $CDCl_3$): δ = 148.3 (C_q), 139.6 (C_q), 129.4 (2C, CH), 128.8 (2C, CH), 127.7 (2C, CH), 127.4 (CH), 117.7 (CH), 113.0 (2C, CH), 48.5 (CH_2). HRMS (ESI): *m/z* Calcd for $C_{13}H_{13}N + H^+$ $[M + H]^+$ 184.1121; Found 184.1120. The ^{1}H and $^{13}C\{^{1}H\}$ spectra are consistent with those reported in the literature.[90]

2.4.3 Mechanistic Experiments

Procedure for Deuterium Labelling Experiment. A standard catalytic hydrogenation reaction was performed using CD_3OD as a solvent. After completion of reaction time, the reaction mixture was evaporated under vacuo. Purification by column chromatography provided the semi-hydrogenated product **4**-[D] with 92% deuterium incorporation (Figure 2.5). This finding suggests that one of the hydrogen sources for hydrogenation is the methanol, and incorporation of two D at α-position supports the reversible protonation of substrate.

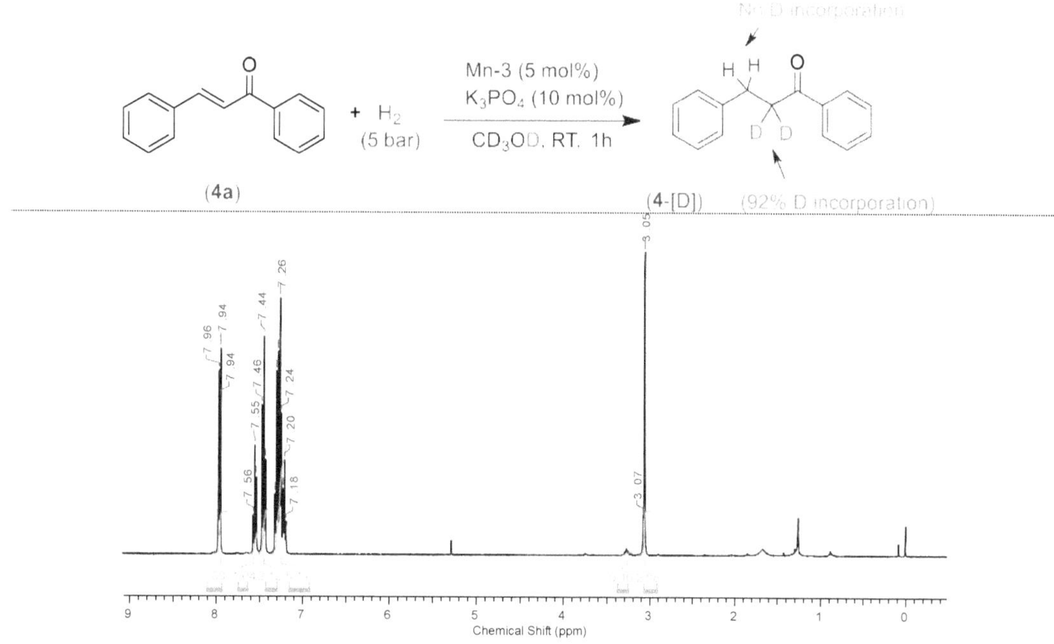

Figure 2.5. ^{1}H NMR of Compound **4**-[D].

Experiment to Understand Non-innocent Behaviour of Ligand *NH*. A standard hydrogenation reaction was performed using **Mn-3^{Me}** as catalyst. After 5 h of the reaction time, the reaction mixture was analyzed by GC that ensured no formation of hydrogenated product. This experiment suggests the *NH* moiety in the manganese catalyst is necessary to generate active de-aromatized catalyst in the presence of base (K_3PO_4) that would activate the H_2 molecule.

(Mn-3^{Me})
(5 mol%)
+ H_2 (5 bar) — K_3PO_4 (10 mol%), MeOH, rt (27 °C), 5 h → No Reaction

Attempted Hydrogenation of (*E*)-1,3-Diphenylprop-2-en-1-ol: Under standard hydrogenation conditions, the compound (*E*)-1,3-diphenylprop-2-en-1-ol was attempted for hydrogenation. The GC analysis of the reaction mixture does not show the formation of alkene hydrogenation product. This experiment suggests that the hydrogenation of (*E*)-chalcone (**4**) does not proceed *via* the 1,2-addition. Hence this reaction proceeds *via* the 1,4-addition of hydrogen.

OH + H_2 (5 bar) —[Mn-3 (5 mol%), K_3PO_4 (10 mol%), MeOH, rt (27 °C), 1 h]→ No Reaction

2.4.4 DFT Calculations

All the calculations in this study have been performed with density functional theory (DFT), with the aid of the Turbomole 7.5 suite of programs,[96] using the PBE functional,[97] along with dispersion correction (DFT-D3).[98] The def2-SVP basis set[99] for Mn and the TZVP basis set for all other atoms have been employed. The resolution of identity (RI),[100] along with the multipole accelerated resolution of identity (marij)[101] approximations have been employed for an accurate and efficient treatment of the electronic Coulomb term in the DFT calculations. A solvent correction was incorporated with optimization calculations using the COSMO model,[102] with methanol ($\varepsilon = 32.7$) as the solvent. The values reported are ΔG values, with zero-point energy corrections, internal energy, and entropic contributions included through frequency calculations on the optimized minima, with the temperature is taken to be 298.15 K. Harmonic frequency calculations were performed for all stationary points to confirm them as local minima or transition state structures.

Energy span model (ESM)- The Turnover frequency (TOF) of the catalytic cycle can be calculated through the Energetic Span Model (ESM) and put into practical use by Shaik and co-workers.[103-105] The ESM imparts an easy method to determine the turnover frequencies (TOFs) of catalytic cycles based on their computed energy profiles. In most cases, the TOF is calculated by the TOF-determining transition state (TDTS), the TOF-determining intermediate (TDI), and by the reaction energy, ΔG_r, as shown below

$$\mathrm{TOF} = \frac{k_B T}{h} e^{-\delta E/RT}$$

where δE is the energy span and is defined as the Gibbs energy difference between the TDTS and the TDI, with the addition of the ΔG_r when the TDTS appears before the TDI. δE is the effective activation energy barrier of the global reaction. The TDTS and TDI are the intermediate and the transition state, respectively, that maximize δE, according to equation 1.

$$\delta E = \begin{cases} TDTS - TDTI, & \text{if TDTS appears after TDI} \\ TFTS - TDI + \Delta G_r, & \text{if TDTS appears before TDI} \end{cases} \quad (1)$$

This model has been used to calculate the TOFs (at 298.15 K). The ESM can be applied in a user-friendly way with the recently developed AUTOF computer program.[35,36,37]

2.4.5 X-ray Structural Data

X-ray intensity data measurements of compound **Mn-2** was carried out on a Bruker D8 VENTURE Kappa Duo PHOTON II CPAD diffractometer equipped with Incoatech multilayer mirrors optics. The intensity measurements were carried out with Mo micro-focus sealed tube diffraction source (MoK_{α}= 0.71073 Å) at 100(2) K temperature. The X-ray generator was operated at 50 kV and 1.4 mA. A preliminary set of cell constants and an orientation matrix were calculated from three matrix sets of 36 frames (each matrix run consists of 12 frames). Data were collected with ω scan width of 0.5° at different settings of φ and 2θ with a frame time of 10-20 sec depending on the diffraction power of the crystals keeping the sample-to-detector distance fixed at 5.00 cm. The X-ray data collection was monitored by APEX3 program (Bruker, 2016).[106] All the data were corrected for Lorentzian, polarization and absorption effects using SAINT and SADABS programs (Bruker, 2016). Using the APEX3 (Bruker) program suite, the structure was solved with the ShelXS-97 (Sheldrick, 2008)[107] structure solution program, using direct methods. The model was refined with a version of ShelXL-2018/3 (Sheldrick, 2015)[108] using Least Squares minimization. All the hydrogen atoms were placed in a geometrically idealized position and constrained to ride on their parent atoms. An *ORTEP* III[109] view of the compounds was drawn with 50% probability displacement ellipsoids, and H atoms are shown as small spheres of arbitrary radii.

Procedure of crystallization: Equimolar ratio of L2 (0.005 g) and $Mn(CO)_5Br$ (0.005 g) was taken in a dry NMR tube and added dry THF (0.5 mL). Then the tube was closed with cap and kept inside the glove box for three days to form orange crystal on the wall of the tube.

Table 2.2. Crystal Data of Compound **Mn-2**.

Crystal Data	**Mn-2**
Formula	$C_{16}H_{21}BrMnN_4O_2P$
Molecular weight	467.19
Crystal Size, mm^3	0.240 × 0.260 × 0.360
Temp. (K)	100(2)
Wavelength (Å)	0.71073
Crystal Syst.	orthorhombic
Space Group	*Pbca*
a/Å	9.1016(5)
b/Å	14.4759(6)
c/Å	28.6257(14)
V/Å^3	3771.5(3)
Z	8
D_{calc}/g cm^{-3}	1.642
μ/mm^{-1}	2.921
F(000)	1880
Ab. Correct.	multi-scan
T_{min}/ T_{max}	0.419/0.541
$2\theta_{max}$	56
Total reflns.	73416
Unique reflns.	4688
Obs. reflns.	3291
h, k, l (min, max)	(-12, 11), (-19, 19), (-38, 38)
R_{int} / R_{sig}	0.1649/ 0.0655
No. of parameters	230
$R1$ [I> $2\sigma(I)$]	0.0494
$wR2$[I> $2\sigma(I)$]	0.0889
$R1$ [all data]	0.0882
$wR2$ [all data]	0.1012
goodness-of-fit	1.051
$\Delta\rho_{max}$, $\Delta\rho_{min}$(eÅ^{-3})	+0.798, -0.589
CCDC No.	2194541

Table 2.3. Bond Lengths (Å) for Compound **Mn-2**.

Br1-Mn1	2.5916(7)	Mn1-C9	1.773(4)
Mn1-C10	1.797(4)	Mn1-N3	2.011(3)
Mn1-N1	2.012(3)	Mn1-P1	2.2462(12)
P1-N4	1.732(3)	P1-C11	1.846(4)
P1-C14	1.849(4)	O1-C9	1.157(5)
O2-C10	1.149(5)	N1-C1	1.344(5)
N1-C5	1.345(5)	N2-N3	1.362(5)
N2-C6	1.368(6)	N2-C5	1.401(6)
N3-C8	1.323(5)	N4-C1	1.375(5)
N4-H4N	0.88		
C1-C2	1.404(5)	C2-C3	1.376(6)
C2-H2	0.95	C3-C4	1.389(7)
C3-H3	0.95	C4-C5	1.377(6)
C4-H4	0.95	C6-C7	1.356(7)
C6-H6	0.95	C7-C8	1.394(6)
C7-H7	0.95	C8-H8	0.95
C11-C12	1.516(6)	C11-C13	1.530(5)
C11-H11	1.0	C12-H12A	0.98
C12-H12B	0.98	C12-H12C	0.98
C13-H13A	0.98	C13-H13B	0.98
C13-H13C	0.98	C14-C16	1.528(5)
C14-C15	1.535(6)	C14-H14	1.0
C15-H15A	0.98	C15-H15B	0.98
C15-H15C	0.98	C16-H16A	0.98
C16-H16B	0.98	C16-H16C	0.98

Table 2.4. Bond Angles (°) for Compound **Mn-2**.

C9-Mn1-C10	87.40(19)	C9-Mn1-N3	91.59(17)
C10-Mn1-N3	99.97(17)	C9-Mn1-N1	96.23(16)
C10-Mn1-N1	175.88(16)	N3-Mn1-N1	78.02(14)
C9-Mn1-P1	92.06(14)	C10-Mn1-P1	99.34(14)

N3-Mn1-P1	160.49(11)	N1-Mn1-P1	82.53(10)
C9-Mn1-Br1	175.54(14)	C10-Mn1-Br1	91.66(13)
N3-Mn1-Br1	84.28(10)	N1-Mn1-Br1	84.57(9)
P1-Mn1-Br1	92.39(3)	N4-P1-C11	103.72(17)
N4-P1-C14	100.77(17)	C11-P1-C14	104.30(19)
N4-P1-Mn1	99.98(12)	C11-P1-Mn1	121.95(14)
C14-P1-Mn1	122.10(14)	C1-N1-C5	119.5(4)
C1-N1-Mn1	122.5(3)	C5-N1-Mn1	117.9(3)
N3-N2-C6	110.9(4)	N3-N2-C5	117.2(3)
C6-N2-C5	131.7(4)	C8-N3-N2	105.3(3)
C8-N3-Mn1	140.2(3)	N2-N3-Mn1	114.5(3)
C1-N4-P1	118.6(3)	N1-C1-N4	116.1(3)
C1-N4-H4N	120.2	P1-N4-H4N	120.2
N1-C1-C2	120.6(4)	N4-C1-C2	123.3(4)
C3-C2-C1	118.4(4)	C3-C2-H2	120.8
C1-C2-H2	120.8	C2-C3-C4	121.3(4)
C2-C3-H3	119.3	C4-C3-H3	119.3
C5-C4-C3	116.6(4)	C5-C4-H4	121.7
C3-C4-H4	121.7	N1-C5-C4	123.5(4)
N1-C5-N2	112.2(4)	C4-C5-N2	124.3(4)
C7-C6-N2	106.6(4)	C7-C6-H6	126.7
N2-C6-H6	126.7	C6-C7-C8	106.1(4)
C6-C7-H7	126.9	C8-C7-H7	126.9
N3-C8-C7	111.1(4)	N3-C8-H8	124.4
C7-C8-H8	124.4	O1-C9-Mn1	175.3(4)
O2-C10-Mn1	176.7(4)	C12-C11-C13	110.5(3)
C12-C11-P1	112.9(3)	C13-C11-P1	111.0(3)
C12-C11-H11	107.4	C13-C11-H11	107.4
P1-C11-H11	107.4	C11-C12-H12A	109.5
C11-C12-H12B	109.5	H12A-C12-H12B	109.5
C11-C12-H12C	109.5	H12A-C12-H12C	109.5
H12B-C12-H12C	109.5	C11-C13-H13A	109.5

C11-C13-H13B	109.5	H13A-C13-H13B	109.5
C11-C13-H13C	109.5	H13A-C13-H13C	109.5
H13B-C13-H13C	109.5	C16-C14-C15	110.5(3)
C16-C14-P1	111.1(3)	C15-C14-P1	114.4(3)
C16-C14-H14	106.8	C15-C14-H14	106.8
P1-C14-H14	106.8	C14-C15-H15A	109.5
C14-C15-H15B	109.5	H15A-C15-H15B	109.5
C14-C15-H15C	109.5	H15A-C15-H15C	109.5
H15B-C15-H15C	109.5	C14-C16-H16A	109.5
C14-C16-H16B	109.5	H16A-C16-H16B	109.5
C14-C16-H16C	109.5	H16A-C16-H16C	109.5
H16B-C16-H16C	109.5		

Table 2.5. Torsion Angles (°) for Compound **Mn-2**.

C6-N2-N3-C8	-0.1(5)	C5-N2-N3-C8	-176.5(4)
C6-N2-N3-Mn1	178.5(3)	C5-N2-N3-Mn1	2.1(4)
C11-P1-N4-C1	132.7(3)	C14-P1-N4-C1	-119.5(3)
Mn1-P1-N4-C1	6.2(3)	C5-N1-C1-N4	177.3(3)
Mn1-N1-C1-N4	1.3(5)	C5-N1-C1-C2	-1.1(6)
Mn1-N1-C1-C2	-177.1(3)	P1-N4-C1-N1	-5.4(5)
P1-N4-C1-C2	173.0(3)	N1-C1-C2-C3	1.8(6)
N4-C1-C2-C3	-176.5(4)	C1-C2-C3-C4	-0.1(6)
C2-C3-C4-C5	-2.0(6)	C1-N1-C5-C4	-1.3(6)
Mn1-N1-C5-C4	174.9(3)	C1-N1-C5-N2	179.2(3)
Mn1-N1-C5-N2	-4.6(4)	C3-C4-C5-N1	2.8(6)
C3-C4-C5-N2	-177.7(4)	N3-N2-C5-N1	1.5(5)
C6-N2-C5-N1	-174.0(4)	N3-N2-C5-C4	-178.0(4)
C6-N2-C5-C4	6.5(7)	N3-N2-C6-C7	0.5(5)
C5-N2-C6-C7	176.2(4)	N2-C6-C7-C8	-0.7(6)
N2-N3-C8-C7	-0.3(5)	Mn1-N3-C8-C7	-178.4(4)
C6-C7-C8-N3	0.7(6)	N4-P1-C11-C12	-37.8(3)

C14-P1-C11-C12	-142.9(3)	Mn1-P1-C11-C12	73.5(3)
N4-P1-C11-C13	-162.5(3)	C14-P1-C11-C13	92.4(3)
Mn1-P1-C11-C13	-51.2(3)	N4-P1-C14-C16	178.5(3)
C11-P1-C14-C16	-74.2(3)	Mn1-P1-C14-C16	69.4(3)
N4-P1-C14-C15	-55.5(3)	C11-P1-C14-C15	51.9(3)
Mn1-P1-C14-C15	-164.6(2)		

2.4.6 ^1H and $^{13}C\{^1H\}$ NMR and IR Spectra of Complexes

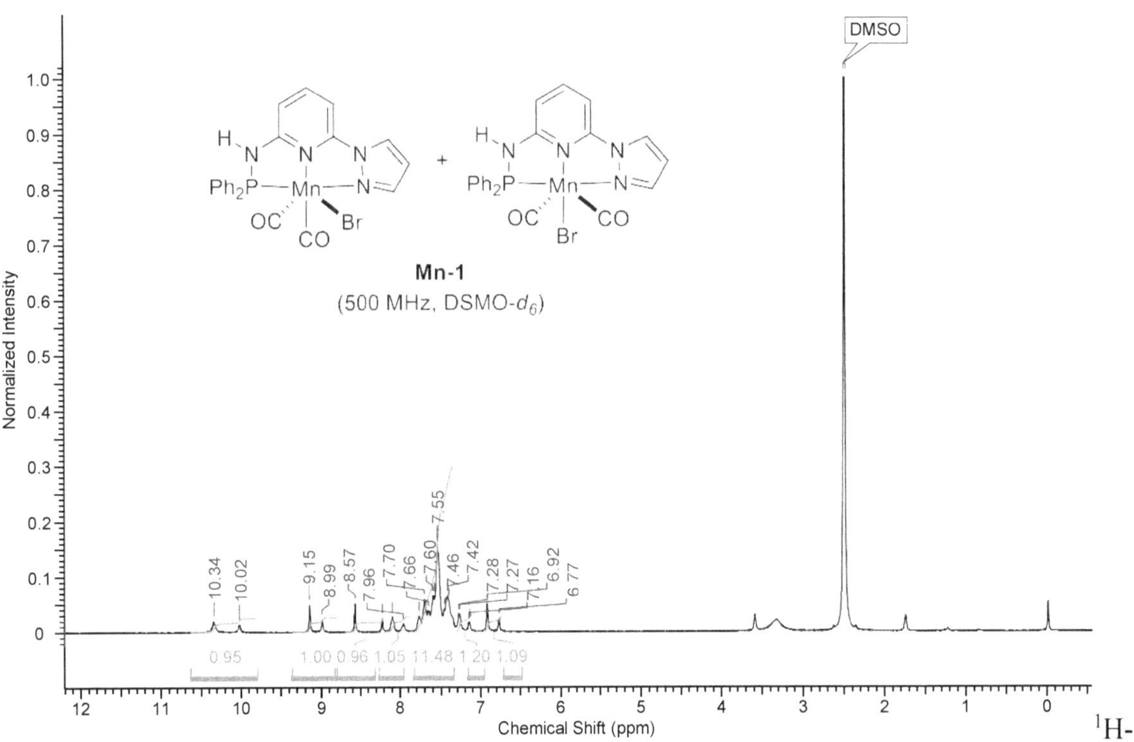

^1H-NMR spectrum of **Mn-1** complex (mixture of two isomers).

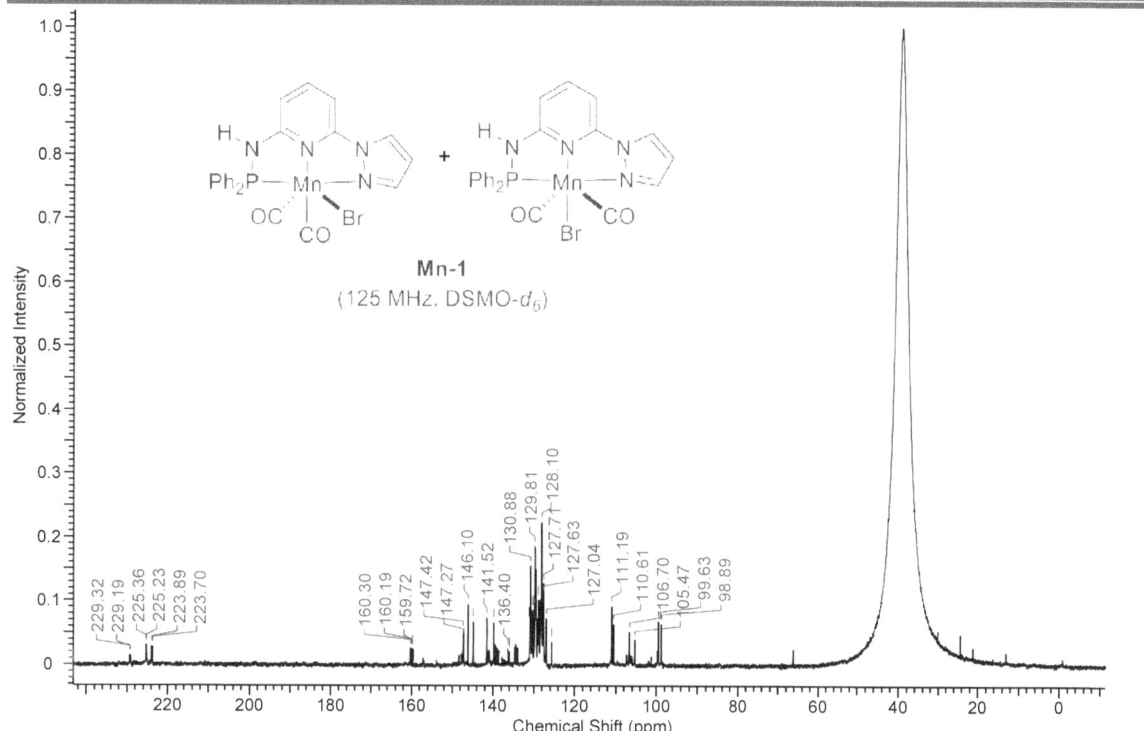

$^{13}C\{^1H\}$-NMR spectrum of **Mn-1** complex (mixture of two isomers).

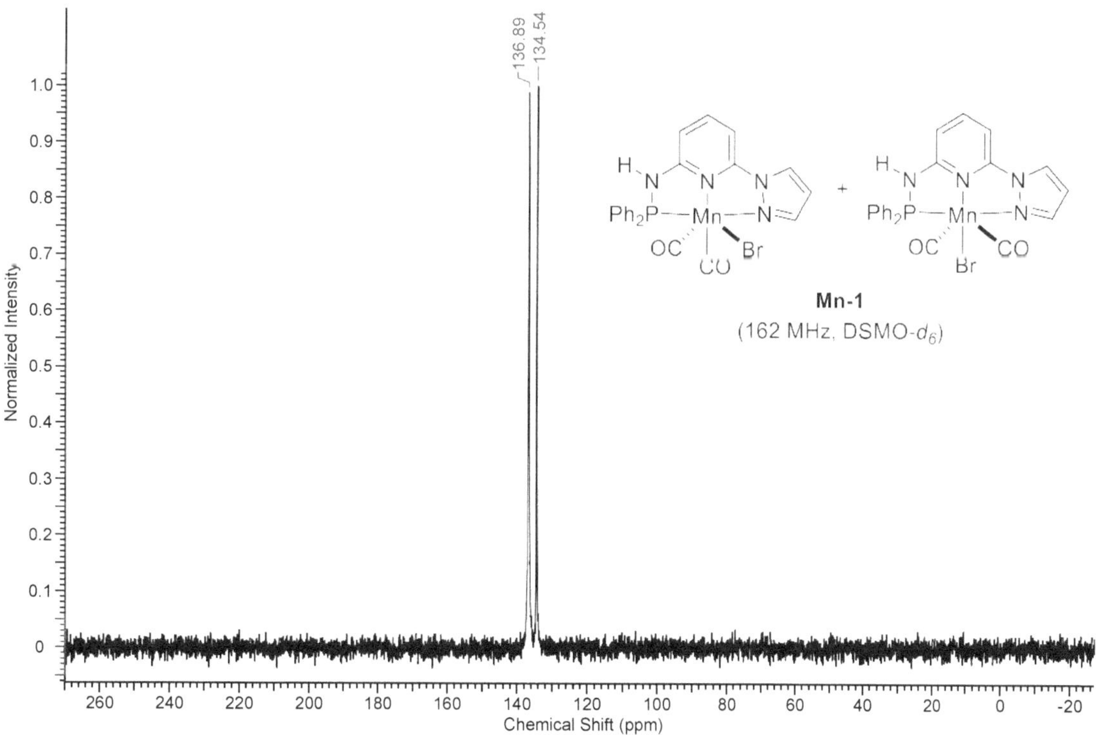

$^{31}P\{^1H\}$-NMR spectrum of **Mn-1** complex (mixture of two isomers).

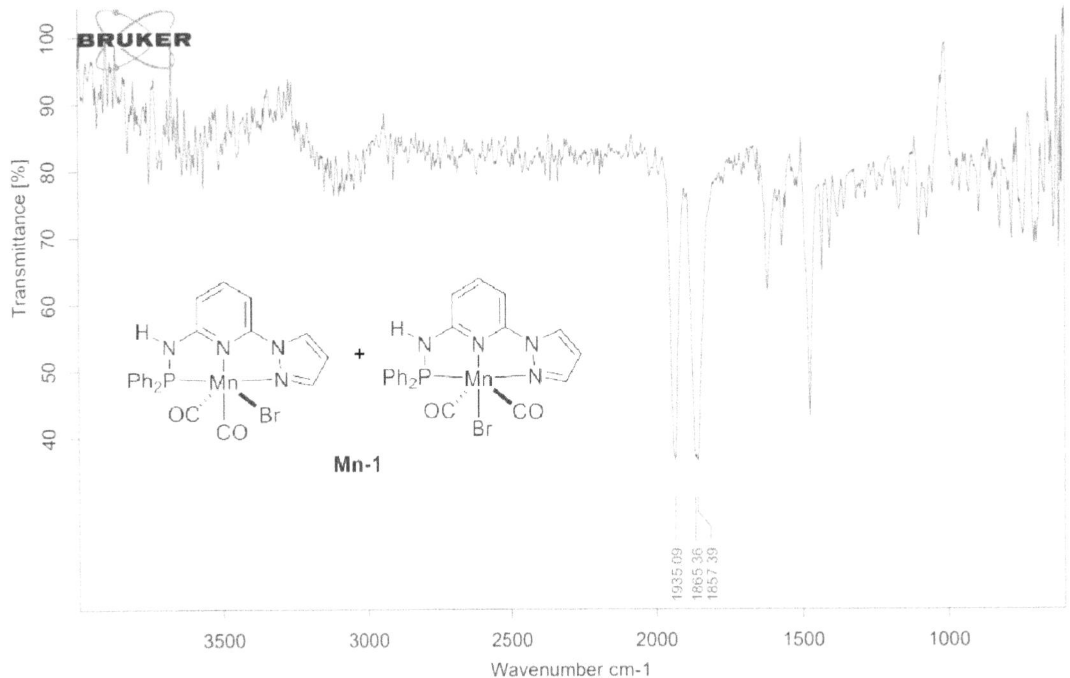

IR spectrum of **Mn-1** complex.

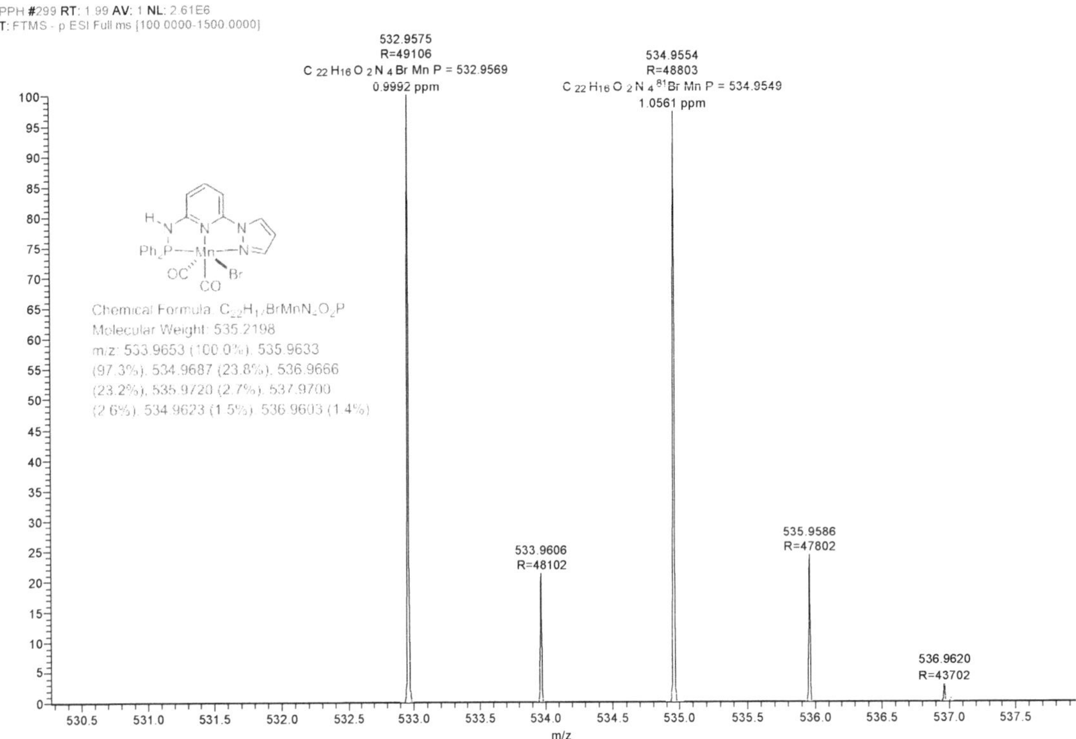

ESI-MS (–ve mode) of **Mn-1** complex.

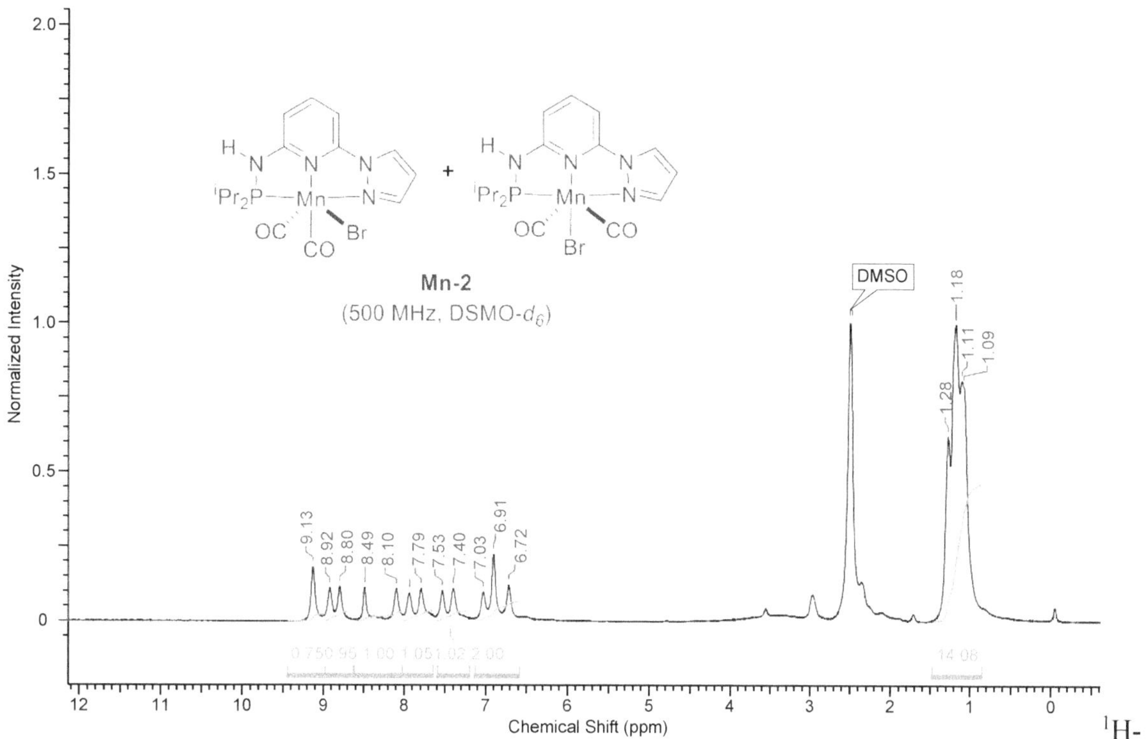

^{1}H-NMR spectrum of **Mn-2** complex (mixture of two isomers).

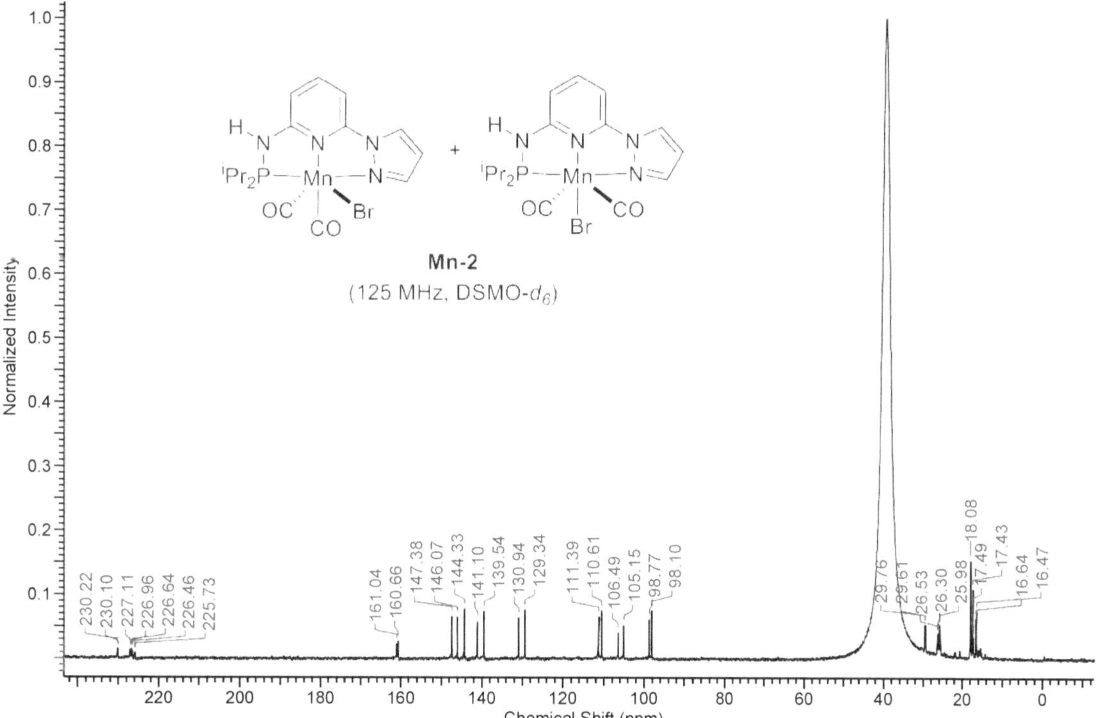

$^{13}C\{^1H\}$-NMR spectrum of **Mn-2** complex (mixture of two isomers).

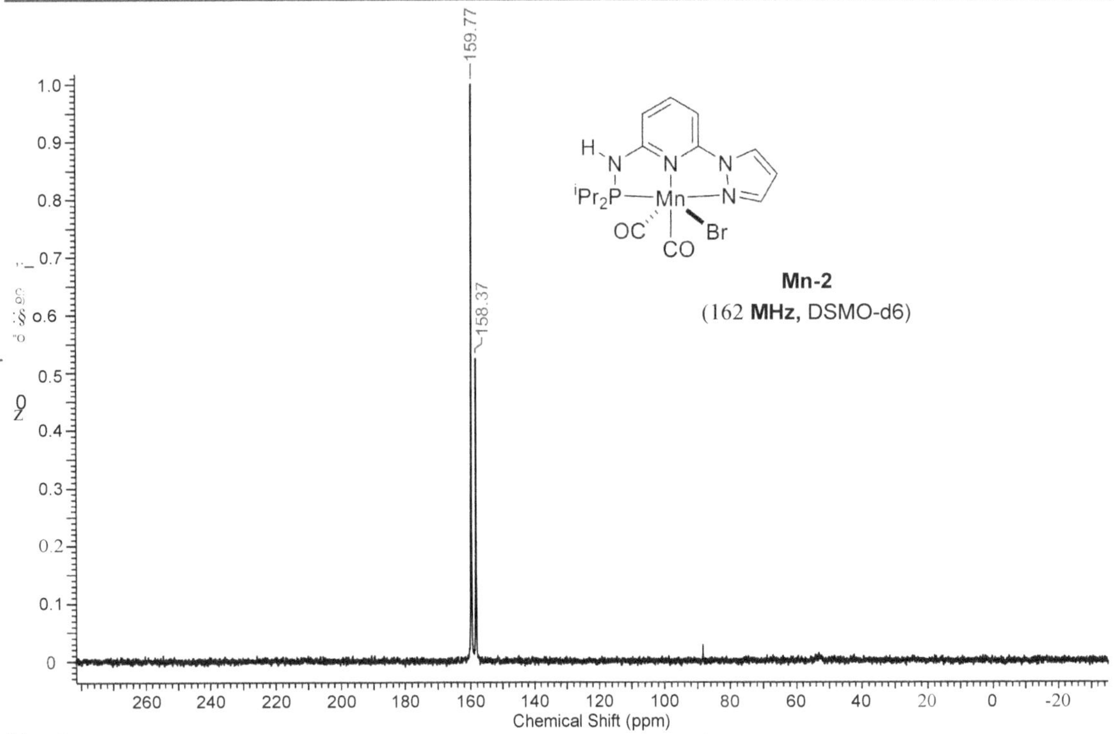

$^{31}P\{^1H\}$-NMR spectrum of **Mn-2** complex (mixture of two isomers).

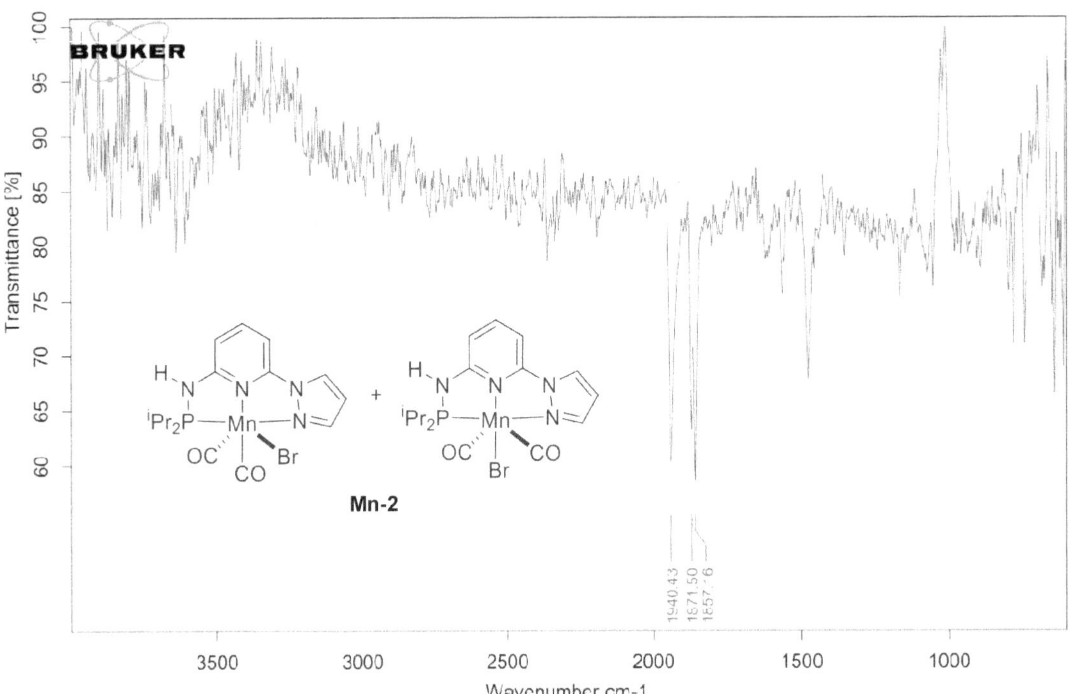

IR spectrum of **Mn-2** complex.

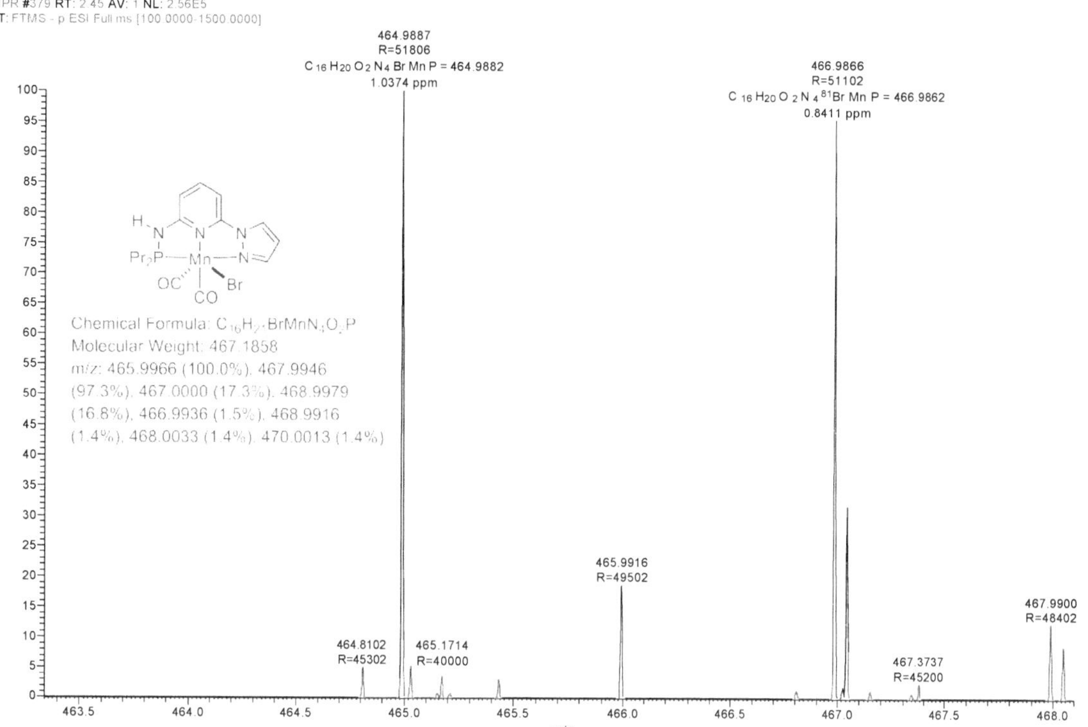

ESI-MS (–ve mode) of **Mn-2** complex.

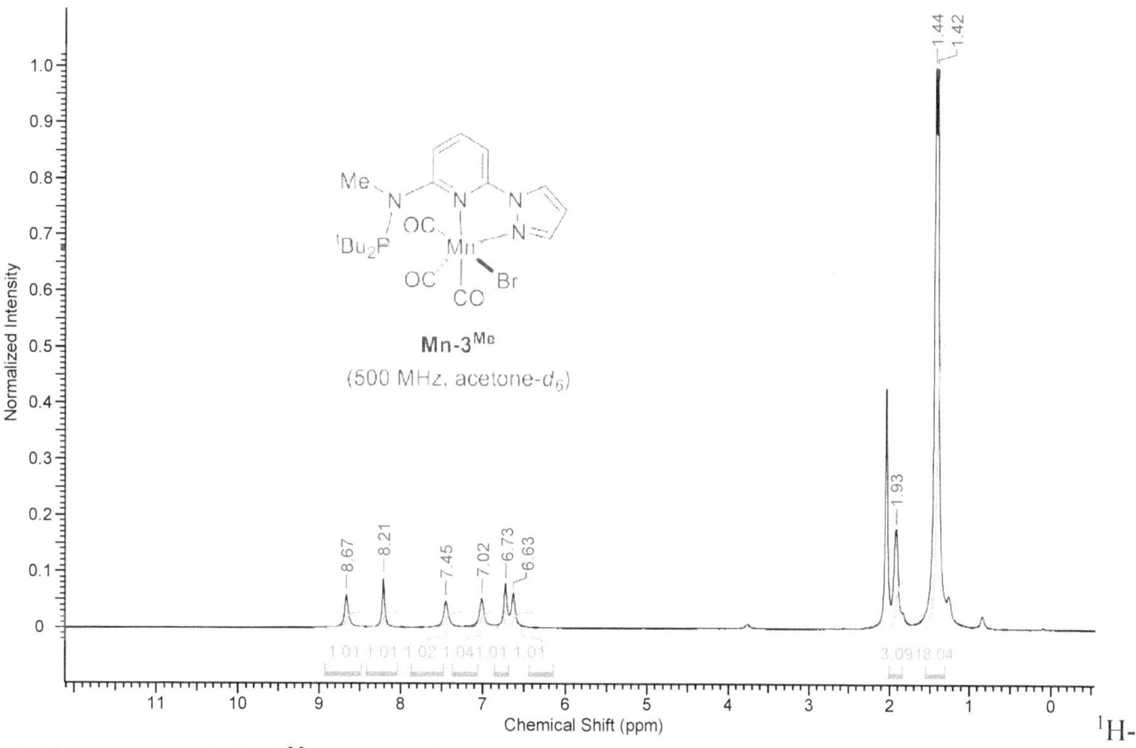

^{1}H-NMR spectrum of **Mn-3^{Me}** complex in acetone-d_6.

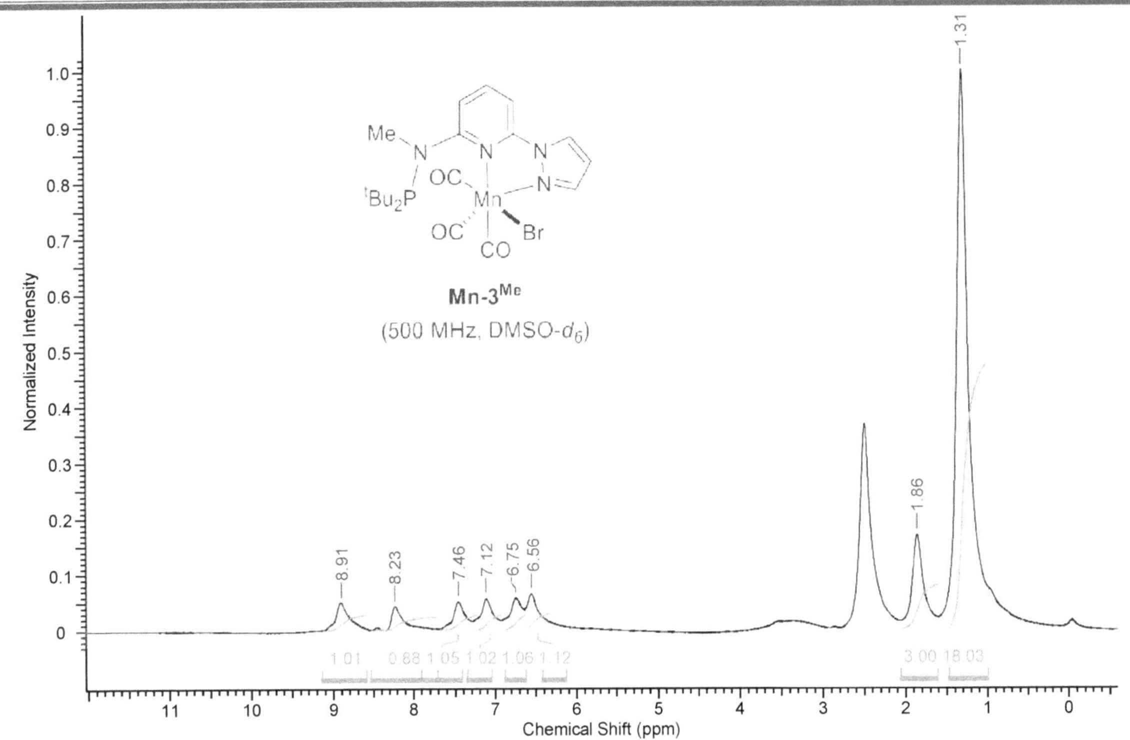

^{1}H-NMR spectrum of **Mn-3^{Me}** complex in DMSO-d_6.

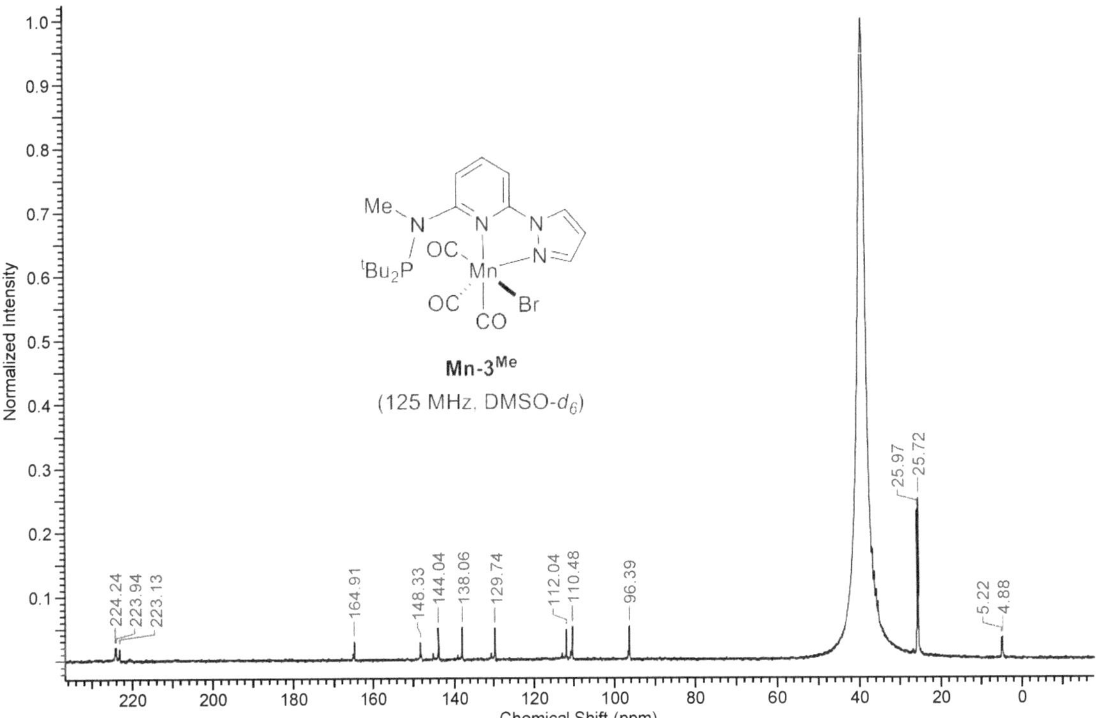

^{13}C{^{1}H}-NMR spectrum of **Mn-3^{Me}** complex

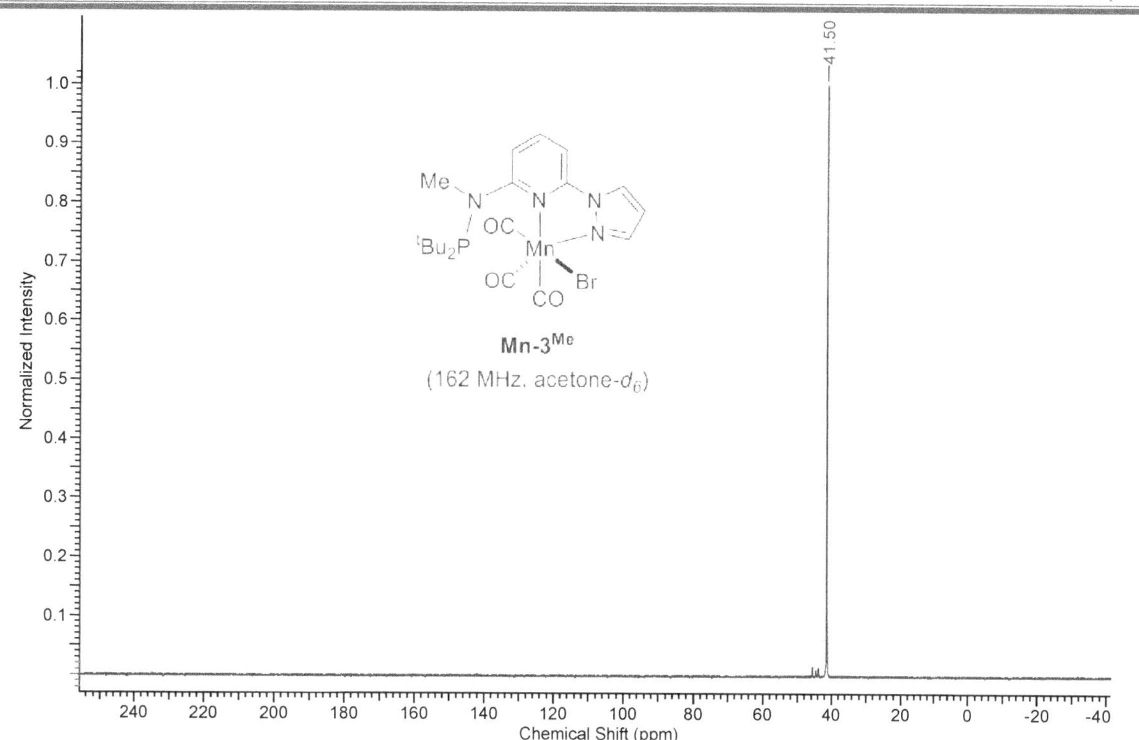

$^{31}P\{^1H\}$-NMR spectrum of **Mn-3^{Me}** complex.

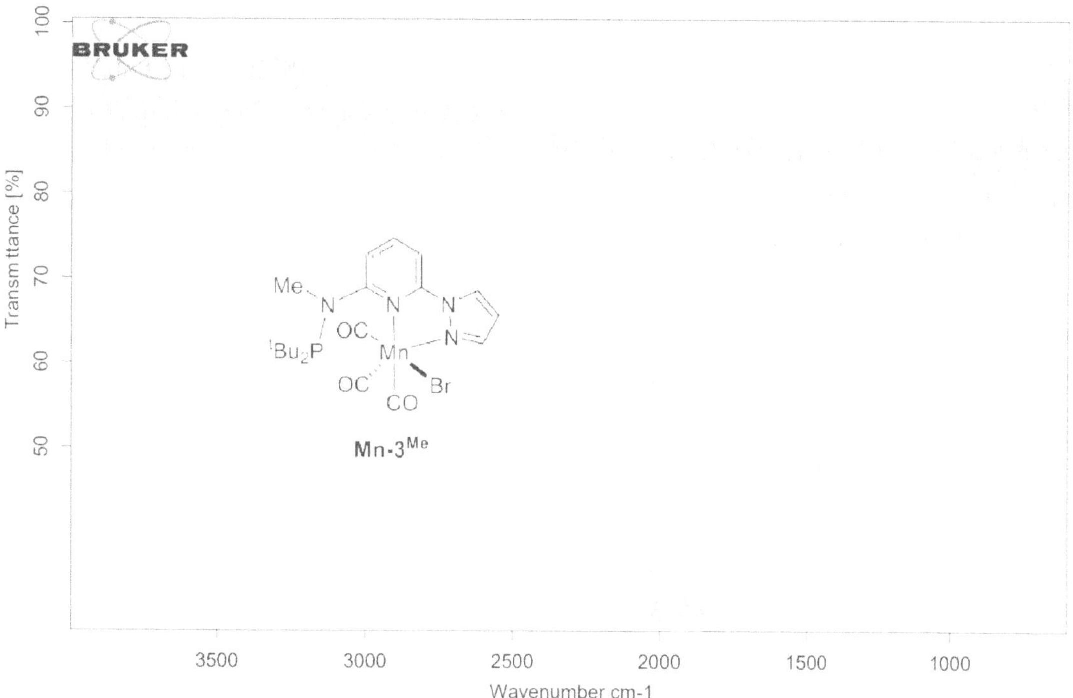

IR spectrum of **Mn-3^{Me}** complex.

2.4.7 ^{1}H and ^{13}C NMR Spectra of Selected Hydrogenated Compounds

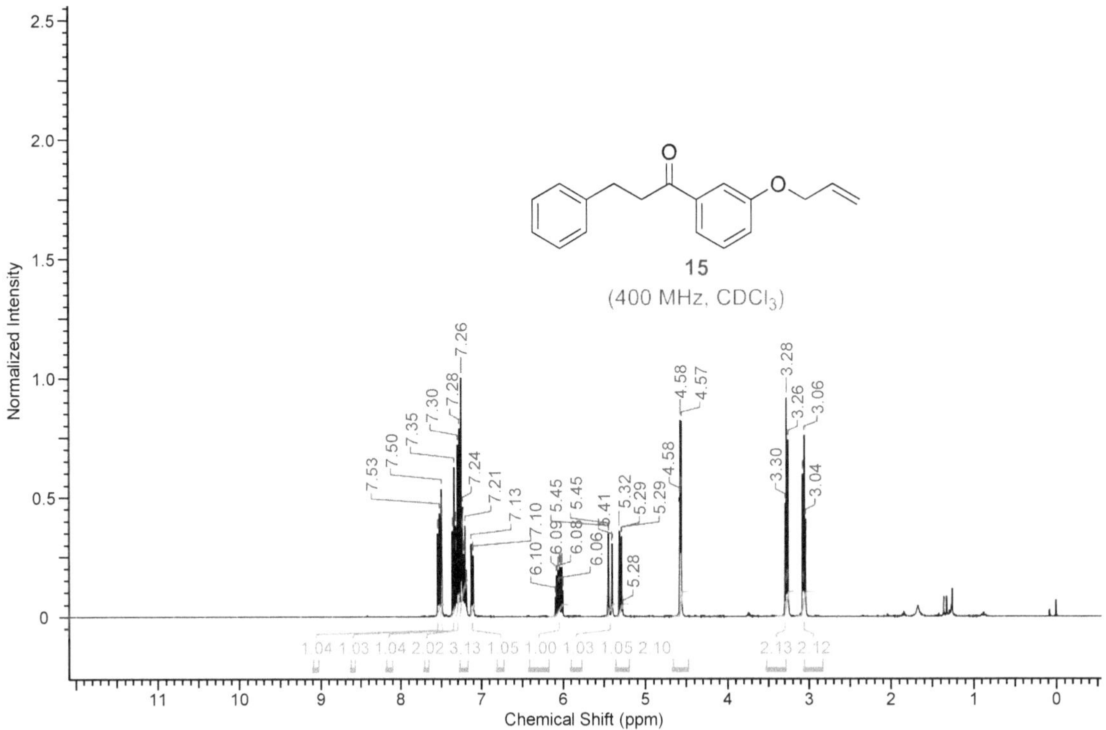

^{1}H-NMR spectrum of **15**.

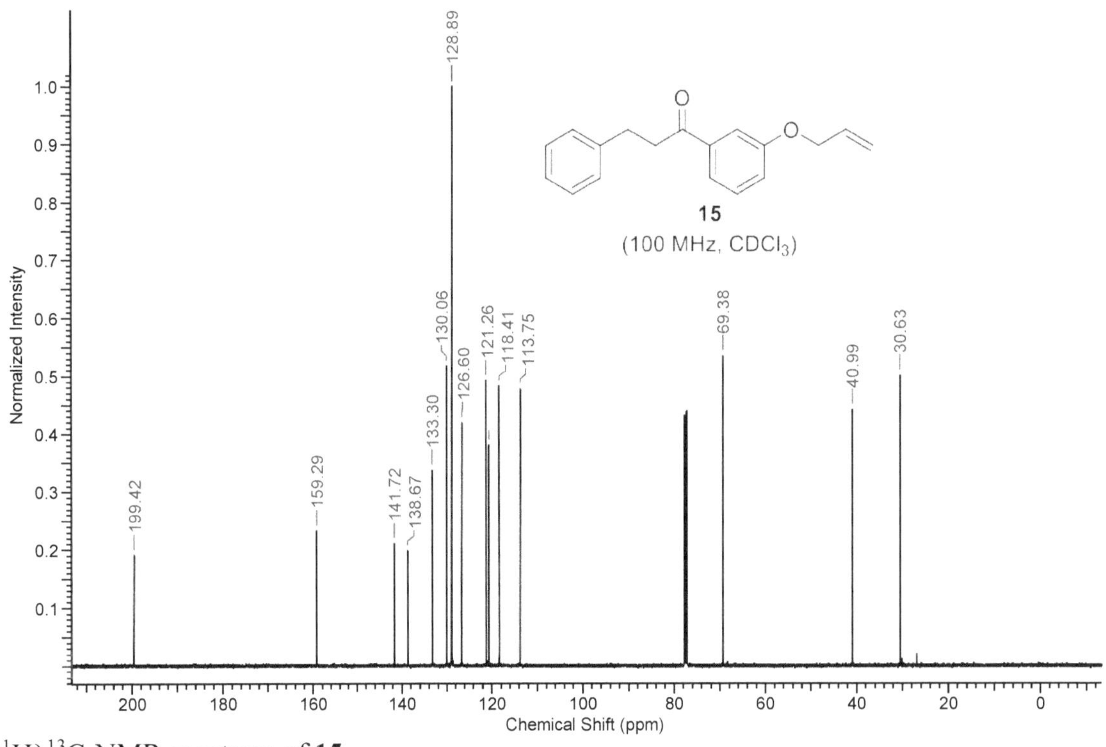

{^{1}H}^{13}C-NMR spectrum of **15**.

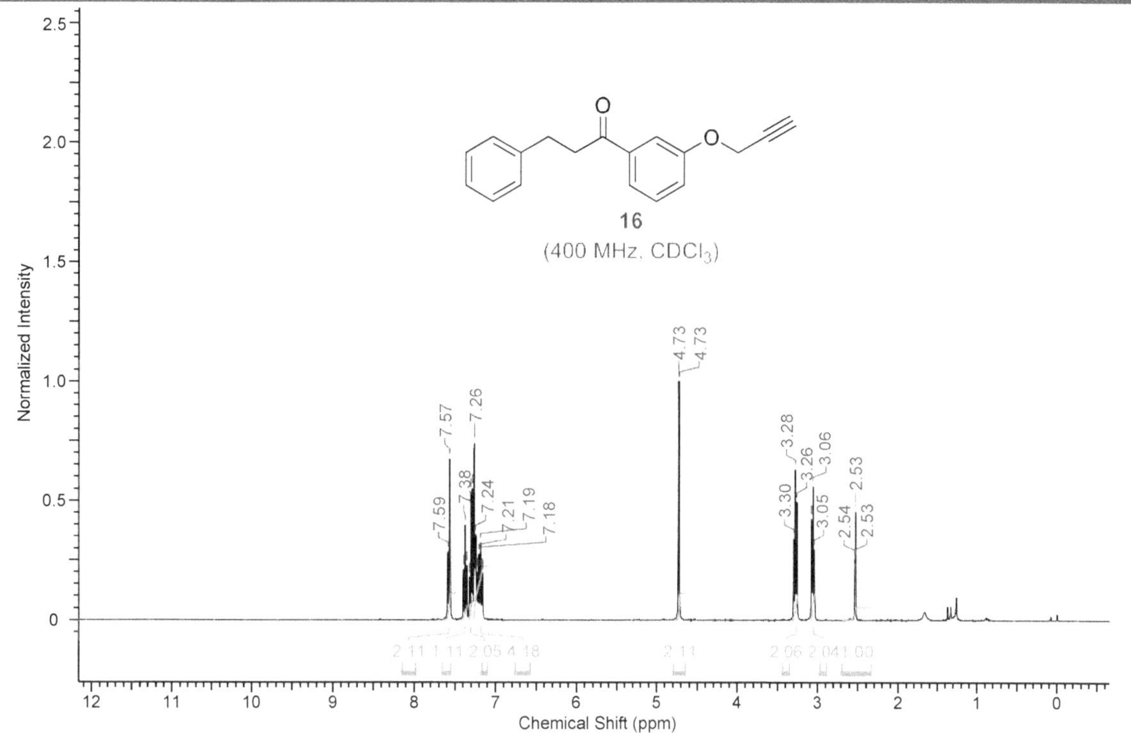

^{1}H-NMR spectrum of **16**.

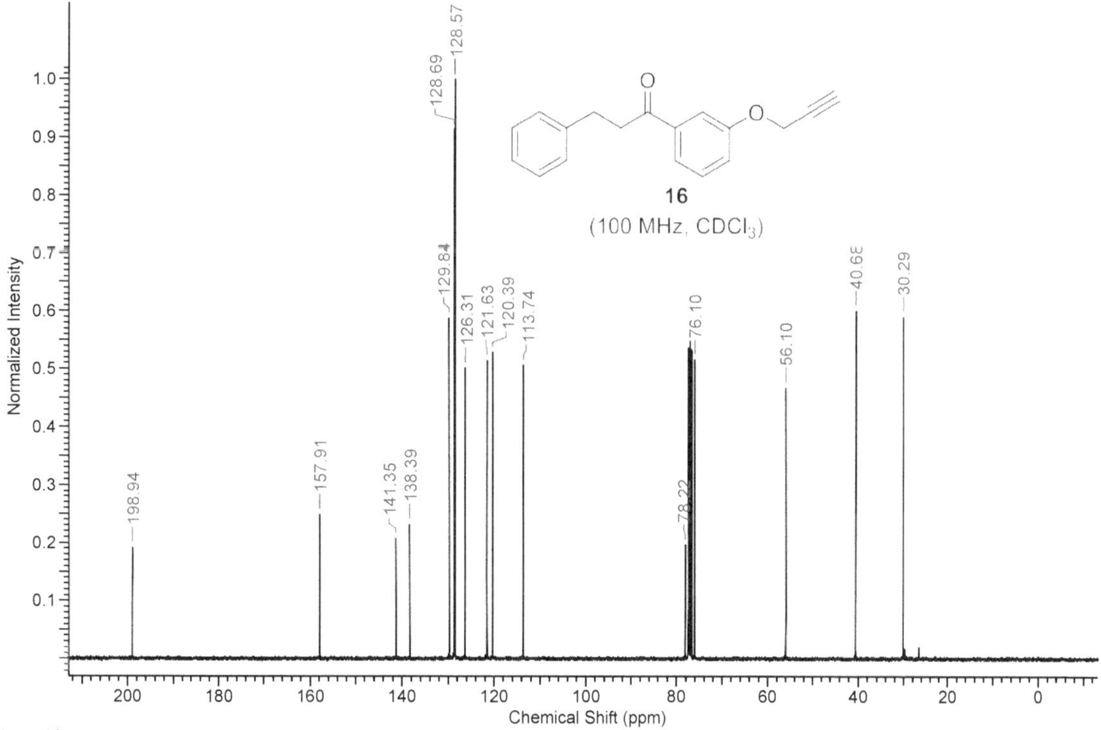

{^{1}H}^{13}C-NMR spectrum of **16**.

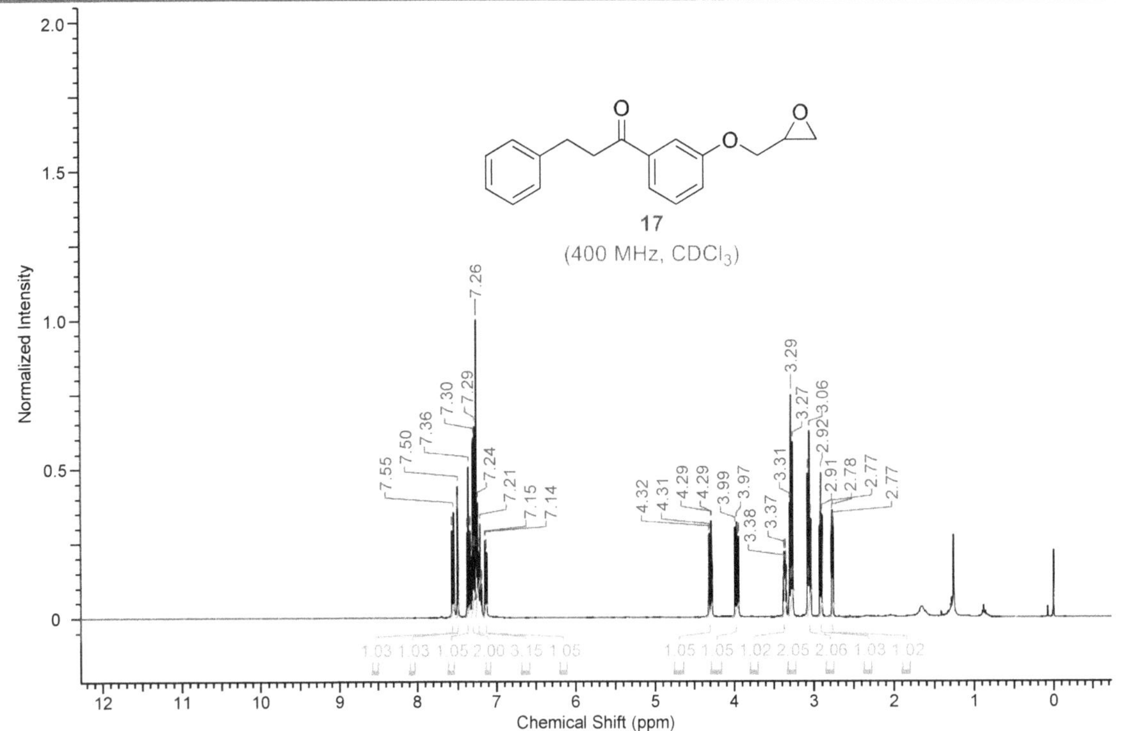

^{1}H-NMR spectrum of **17**.

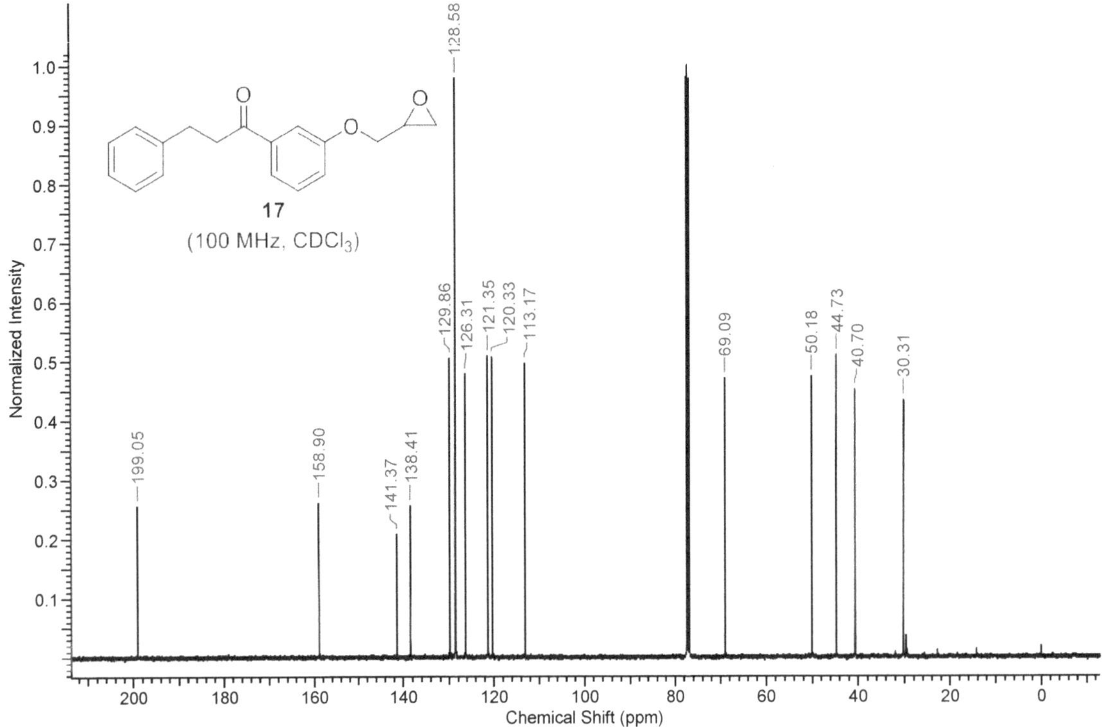

{^{1}H}^{13}C-NMR spectrum of **17**.

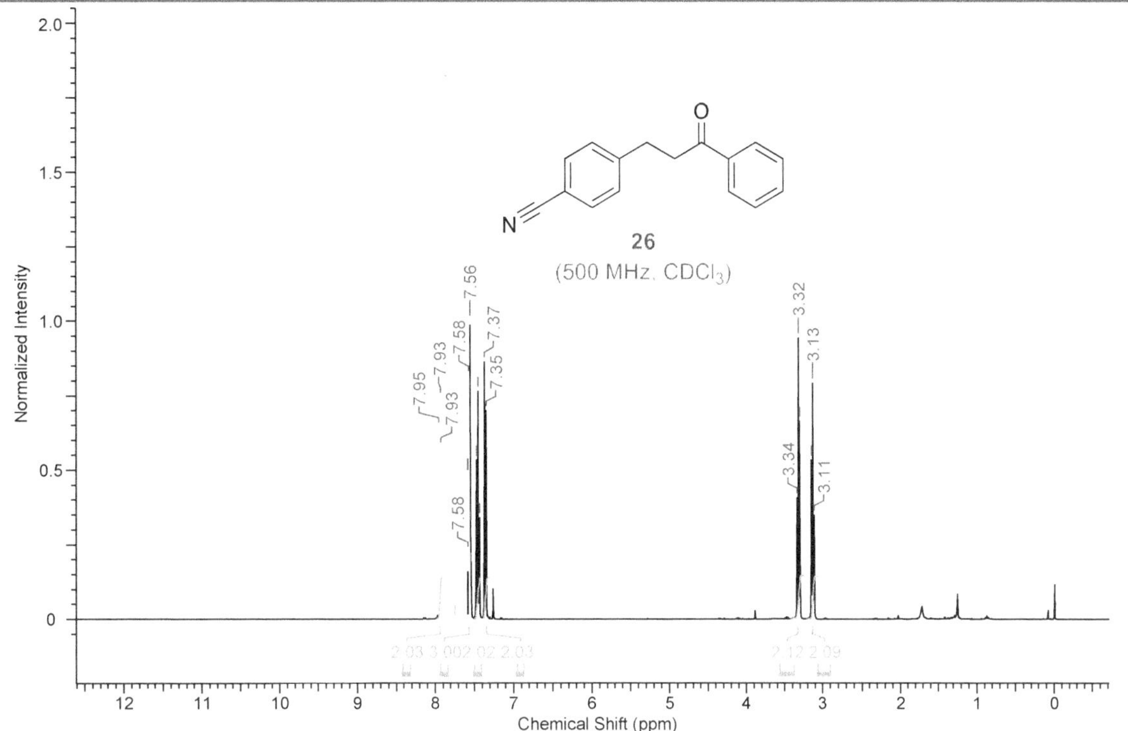

^{1}H-NMR spectrum of **26**.

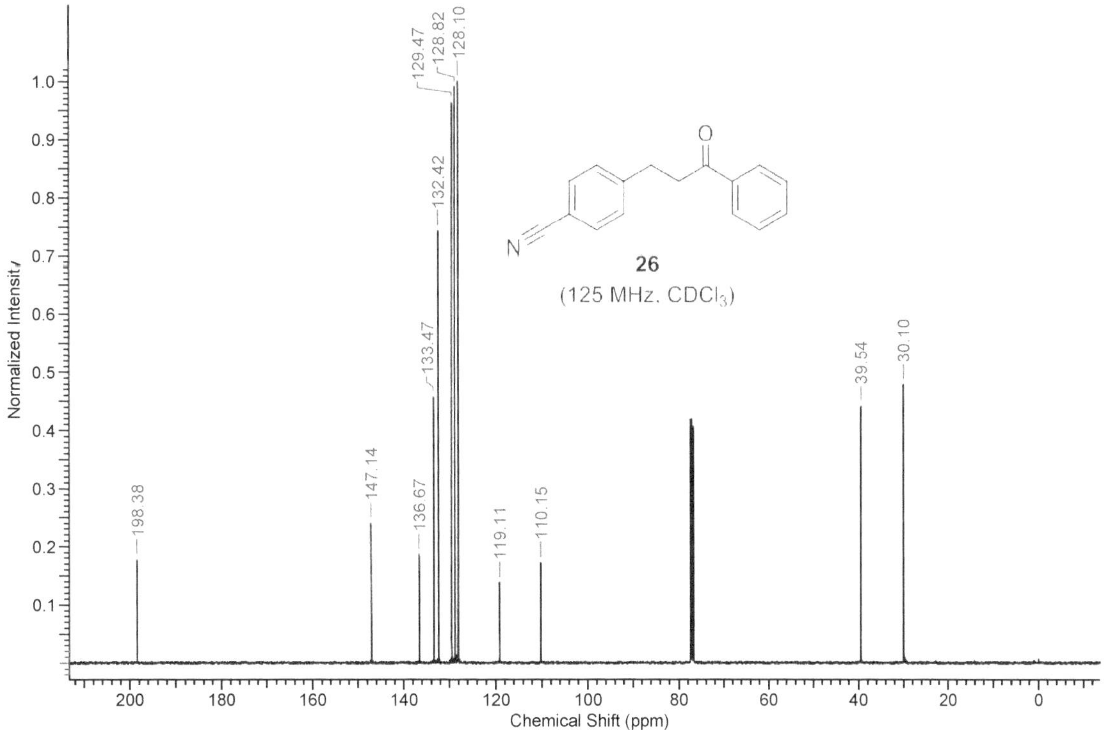

{^{1}H}^{13}C-NMR spectrum of **26**.

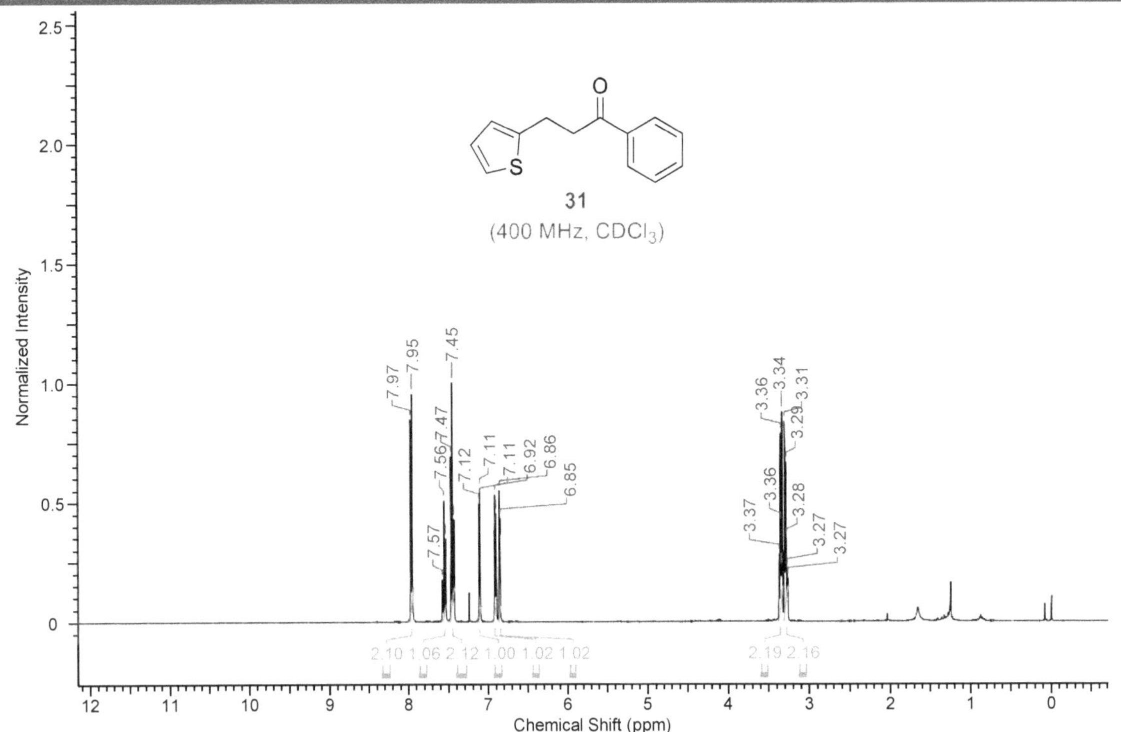

^{1}H-NMR spectrum of **31**.

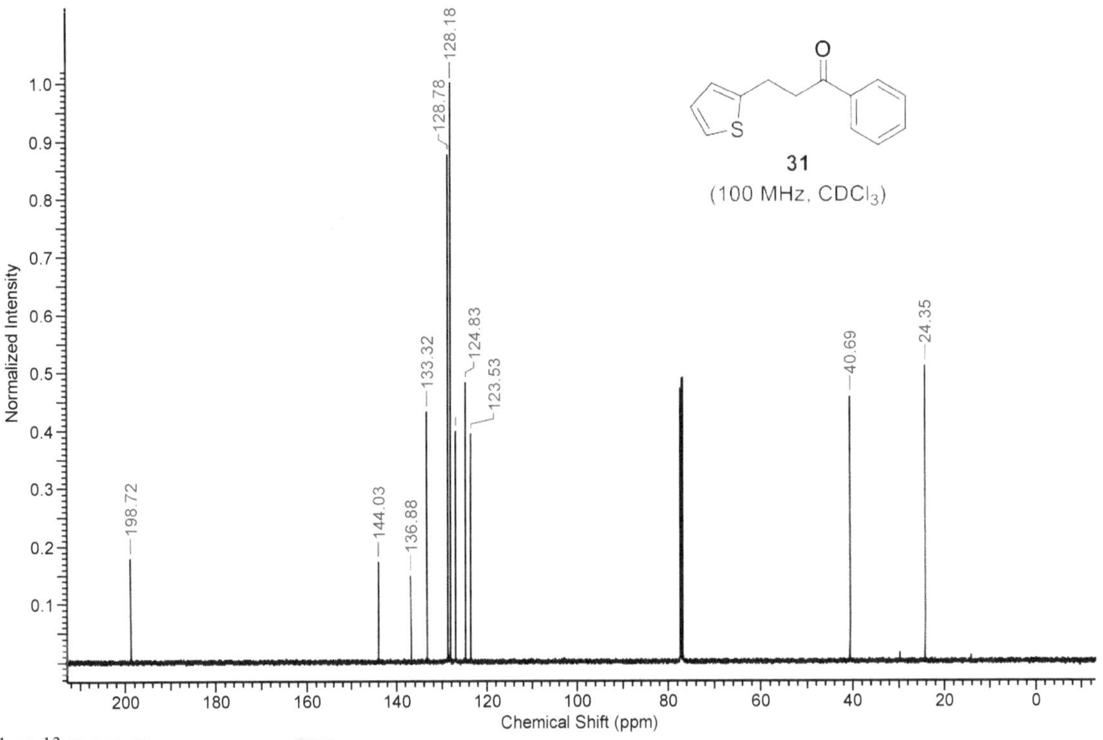

{^{1}H}^{13}C-NMR spectrum of **31**.

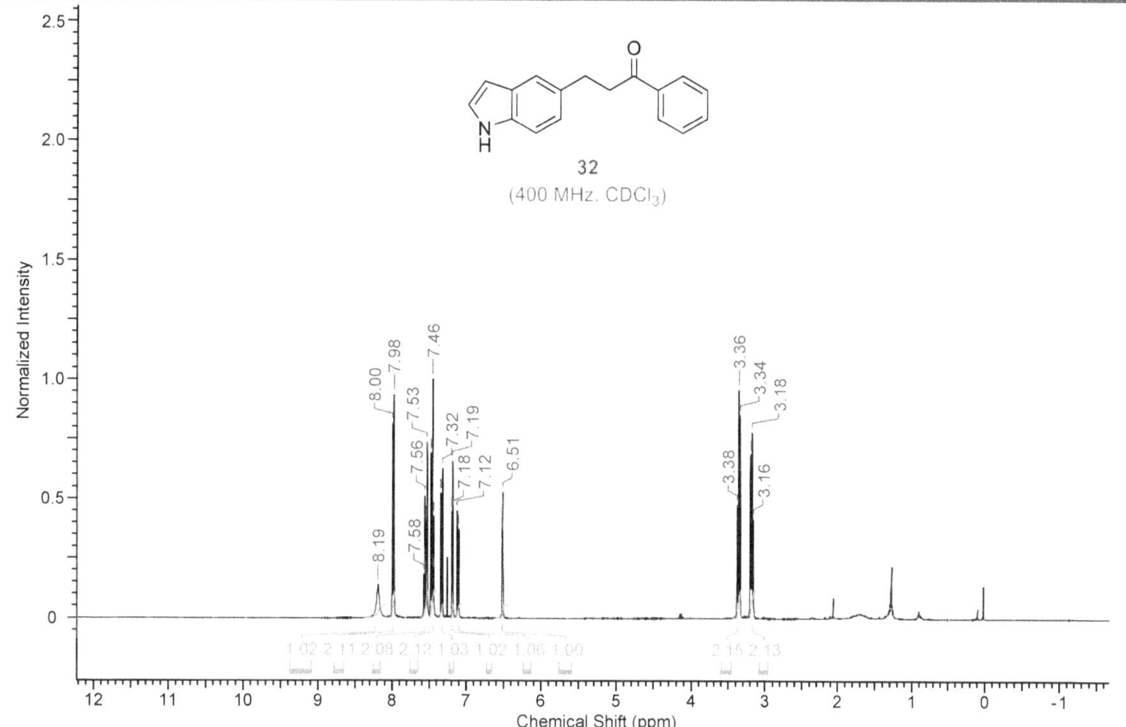

^{1}H-NMR spectrum of **32**.

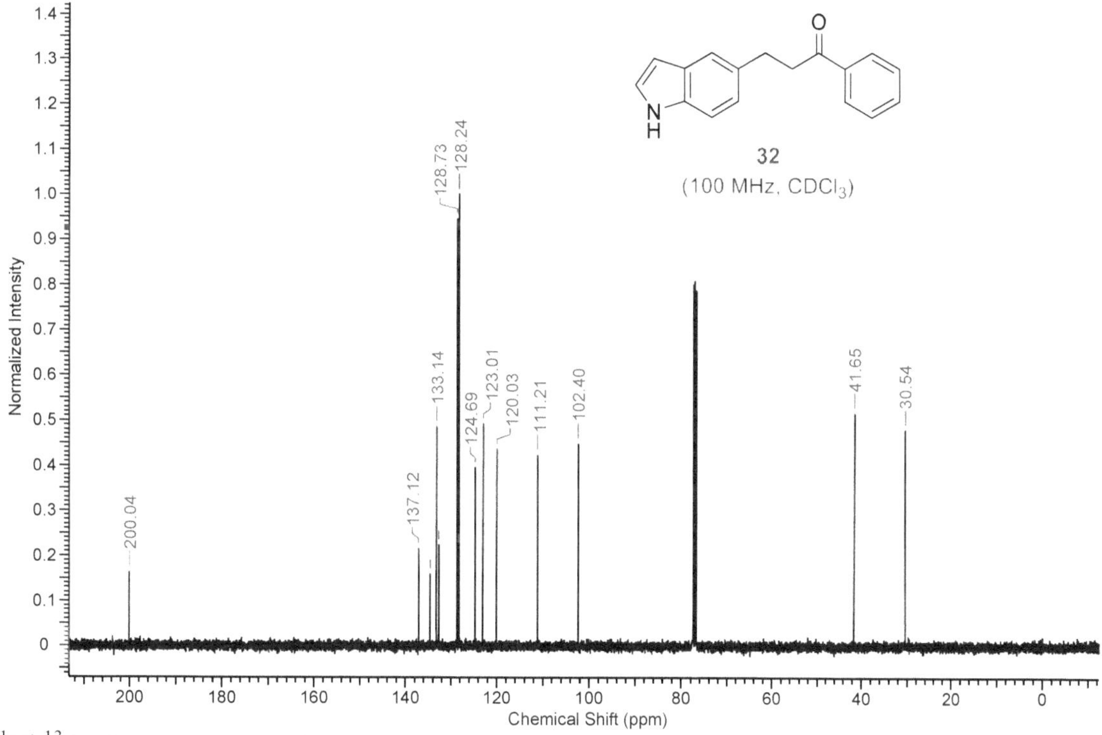

{^{1}H}^{13}C-NMR spectrum of **32**.

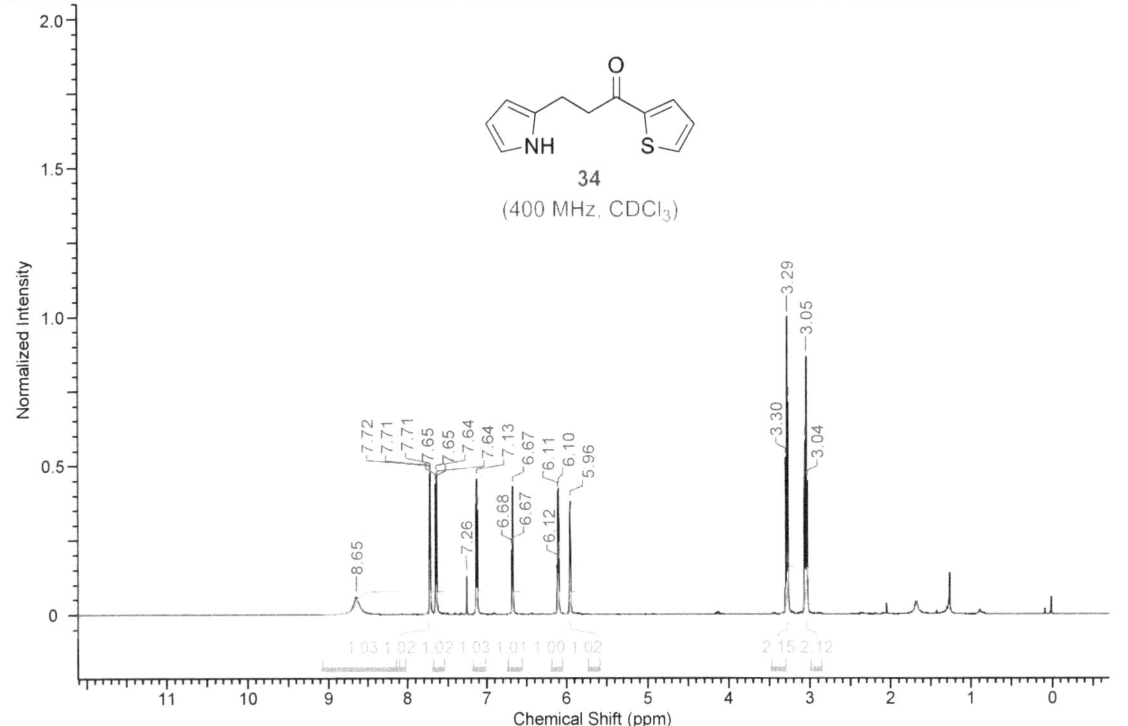

^{1}H-NMR spectrum of **34**.

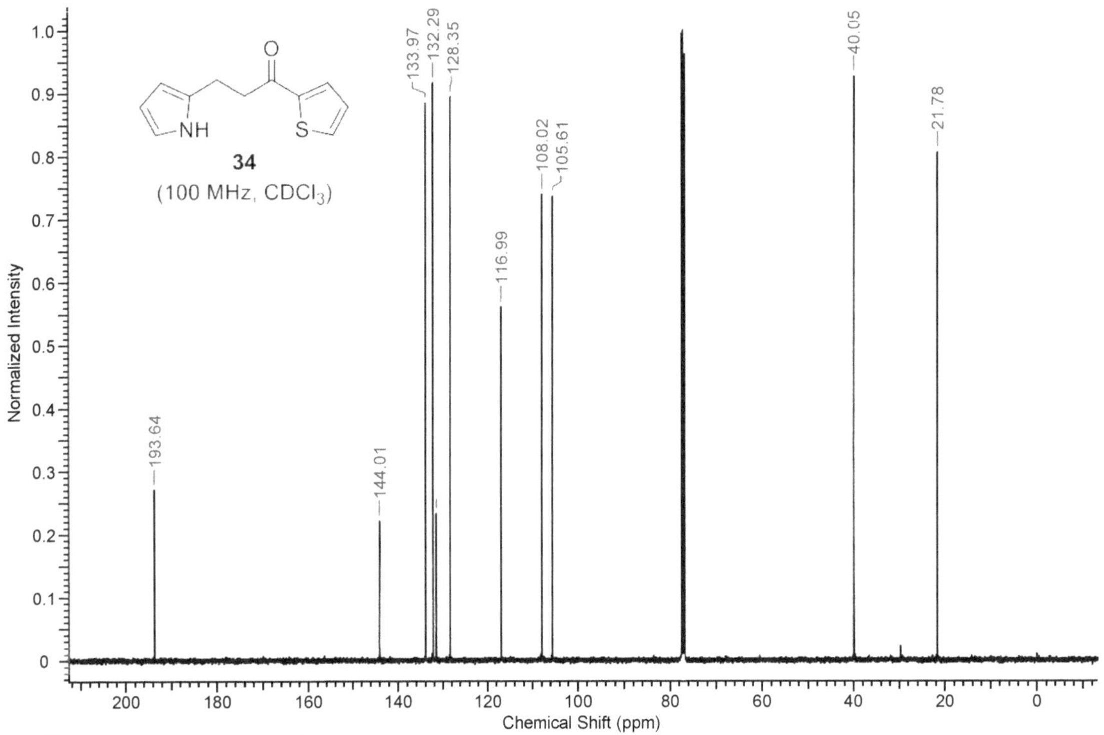

{^{1}H}^{13}C-NMR spectrum of **34**.

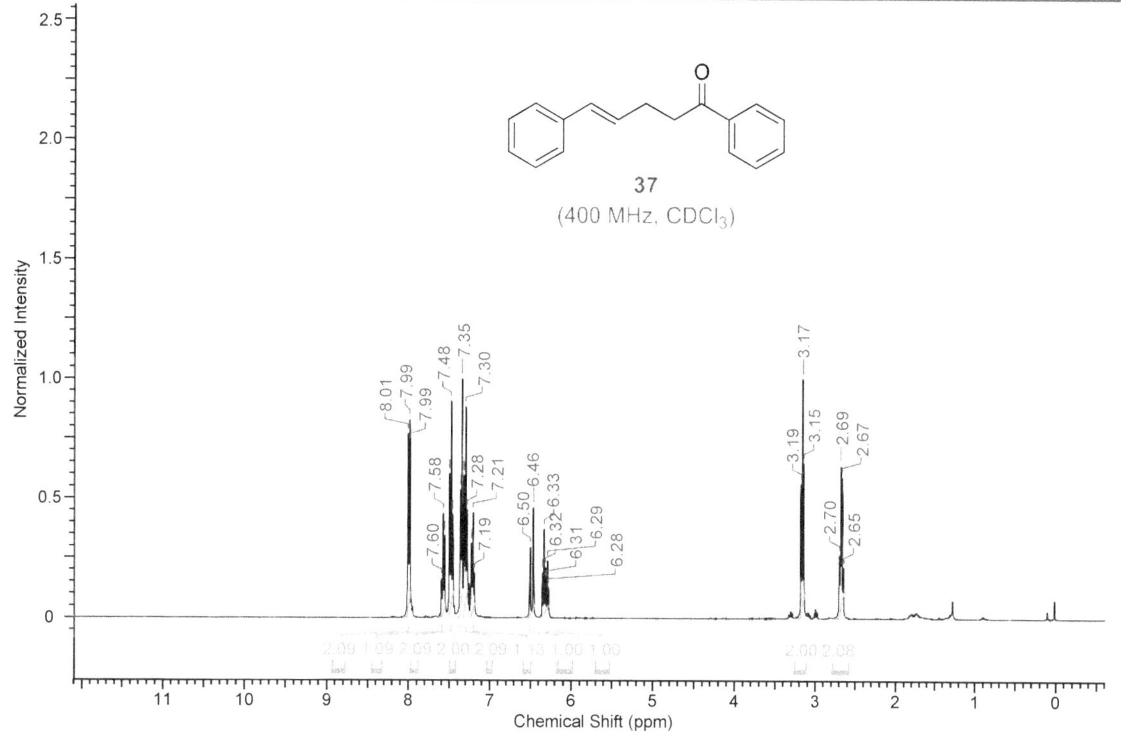

^{1}H-NMR spectrum of **37**.

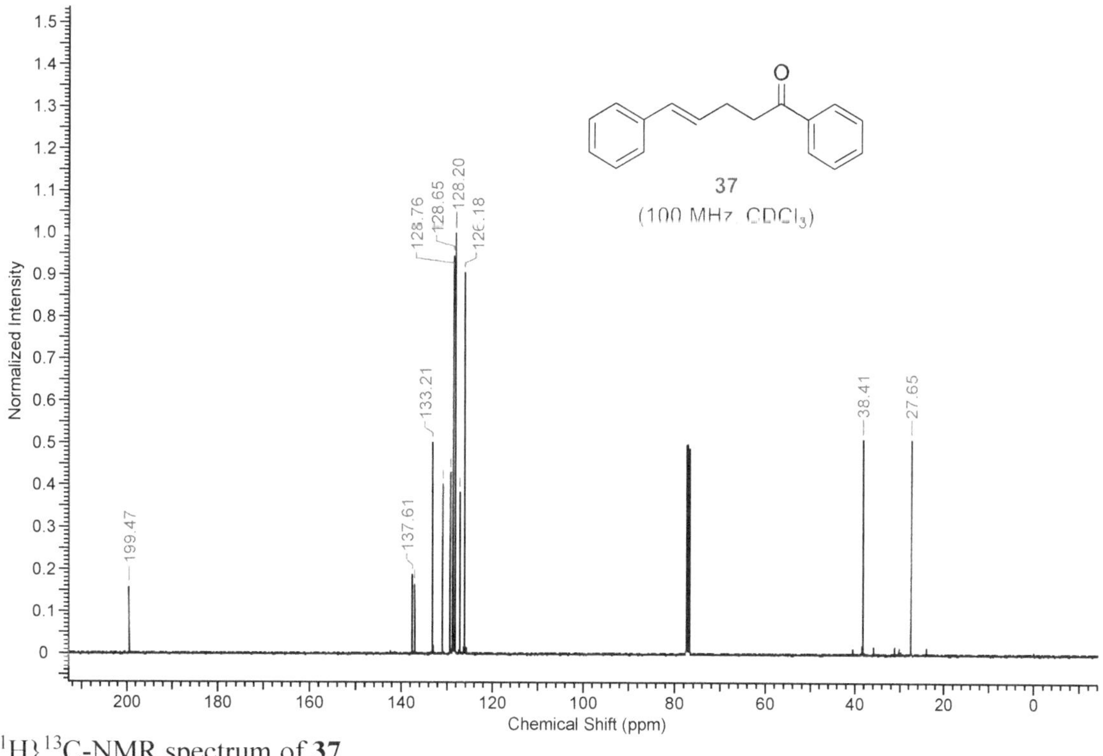

{^{1}H}^{13}C-NMR spectrum of **37**.

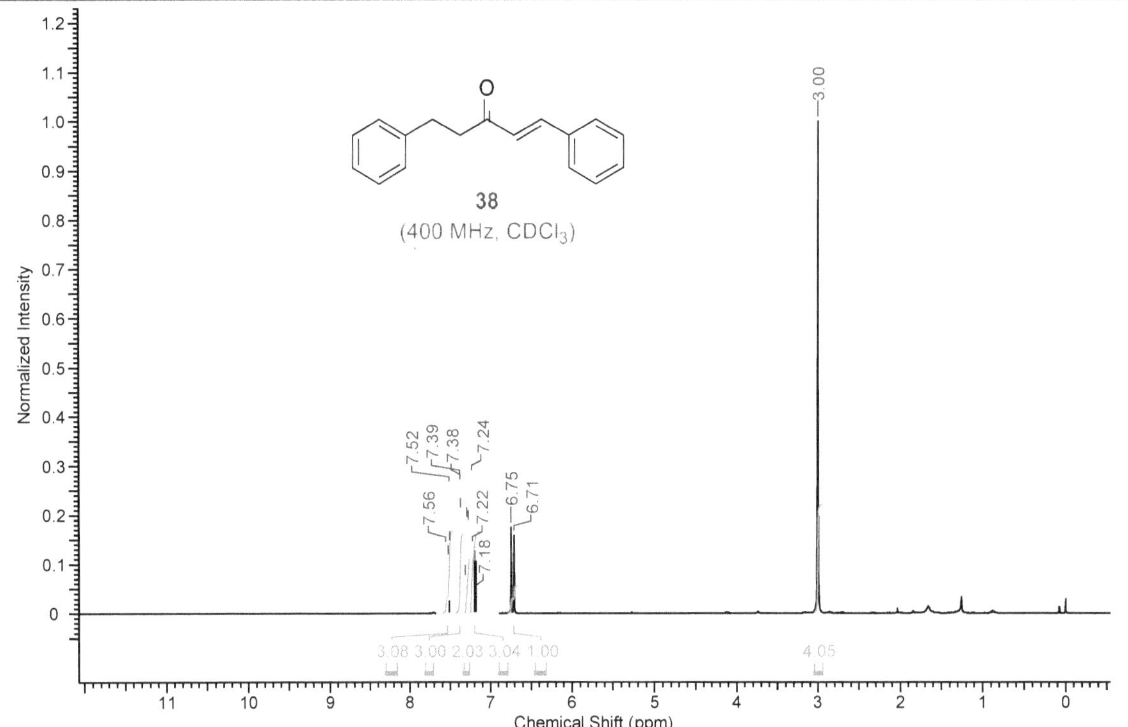

^{1}H-NMR spectrum of **38**.

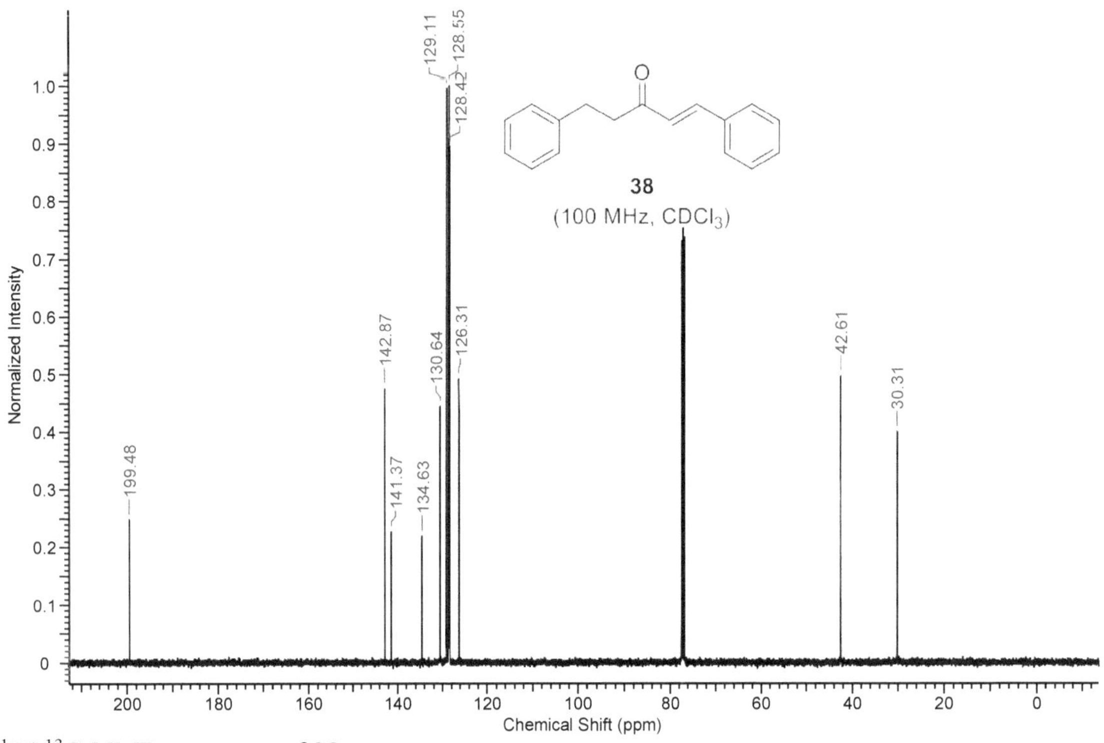

{^{1}H}^{13}C-NMR spectrum of **38**.

2.5 REFERENCES

(1) Rylander, P. N., Academic Press, New York. Academic Press: **1979**; pp 1-12.

(2) Andersson, P. G.; Munslow, I. J., Wiley-VCH Verlag GmbH & Co. **2008**.

(3) Pritchard, J.; Filonenko, G. A.; van Putten, R.; Hensen, E. J. M.; Pidko, E. A., *Chem. Soc. Rev.* **2015**, *44*, 3808-3833.

(4) Tamura, M.; Nakagawa, Y.; Tomishige, K., *J. Jpn. Petroleum Inst.* **2019**, *62*, 106-119.

(5) Wang, Y.; Wang, M.; Li, Y.; Liu, Q., Chem, 2021, 7, 1180–1223.

(6) De Vries, J. G.; Elsevier, C. J., Weinheim Wiley-VCH: **2007**.

(7) Noyori, R.; Ohkuma, T., *Angew. Chem., Int. Ed.* **2001**, *40*, 40-73.

(8) Xie, J.-H.; Zhu, S.-F.; Zhou, Q.-L., *Chem. Rev.* **2011**, *111*, 1713-1760.

(9) Clarke, M. L., *Catal. Sci. Technol.* **2012**, *2*, 2418-2423.

(10) Wang, D.-S.; Chen, Q.-A.; Lu, S.-M.; Zhou, Y.-G., *Chem. Rev.* **2012**, *112*, 2557-2590.

(11) Werkmeister, S.; Junge, K.; Beller, M., *Org. Pro. Res. Dev.* **2014**, *18*, 289-302.

(12) He, Y.-M.; Feng, Y.; Fan, Q.-H., *Acc. Chem. Res.* **2014**, *47*, 2894-2906.

(13) Filonenko, G. A.; van Putten, R.; Hensen, E. J. M.; Pidko, E. A., *Chem. Soc. Rev.* **2018**, *47*, 1459-1483.

(14) Zell, T.; Langer, R., *ChemCatChem* **2018**, *10*, 1930-1940.

(15) Alig, L.; Fritz, M.; Schneider, S., *Chem. Rev.* **2019**, *119*, 2681-2751.

(16) Irrgang, T.; Kempe, R., *Chem. Rev.* **2019**, *119*, 2524-2549.

(17) Bauer, I.; Knölker, H.-J., *Chem. Rev.* **2015**, *115*, 3170-3387.

(18) Zell, T.; Milstein, D., *Acc. Chem. Res.* **2015**, *48*, 1979-1994.

(19) Chakraborty, S.; Bhattacharya, P.; Dai, H.; Guan, H., *Acc. Chem. Res.* **2015**, *48*, 1995-2003.

(20) Fürstner, A., *ACS Cent. Sci.* **2016**, *2*, 778-789.

(21) Mukherjee, A.; Milstein, D., *ACS Catal.* **2018**, *8*, 11435-11469.

(22) Liu, W.; Sahoo, B.; Junge, K.; Beller, M., *Acc. Chem. Res.* **2018**, 51, 1858-1869.

(23) Ai, W.; Zhong, R.; Liu, X.; Liu, Q., *Chem. Rev.* **2019**, *119*, 2876-2953.

(24) Bullock, R. M., *Science* **2013**, *342*, 1054-1055.

(25) Zuo, W.; Lough, A. J.; Li, Y. F.; Morris, R. H., *Science* **2013**, *342*, 1080-1083.

(26) Friedfeld, M. R.; Shevlin, M.; Hoyt, J. M.; Krska, S. W.; Tudge, M. T.; Chirik, P. J., *Science* **2013**, *342*, 1076-1080.

(27) Friedfeld, M. R.; Zhong, H.; Ruck, R. T.; Shevlin, M.; Chirik, P. J., *Science* **2018**, *360*, 888-893.

(28) Elangovan, S.; Topf, C.; Fischer, S.; Jiao, H.; Spannenberg, A.; Baumann, W.;

Ludwig, R.; Junge, K.; Beller, M., *J. Am. Chem. Soc.* **2016**, *138*, 8809-8814.

(29) Kallmeier, F.; Irrgang, T.; Dietel, T.; Kempe, R., *Angew. Chem., Int. Ed.* **2016**, *55*, 11806-11809.

(30) Bruneau-Voisine, A.; Wang, D.; Roisnel, T.; Darcel, C.; Sortais, J.-B., *Catal. Commun.* **2017**, *92*, 1-4.

(31) Wei, D.; Bruneau-Voisine, A.; Chauvin, T. o.; Dorcet, V.; Roisnel, T.; Valyaev, D. A.; Lugan, N. l.; Sortais, J.-B., *Adv. Synth. Catal.* **2018**, *360*, 676-681.

(32) Buhaibeh, R.; Filippov, O. A.; Bruneau-Voisine, A.; Willot, J.; Duhayon, C.; Valyaev, D. A.; Lugan, N.; Canac, Y.; Sortais, J.-B., *Angew. Chem., Int. Ed.* **2019**, *58*, 6727-6731.

(33) Buhaibeh, R.; Duhayon, C.; Valyaev, D. A.; Sortais, J.-B.; Canac, Y., *Organometallics* **2021**, *40*, 231-241.

(34) Glatz, M.; Stöger, B.; Himmelbauer, D.; Veiros, L. F.; *ACS Catal.* **2018**, *8*, 4009-4016.

(35) Weber, S.; Stöger, B.; Kirchner, K., *Org. Lett.* **2018**, *20*, 7212-7215.

(36) Weber, S.; Brünig, J.; Veiros, L. F.; Kirchner, K., *Organometallics* **2021**, *40*, 1388-1394.

(37) Pan, H.-J.; Hu, X., *Angew. Chem., Int. Ed.* **2020**, *59*, 4942-4946.

(38) Yang, W.; Chernyshov, I. Y.; van Schendel, R. K. A.; Weber, M.; Müller, C.; Filonenko, G. A.; Pidko, E. A., *Nat. Commun.* **2021**, *12*, 12.

(39) Vielhaber, T.; Topf, C., *Appl. Catal. A: Gen.* **2021**, *623*, 118280.

(40) Maji, B.; Barman, M. K., *Synthesis* **2017**, *49*, 3377-3393.

(41) Garbe, M.; Junge, K.; Beller, M., *Eur. J. Org. Chem.* **2017**, *2017*, 4344-4362.

(42) Kallmeier, F.; Kempe, R., *Angew. Chem. Int. Ed.* **2018**, *57*, 46-60.

(43) Gorgas, N.; Kirchner, K., *Acc. Chem. Res.* **2018**, *51*, 1558-1569.

(44) Wei, D.; Bruneau-Voisine, A.; Valyaev, D. A.; Lugan, N.; Sortais, J.-B., *Chem. Commun.* **2018**, *54*, 4302-4305.

(45) Wang, Y.; Zhu, L.; Shao, Z.; Li, G.; Lan, Y.; Liu, Q., *J. Am. Chem. Soc.* **2019**, *141*, 17337-17349.

(46) Freitag, F.; Irrgang, T.; Kempe, R., *J. Am. Chem. Soc.* **2019**, *141*, 11677-11685.

(47) Wang, Z.; Chen, L.; Mao, G.; Wang, C., *Chin. Chem. Lett.* **2020**, *31*, 1890-1894.

(48) Wang, Y.; Liu, S.; Yang, H.; Li, H.; Lan, Y.; Liu, Q., *Nat. Chem.* **2022**, DOI: 10.1038/s41557-022-01036-6.

(49) Widegren, M. B.; Harkness, G. J.; Slawin, A. M. Z.; Cordes, D. B.; Clarke, M. L.,

Angew. Chem., Int. Ed. **2017**, *56*, 5825-5828.

(50) Garbe, M.; Junge, K.; Walker, S.; Wei, Z.; Jiao, H.; Spannenberg, A.; Bachmann, S.; Scalone, M.; Beller, M., *Angew. Chem., Int. Ed.* **2017**, *56*, 11237-11241.

(51) Garbe, M.; Wei, Z.; Tannert, B.; Spannenberg, A.; Jiao, H.; Bachmann, S.; Scalone, M.; Junge, K.; Beller, M., *Adv. Synth. Catal.* **2019**, *361*, 1913-1920.

(52) Zhang, L.; Tang, Y.; Han, Z.; Ding, K., *Angew. Chem., Int. Ed.* **2019**, *58*, 4973-4977.

(53) Zhang, L.; Wang, Z.; Han, Z.; Ding, K., *Angew. Chem., Int. Ed.* **2020**, *59*, 15565-15569.

(54) Ling, F.; Hou, H.; Chen, J.; Nian, S.; Yi, X.; Wang, Z.; Song, D.; Zhong, W., *Org. Lett.* **2019**, *21*, 3937-3941.

(55) Zeng, L.; Yang, H.; Zhao, M.; Wen, J.; Tucker, J. H. R.; Zhang, X., *ACS Catal.* **2020**, *10*, 13794-13799.

(56) Ling, F.; Chen, J.; Nian, S.; Hou, H.; Yi, X.; Wu, F.; Xu, M.; Zhong, W., *Synlett* **2020**, *31*, 285-289.

(57) Seo, C. S. G.; Tsui, B. T. H.; Gradiski, M. V.; Smith, S. A. M.; Morris, R. H., *Catal. Sci. Technol.* **2021**, *11*, 3153-3163.

(58) Liu, C.; Wang, M.; Liu, S.; Wang, Y.; Peng, Y.; Lan, Y.; Liu, Q., *Angew. Chem., Int. Ed.* **2021**, *60*, 5108-5113;

(59) Liu, C.; Wang, M.; Xu, Y.; Li, Y.; Liu, Q., *Angew. Chem. Int. Ed.* **2022**, *61*, DOI: 10.1002/anie.202202814.

(60) Weber, S.; Stöger, B.; Veiros, L. F.; Kirchner, K., *ACS Catal.* **2019**, *9*, 9715-9720.

(61) Rahaman, S. M. W.; Pandey, D. K.; Rivada-Wheelaghan, O.; Dubey, A.; Fayzullin, R. R.; Khusnutdinova, J. R., *ChemCatChem* **2020**, *12*, 5912-5918.

(62) Hu, Q.; Hu, Y.; Liu, Y.; Zhang, Z.; Liu, Y.; Zhang, W., *Chem. Eur. J.* **2017**, *23*, 1040-1043.

(63) Chang, W.; Gong, X.; Wang, S.; Xiao, L.-P.; Song, G., *Org. Biomol. Chem.* **2017**, *15*, 3466-3471.

(64) Soto, M.; Soengas, R. G.; Rodríguez-Solla, H., *Adv. Synth. Catal.* **2020**, *362*, 5422-5431.

(65) Gu, Y.; Norton, J. R.; Salahi, F.; Lisnyak, V. G.; Zhou, Z.; Snyder, S. A., *J. Am. Chem. Soc.* **2021**, *143*, 9657-9663.

(66) Gong, D.; Liu, W.; Chen, T.; Chen, Z.-R.; Huang, K.-W., *J. Mol. Catal. A Chem.* **2014**, *395*, 100-107.

(67) Gong, D.; Zhang, X.; Huang, K.-W., *Dalton Trans.* **2016**, *45*, 19399-19407.

(68) Chen, H.; Pan, W.; Huang, K.-W.; Zhang, X.; Gong, D., *Polym. Chem.* **2017**, *8*, 1805-1814.

(69) Gonçalves, T. P.; Huang, K.-W., *J. Am. Chem. Soc.* **2017**, *139*, 13442-13449.

(70) Shimbayashi, T.; Fujita, K.-i., *Catalysts* **2020**, *10*, 635.

(71) Genç, S.; Günnaz, S.; Çetinkaya, B.; Gü lcemal, S. l.; Gü lcemal, D., *J. Org. Chem.* **2018**, *83*, 2875-2881.

(72) Cao, X.-N.; Wan, X.-M.; Yang, F.-L.; Li, K.; Hao, X.-Q.; Shao, T.; Zhu, X.; Song, M.-P., *J. Org. Chem.* **2018**, *83*, 3657-3668.

(73) Nallagangula, M.; Sujatha, C.; Bhat, V. T.; Namitharan, K., *Chem. Commun.* **2019**, *55*, 8490-8493.

(74) Cao, X.-N.; Wan, X.-M.; Yang, F.-L.; Li, K.; Hao, X.-Q.; Shao, T.; Zhu, X.; Song, M.-P., *J. Org. Chem.,* **2018**, *83*, 3657-3668.

(75) Lan, X.-B.; Ye, Z.; Huang, M.; Liu, J.; Liu, Y.; Ke, Z., *Org. Lett.* **2019**, *21*, 8065-8070.

(76) Genç, S.; Gülcemal, S.; Günnaz, S.; Çetinkaya, B.; Gülcemal, D., *Org. Lett.* **2021**, *23*, 5229-5234.

(77) Rakers, L.; Schäfers, F.; Glorius, F., *Chem. Eur. J.,* **2018**, *24*, 15529-15532.

(78) Lan, X.-B.; Ye, Z.; Huang, M.; Liu, J.; Liu, Y.; Ke, Z., *Org. Lett.,* **2019**, *21*, 8065-8070.

(79) Teja, C.; Nawaz Khan, F. R., *ACS Omega* **2019**, *4*, 8046-8055.

(80) Bhattacharyya, D.; Sarmah, B. K.; Nandi, S.; Srivastava, H. K.; Das, A., *Org. Lett.* **2021**, *23*, 869-875.

(81) Lan, X.-B.; Ye, Z.; Liu, J.; Huang, M.; Shao, Y.; Cai, X.; Liu, Y.; Ke, Z., *ChemSusChem*, **2020**, *13*, 2557-2563.

(82) Ding, Z.-C.; Li, C.-Y.; Chen, J.-J.; Zeng, J.-H.; Tang, H.-T.; Ding, Y.-J.; Zhan, Z.-P., *Adv. Synth. Catal.* **2017**, *359*, 2280-2287.

(83) Verma, A.; Hazra, S.; Dolui, P.; Elias, A. J., *Asian J. Org. Chem.* **2021**, *10*, 626-633.

(84) Jiang, Q.; Guo, T.; Wang, Q.; Wu, P.; Yu, Z., *Adv. Synth. Catal.* **2013**, *355*, 1874-1880.

(85) Genç, S.; Günnaz, S.; Çetinkaya, B.; Gülcemal, S.; Gülcemal, D., *J. Org. Chem.*, **2018**, *83*, 2875-2881.

(86) Jung, S.; Lee, H.; Moon, Y.; Jung, H.-Y.; Hong, S., *ACS Catal.* **2019**, *9*, 9891-9896.

(87) Guven, S.; Kundu, G.; Weßels, A.; Ward, J. S.; Rissanen, K.; Schoenebeck, F., *J. Am. Chem. Soc.* **2021**, *143*, 8375-8380.

(88) Fu, Y.-H.; Bergbreiter, D. E., *ChemCatChem* **2020**, *12*, 6050-6058.

(89) Alanthadka, A.; Bera, S.; Banerjee, D., *J. Org. Chem.* **2019**, *84*, 11676-11686.

(90) Zhou, J.; Jia, M.; Song, M.; Huang, Z.; Steiner, A.; An, Q.; Ma, J.; Guo, Z.; Zhang, Q.; Sun, H.; Robertson, C.; Bacsa, J.; Xiao J.; Li, C., *Angew. Chem. Int. Ed.,* **2022**, *61*, DOI: 10.1002/anie.202205983.

(91) Duan, Y.-N.; Du, X.; Cui, Z.; Zeng, Y.; Liu, Y.; Yang, T.; Wen, J.; Zhang, X., *J. Am. Chem. Soc.* **2019**, *141*, 20424-20433.

(92) Elangovan, S.; Topf, C.; Fischer, S.; Jiao, H.; Spannenberg, A.; Baumann, W.; Ludwig, R.; Junge, K.; Beller, M., *J. Am. Chem. Soc.* **2016**, *138*, 8809-8814.

(93) Zimmermann, B. M.; Ngoc, T. T.; Tzaras, D.-I.; Kaicharla, T.; Teichert, J. F., *J. Am. Chem. Soc.* **2021**, *143*, 16865-16873.

(94) Zubar, V.; Lichtenberger, N.; Schelwies, M.; Oeser, T.; Hashmi, A. S. K.; Schaub, T., *ChemCatChem* **2022**, *14*, e202101443.

(95) Bhawar, R.; Patil, K. S.; Bose, S. K., *New. J. Chem.* **2021**, *45*, 15028-15034.

(96) Liu, Y.; He, S.; Quan, Z.; Cai, H.; Zhao, Y.; Wang, B., *Green Chem.* **2019**, *21*, 830-838.

(97) TURBMOLE 7.5, V., A Development of University of Karlsruhe and Forschungszentrum Karlsruhe GmbH, 1989-2007, TURBOMOLE GmbH, since 2007. **2020**.

(98) Perdew, J. P.; Burke, K.; Ernzerhof, M., *Phy. Rev. Lett.* **1996**, *77*, 3865-3868.

(99) Grimme, S.; Antony, J.; Ehrlich, S.; Krieg, H., *J. Chem. Phys.* **2010**, *132*, 154104-1.

(100) Eichkorn, K.; Weigend, F.; Treutler, O.; Ahlrichs, R., *Theor. Chem. Acta.* **1997**, *97*, 119-124.

(101) Eichkorn, K.; Treutler, O.; Öhm, H.; Häser, M.; Ahlrichs, R., *Chem. Phy. Lett.* **1995**, *240*, 283-290.

(102) Sierka, M.; Hogekamp, A.; Ahlrichs, R., *J. Chem. Phy.* **2003**, *118*, 9136-9148.

(103) Klamt, A.; Schüürmann, G., *J. Chem. Soc. Perkin Trans.* **1993**, 799-805.

(104) Kozuch, S.; Shaik, S., *J. Am. Chem. Soc.* **2006**, *128*, 3355-3365.

(105) Kozuch, S.; Shaik, S., *J. Phys. Chem. A* **2008**, *112*, 6032-6041.

(106) Uhe, A.; Kozuch, S.; Shaik, S., *J. Comput. Chem.* **2011**, *32*, 978-985.

(107) Bruker, *APEX3, SAINT and SADABS. Bruker AXS Inc.*, Madison, Wisconsin, USA. **2016**.

(108) Sheldrick, G. M., *Acta Crystallogr.* **2008**, *A64*, 112-122.

(109) Sheldrick, G. M., *Acta Crystallogr.* **2015**, *C71*, 3-8.

(110) Sharma, D. M.; Punji, B., Synthesis and Implication of 3d Transition Metal Catalysts for the Hydrogenative Transformations of Alkynes Chalcones and Nitriles, CSIR-National Chemical Laboratory, Pune, 2022. http://hdl.handle.net/10603/483805.

Chapter 3

Chemoselective Hydrogenation of α,β-Epoxy Ketones to α-Hydroxy Epoxides by a Well-Defined Manganese Catalyst

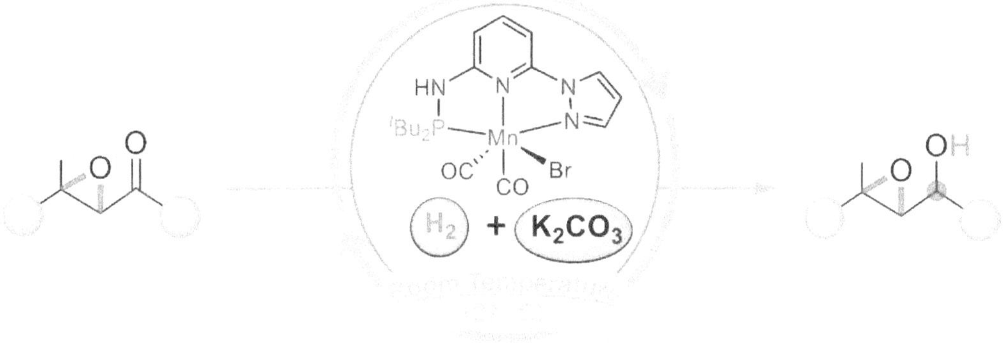

HN
$^{t}Bu_2P$
Mn
N
N
N
OC
CO
Br
H_2 + K_2CO_3
O
O
OH

This chapter has been partly adapted from the publication "Manganese-Catalyzed Chemoselective Direct Hydrogenation of α,β-Epoxy Ketones and α-Ketoamides at Room Temperature" **Shabade, A. B.**; Singh, R. K.; Gonnade, R. G.; and Punji, B. *Adv. Synth. Catal.*, **2024**, doi.org/10.1002/adsc.202400267.

3.1 INTRODUCTION

The α-hydroxy epoxides are key starting materials in a variety of organic transformations as well as in the production of medicines and agrochemicals.[1-3] Further, they are crucial structural motifs in diverse biologically active molecules and drug candidates (Scheme 3.1a).[4] The α-hydroxy epoxides can be synthesized by epoxidation of allyl alcohols, coupling of organometallic reagent to α,β-epoxy ketones, and chemoselective hydrogenation of α,β-epoxy ketones.[5,6] In particular, selective hydrogenation of α,β-epoxy ketones is traditionally followed, wherein borohydrides, silanes, and tin-hydrides are used as hydrogen sources (Scheme 3.1b).[7-10] These protocols suffer from poor chemoselectivity, require extreme conditions, and are accompanied by toxic waste production. Therefore, the transition metal-catalyzed protocol using a mild and environment-friendly hydrogen source is given significant attention. In that direction, the catalysts based on cost-effective and abundant 3d metals are substantially explored for the independent hydrogenation of carbonyl, imine or epoxide functionalities using molecular hydrogen.[11-16] However, the selective hydrogenation of C=O moiety in presence of other reducible functionality is highly challenging and extremely rare using base metal catalysts.[17-25] Notably, the diastereoselective reduction of ketone in α,β-epoxy ketones is developed by noble ruthenium-based metal catalysts using isopropanol or Et_3N/HCOOH as sacrificial hydrogen sources (Scheme 3.1c).[26,27]

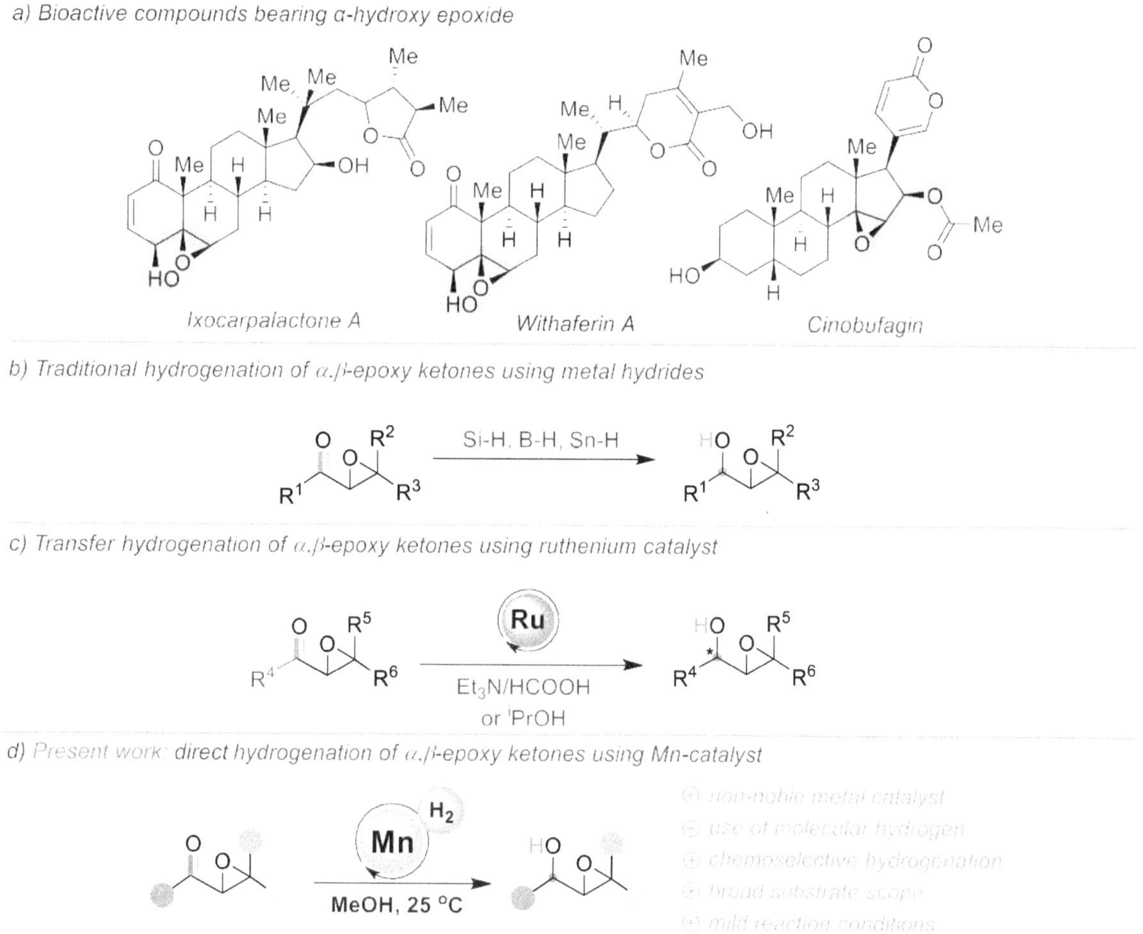

a) Bioactive compounds bearing α-hydroxy epoxide

Ixocarpalactone A   Withaferin A   Cinobufagin

b) Traditional hydrogenation of α,β-epoxy ketones using metal hydrides

c) Transfer hydrogenation of α,β-epoxy ketones using ruthenium catalyst

d) Present work: direct hydrogenation of α,β-epoxy ketones using Mn-catalyst

Scheme 3.1 Bioactive Compounds Containing *α*-Hydroxy Epoxide and Selective Reduction of *α,β*-Epoxy Ketones.

The direct hydrogenation of alkenes, carbonyls, imines, and nitriles has been studied by several research groups using manganese-based catalysts, as it is the third most prevalent transition metal with reduced toxicity.[28-37] However, the chemoselective reduction of an unsaturated bond using manganese catalysts and molecular hydrogen is scarce,[38-42] and the reactions were performed at high H_2 pressure and elevated temperature, which is a significant drawback. Unfortunately, chemoselective hydrogenation of *α,β*-epoxy ketones employing cost-effective and sustainable 3d metal catalysts and industrial-friendly molecular H_2 has not been precedented. In this chapter, we reveal an effective protocol for the chemoselective hydrogenation of C=O bonds in *α,β*-epoxy ketones to produce valuable *α*-hydroxy epoxides at room temperature (25 °C). This process uses Mn(I) catalyst and moderate H_2. This protocol takes into account the synthetic importance of epoxy alcohols and is a step towards sustainable metal catalysis (Scheme 3.1d).

3.2 RESULTS AND DISCUSSION

3.2.1 Optimization of Reaction Conditions

Given the unavailability of a cost-effective and direct hydrogenative approach for the synthesis *α,β*-epoxy alcohols, we have investigated the chemoselective C=O bond hydrogenation in phenyl(3-phenyloxiran-2-yl)methanone (**1a**) using recently developed manganese catalysts (**Mn-1** to **Mn-3**; Table 3.1). Thus, using 30 bar of molecular H_2 as a hydrogen source at 50 °C, compound **1a** was chemoselectively converted to phenyl(3-phenyloxiran-2-yl)methanol (**1**) under Mn-catalyst and a catalytic amount of KO^{t}Bu in MeOH (Table 3.1, entries 1-3). Notably, the hydrogenation provided quantitative conversion to desired product **1** in the presence of catalysts **Mn-2** and **Mn-3**, whereas **Mn-1** was less effective. Performing the reaction at 5 or 2 bar H_2 employing **Mn-3** catalyst also provided quantitative hydrogenated product **1**, whereas a further decrease in H_2 pressure resulted in low yield (Table 3.1, entries 4-7). The hydrogenation proceeded even at room temperature (25 °C) using the **Mn-3**/KO^{t}Bu system and 2 bar H_2 to afford compound **1** in 92% yield (entry 6). Screening various bases as catalytic additives indicates that the mild bases K_3PO_4 and K_2CO_3 are as effective as KO^{t}Bu (Table 3.1, entries 10, 11). Notably, the effectiveness of a catalytic mild base in this Mn-catalyzed hydrogenation is notable, as many Mn-catalyzed processes generally employ a strong base. The hydrogenation in EtOH or ^{i}PrOH provided a low yield, whereas the hydrogenation was not observed under aprotic solvents such as 1,4-dioxane, 2-MeTHF, or toluene (Table 3.1, entries 17-20). Lowering the catalyst loading or reaction time led to a decline in reaction yield (entry 24). Under the optimized condition (entry 11), the catalysts **Mn-1** and **Mn-2** afforded 10% and 76% of **1**, respectively. Using $Mn(CO)_5Br$ as a catalyst with or without ligand **L3** gave a trace or low yield of **1**, indicating the significance of structurally defined Mn(I) complex **Mn-3**. The attempted hydrogenation using $MnCl_2$ or $MnBr_2$ and other metal salts (Fe, Co, Ni) failed to perform the hydrogenation (Table 3.1, entries 29-37). The use of **Mn-3** catalyst is essential for hydrogenation, and the reaction failed in its absence (entry 38). Thus, upon optimization, the reaction parameters for chemoselective hydrogenation of **1a** to **1** are as follows; **1a** (0.2 mmol), catalyst **Mn-3** (5 mol%), 10 mol% of K_2CO_3 and 2 bar H_2 in MeOH (1.0 mL) at room temperature for 16 h (entry 11).

Table 3.1 Optimization of Reaction Conditions. [a]

[**Mn**] (5.0 mol%)
base (10 mol%)
H_2 (bar), MeOH
T (°C), t (h)

(**1a**) (**1**)

Entry	[Mn]	Base	H_2 (bar)	T (°C)/t (h)	Conv (%)[b]
1	**Mn-1**	KO^{t}Bu	30	50/16	28
2	**Mn-2**	KO^{t}Bu	30	50/16	95 (87)
3	**Mn-3**	KO^{t}Bu	30	50/16	98 (90)
4	**Mn-3**	KO^{t}Bu	5	50/16	98 (90)
5	**Mn-3**	KO^{t}Bu	2	50/16	98 (92)
6	**Mn-3**	KO^{t}Bu	2	RT/16	99 (92)
7	**Mn-3**	KO^{t}Bu	1	RT/16	38
8	**Mn-3**	NaO^{t}Bu	2	RT/16	26
9	**Mn-3**	LiO^{t}Bu	2	RT/16	28
10	**Mn-3**	K_3PO_4	2	RT/16	99 (93)
11	**Mn-3**	**K_2CO_3**	**2**	**RT/16**	**99 (93)**
12	**Mn-3**	KOAc	2	RT/16	28
13	**Mn-3**	Na_2CO_3	2	RT/16	19
14	**Mn-3**	NaOAc	2	RT/16	trace
15	**Mn-3**	Li_2CO_3	2	RT/16	16
16	**Mn-3**	LiOAc	2	RT/16	trace
17[c]	**Mn-3**	K_2CO_3	2	RT/16	28
18[d]	**Mn-3**	K_2CO_3	2	RT/16	22
19[e]	**Mn-3**	K_2CO_3	2	RT/16	NR
20[f]	**Mn-3**	K_2CO_3	2	RT/16	NR
21[g]	**Mn-3**	K_2CO_3	2	RT/16	81
22[h]	**Mn-3**	K_2CO_3	2	RT/16	69
23[i]	**Mn-3**	K_2CO_3	2	RT/16	96 (88)
24	**Mn-3**	K_2CO_3	2	RT/8	95 (91)
25	**Mn-1**	K_2CO_3	2	RT/16	10
26	**Mn-2**	K_2CO_3	2	RT/16	76
27	$Mn(CO)_5Br$/**L3**	K_2CO_3	2	RT/16	51
28	$Mn(CO)_5Br$	K_2CO_3	2	RT/16	trace
29	$Mn_2(CO)_{10}$/**L3**	K_2CO_3	2	RT/16	NR
30	$MnCl_2$/**L3**	K_2CO_3	2	RT/16	NR

31	$MnBr_2$/**L3**	K_2CO_3	2	RT/16	NR
32	$FeCl_2$/**L3**	K_2CO_3	2	RT/16	NR
33	$FeCl_3$/**L3**	K_2CO_3	2	RT/16	NR
34	$CoCl_2$/**L3**	K_2CO_3	2	RT/16	NR
35	$Co(OAc)_2$/**L3**	K_2CO_3	2	RT/16	trace
36	(DME)$NiCl_2$/**L3**	K_2CO_3	2	RT/16	NR
37	$Ni(OAc)_2$/**L3**	K_2CO_3	2	RT/16	NR
38	--	K_2CO_3	2	RT/16	NR

(Mn-1) **(Mn-2)** **(Mn-3)** (L3)

[a] Reaction conditions: **1a** (0.042 g, 0.20 mmol), base (0.02 mmol), **[Mn]** catalyst (0.01 mmol, 5 mol%), solvent (1.0 mL). [b] ^{1}H-NMR yield using CH_2Br_2 as standard. Isolated yields are given in parentheses. [c] Using EtOH as solvent. [d] Using ^{i}PrOH as solvent. [e] Using 1,4-dioxane as solvent. [f] Using 2MeTHF as solvent. [g] 0.5 mL MeOH was used. [h] 0.2 mL MeOH was used. [i] Using 3 mol% of **Mn-3** and 6 mol% of K_2CO_3. All three manganese complexes contain a mixture of two geometrical isomers. RT = room temperature (~ 25 °C).

3.2.2 Substrate Scope for Diaryl and Alkyl *α,β*-Epoxy Ketones

Next, we explored the generality and limitations of selective C=O hydrogenation of aryl-substituted *α,β*-epoxy ketones (Scheme 3.2). The *α,β*-epoxy ketones containing electronically distinct substituents on the benzoyl ring reacted smoothly to deliver *α*-hydroxy epoxides, **1-8** in good to excellent yields. The tolerability of halide functionalities, –F, –Br, and –I is notable as they allow late-stage modification to access more complex molecules. The hydrogenation proceeded efficiently even in the presence of electron-withdrawing substituents, $–CF_3$ and $–SO_2Me$ (**7**, **8**), demonstrating the uniqueness of the protocol. The terminal alkenyl and alkynyl moieties remain unaffected, and excellent chemoselective C=O bond reduction was achieved to produce **9** and **10** in 51% and 68%, respectively. The *ortho*-substituted aryl and naphthyl-derived *α,β*-epoxy ketone hydrogenated with excellent chemoselectivity and yields (**11-13**). The synthetically crucial functionalities, such as –Cl, –CF_3, –CN, and alkynyl, were well compatible even at the phenyl ring (**14-17**). Notably, a formyl group (–CHO) is hydrogenated under the condition to deliver epoxy alcohol **18** in 97% yield, which could be due to the high electrophilicity nature of C=O of the aldehyde. The *meta* fluoro-, nitro-substituted aryl epoxy ketones undergo efficient hydrogenation, yielding **20** and **21** in 87% and 84%, respectively. An epoxy cyclic ketone **22a** delivered a 63% yield of epoxy alcohol **22**. Interestingly, higher-scale hydrogenation of **1a** (0.5 g, 2.2 mmol) produced **1** in 92% indicating the potential scalability of the protocol (Scheme 3.2, in parenthesis). Overall, the developed protocol is highly efficient and chemoselective for the reduction of C=O in *α,β*-epoxy ketones to deliver synthetically demanding *α,β*-epoxy alcohols. The compatibility of H_2-sensitive functionalities, such as halides, nitrile, nitro, alkenyl, alkynyl, and epoxy, makes this Mn-catalyzed protocol distinctive. Such chemoselective hydrogenation employing a base metal catalyst has not been precedented.

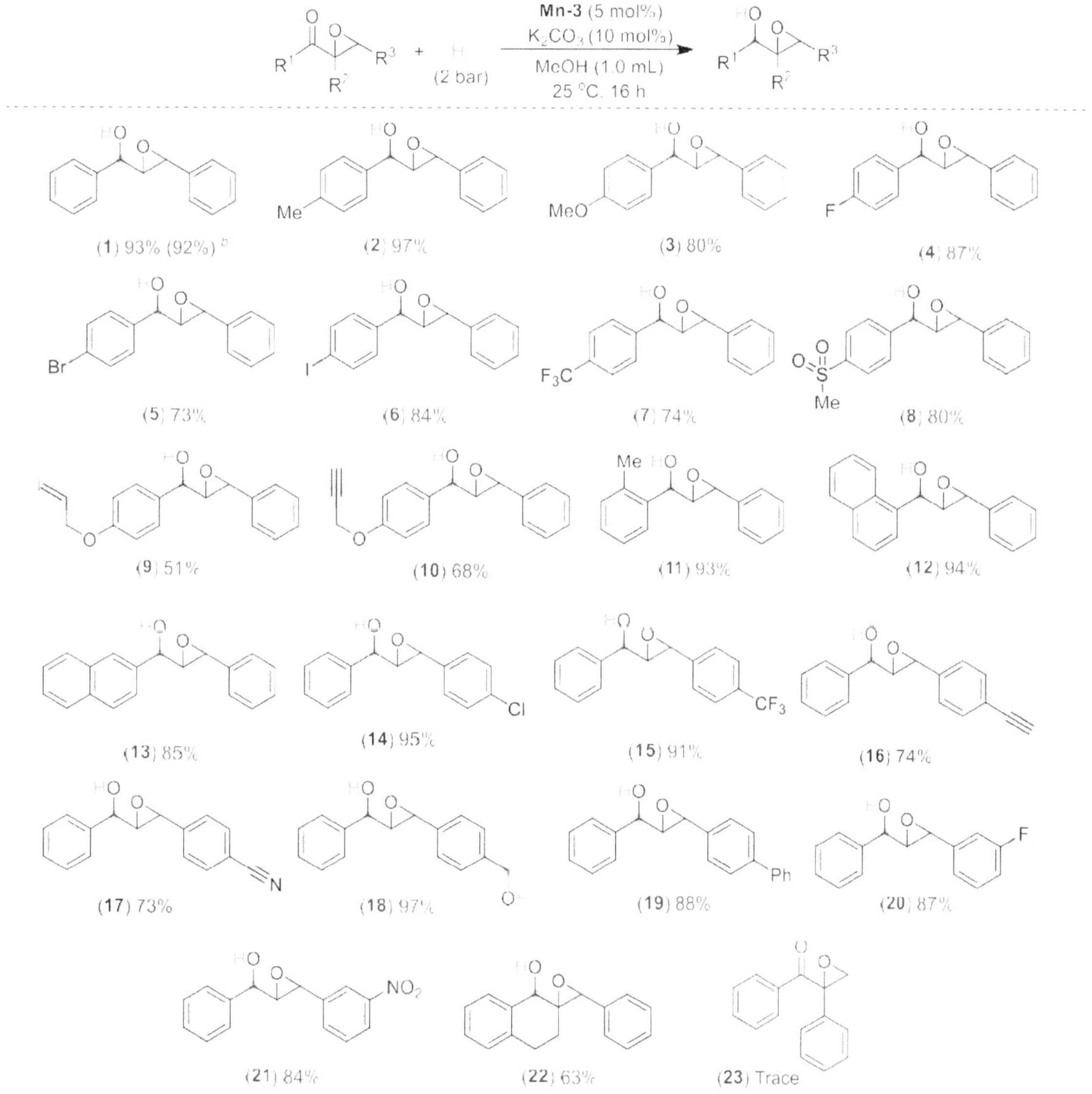

Mn-3 (5 mol%), K_2CO_3 (10 mol%), H_2 (2 bar), MeOH (1.0 mL), 25 °C, 16 h

(1) 93% (92%)[b] (2) 97% (3) 80% (4) 87% (5) 73% (6) 84% (7) 74% (8) 80% (9) 51% (10) 68% (11) 93% (12) 94% (13) 85% (14) 95% (15) 91% (16) 74% (17) 73% (18) 97% (19) 88% (20) 87% (21) 84% (22) 63% (23) Trace

Scheme 3.2 Scope for Chemoselective C=O Bond Hydrogenation of Diaryl *α,β*-Epoxy Ketones. Reaction conditions: *α,β*-Epoxy ketone (0.20 mmol), H_2 (2 bar), **Mn-3** (0.005 g, 0.01 mmol), K_2CO_3

(0.0028 g, 0.02 mmol). Yields are of isolated compounds. [a] Reaction on a 0.5 g scale. All compounds are obtained as two or three isomeric mixtures.

Unfortunately, the 2,2-disubstituted epoxy ketone, **23a** failed to undergo the desired hydrogenation, probably due to the steric crowding. Though excellent chemoselectivity towards C=O bond hydrogenation is observed in this protocol, a trace amount of diol due to epoxy ring opening was unavoidable in some cases.

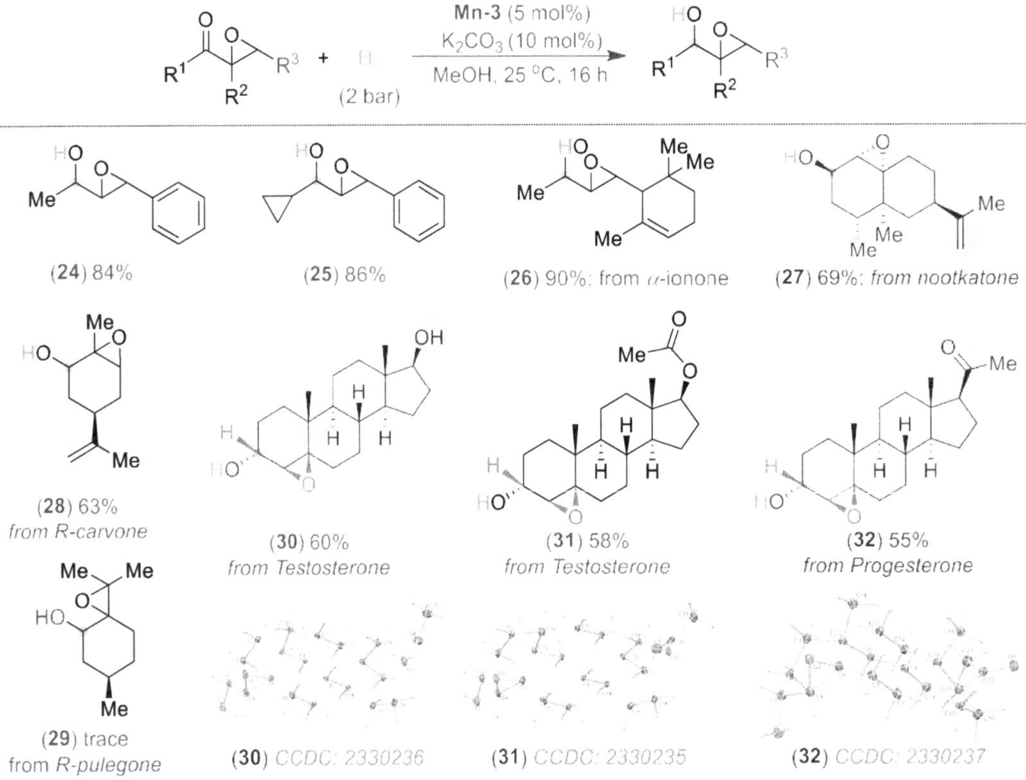

Mn-3 (5 mol%), K_2CO_3 (10 mol%), MeOH, 25 °C, 16 h; H_2 (2 bar)

(24) 84% (25) 86% (26) 90%; from α-ionone (27) 69%; from nootkatone

(28) 63% from R-carvone (30) 60% from Testosterone (31) 58% from Testosterone (32) 55% from Progesterone

(29) trace from R-pulegone

(30) CCDC: 2330236 (31) CCDC: 2330235 (32) CCDC: 2330237

Scheme 3.3 Scope of Alkyl-based α,β-Epoxy Ketones. Reaction conditions: α,β-Epoxy ketone (0.20 mmol), **Mn-3** (0.005 g, 0.01 mmol), K_2CO_3 (0.0028 g, 0.02 mmol), MeOH (1.0 mL), H_2 (2 bar). Yields are of isolated compounds.

The Mn-catalyzed protocol was extended to the hydrogenation of methyl- and cyclopropyl-based α,β-epoxy ketones, wherein compounds **24** and **25** were obtained in 84% and 86% yields, respectively, without compromising the reactivity and selectivity (Scheme 3.3). The epoxy ketones derived from terpenes (α-ionone, nootkatone, and *R*-carvone) delivered the chemoselectively hydrogenated compounds **26-28**, in good yields with untouched internal and terminal alkenyl moieties. Similarly, the steroids (testosterone and progesterone)-derived epoxy ketones delivered moderate to good yields of compounds, **30-32**. The endurance of –OH, –OAc, and –C(O)Me groups under the reaction condition is notable. The structural conformations of **30-32** were studied by a crystallographic analysis, wherein anti-isomers were observed. Unfortunately, the pulegone-derived epoxy ketone afforded a trace of product **29**, probably due to the high steric crowding. The notable reactivity, selectivity, and functional tolerance of this Mn-catalyzed protocol highlight the prospective usefulness of this methodology for late-stage modifications.

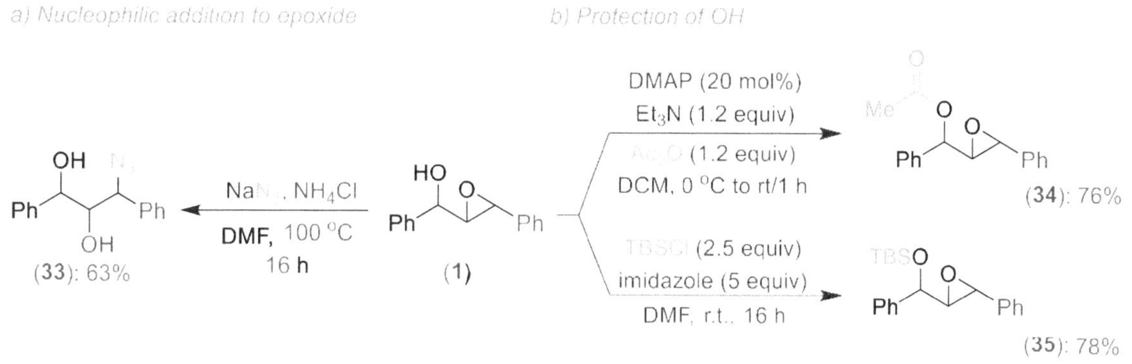

a) Nucleophilic addition to epoxide

b) Protection of OH

DMAP (20 mol%)
Et_3N (1.2 equiv)
Ac_2O (1.2 equiv)
DCM, 0 °C to rt/1 h

NaN_3, NH_4Cl
DMF, 100 °C
16 h

(33): 63%

(1)

(34): 76%

TBSCl (2.5 equiv)
imidazole (5 equiv)
DMF, r.t., 16 h

(35): 78%

Scheme 3.4 Derivatization of α-Hydroxy Epoxides.

Further functionalization of the hydrogenated compounds was demonstrated by performing nucleophilic addition of azide with **1** to achieve azo compound **33** in 63% yield (Scheme 3.4a). The alcoholic –OH can be easily protected with acetyl and silyl groups using Ac_2O and TBSCl, respectively, in good yields of **34** and **35** (Scheme 3.4b). We believe the current protocol would benefit further organic transformation in various fields.

3.2.3 Mechanistic Aspects

The hydrogenation was attempted using the **Mn-3^{Me}** complex, wherein the expected product was not observed (Scheme 3.5a). Interestingly, hydrogenation of **1a** using dearomatized complex **Mn-3'** as catalyst delivered **1** in 94% (Scheme 3.5b). These findings suggest the crucial role of ligand's *NH* in **Mn-3**, and indicate **Mn-3'** as an active catalyst for the activation of H_2 through metal-ligand cooperation (MLC). A standard reaction using CD_3OD as the solvent did not incorporate deuterium in the product (Scheme 3.5c). Moreover, the hydrogenation reaction failed to deliver **1** in aprotic solvents, THF, 1,4-dioxane, or toluene. These observations indicate that the molecular H_2 and MeOH act as the hydrogen (hydride) and proton sources in the reduction. It also rules out methanol as the sole hydrogen source for the reduction process. Unfortunately, we could not avail D_2 gas for additional deuterium labeling study.

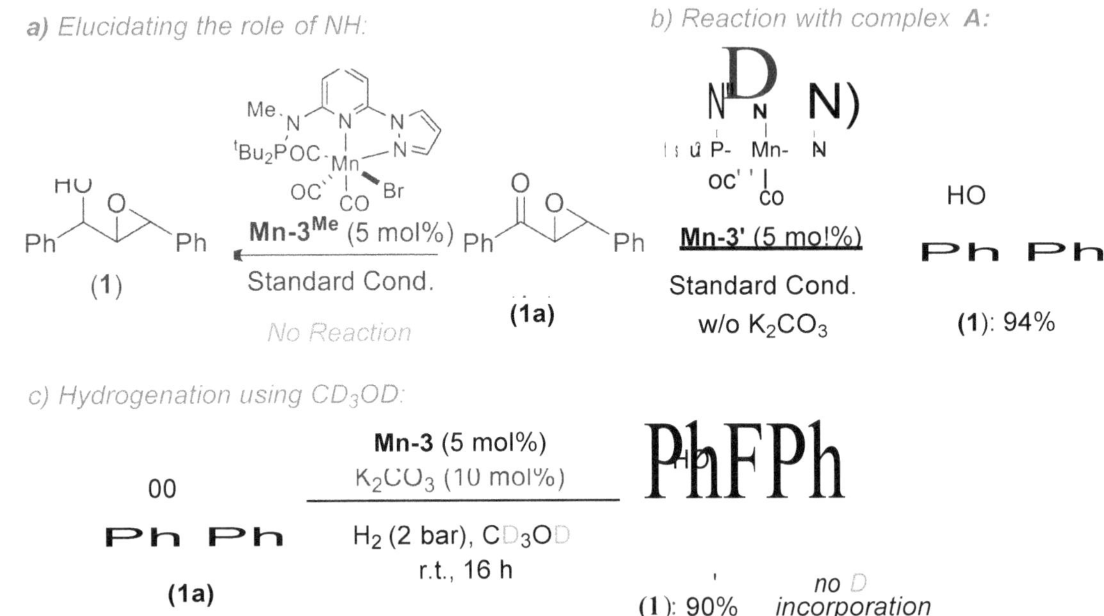

Scheme 3.5 Mechanistic experiments.

3.2.4 Plausible Catalytic Cycle

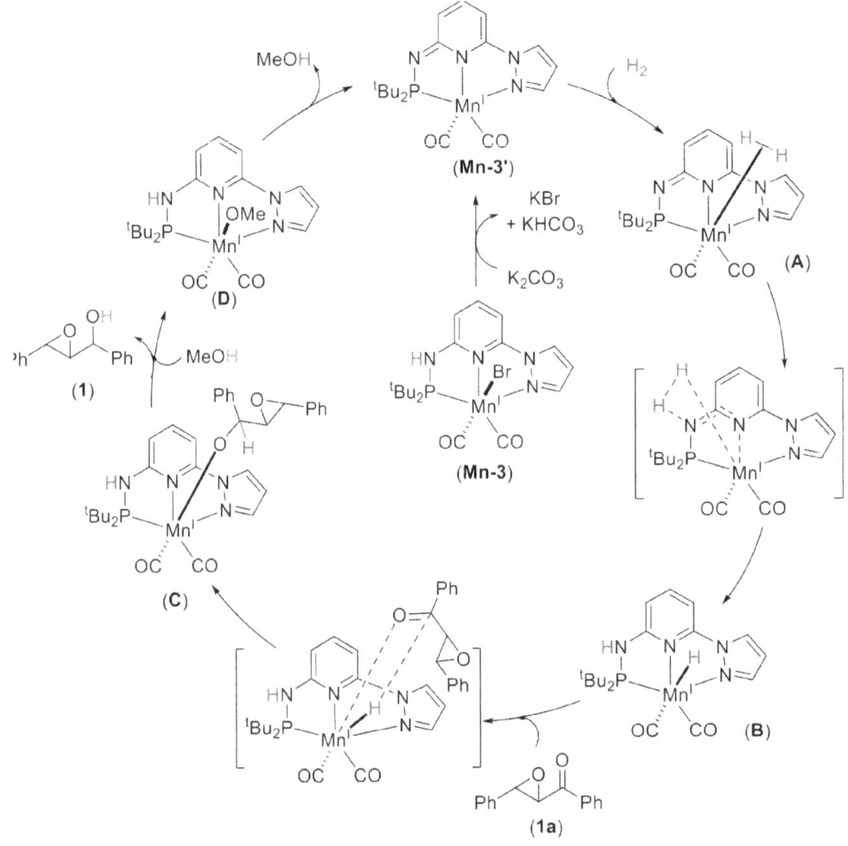

Figure 3.1 A Plausible Mechanism for the Chemoselective Hydrogenation of α,β-Epoxy Ketone.

Based on the experimental outcome and literature,[38,43] we proposed a catalytic cycle that proceeds with the formation of dearomatized intermediate **Mn-3'** from the reaction of **Mn-3** with K_2CO_3 (Figure 3.1). Similar transformations of metal complexes into dearomatized intermediates are well known.[44,45] The molecular H_2 will coordinate to **Mn-3'** followed by H_2 activation via metal-ligand cooperation (MLC) leading to intermediate **B**. The reaction substrate **1a** with **B** followed by 1,2-hydride transfer resulted in species **C**. The semi-hydrogenated species in **C** will undergo protonation in the presence of MeOH to form **1** and intermediate **D**. Finally, active catalyst **Mn-3'** will be regenerated from species **D**.

3.3 CONCLUSION

In summary, we disclosed an efficient and chemoselective hydrogenation protocol for the exclusive synthesis of α-hydroxy epoxides from α,β-epoxy ketones using a cost-effective manganese catalyst and benign H_2 source. The employment of $(^{tBu2}PNN^{Pyz})Mn(I)/K_2CO_3$ system, moderate hydrogen pressure, and friendly MeOH solvent at room temperature is beneficial to the commonly employed noble metal-based catalyst, strong KO^tBu, hazardous hydrogen source, and extreme conditions. A range of alkyl and aryl-substituted α,β-epoxy ketones hydrogenated to α-hydroxy epoxides with the endurance of synthetically vital and sensitive functionalities, such as halides, $-CF_3$, acetyl, nitrile, nitro, hydroxy, alkoxy, alkenyl, and alkynyl groups. The reaction proceeds via the H_2 activation by metal-ligand cooperation and protonation by MeOH. A higher-scale reaction and synthetic modifications highlight the applicability of the protocol.

3.4 EXPERIMENTAL SECTION

All the manipulations were conducted under an argon atmosphere either in a glove box or using standard Schlenk techniques in pre-dried glassware. The catalytic reactions were performed in oven-dried glass vials with magnetic bar by placing them in the pressure reactor. Solvents were dried over Na/benzophenone or Mg and distilled prior to use. Liquid reagents were flushed with argon prior to use. All other chemicals were obtained from commercial sources and were used without further purification. All α,β-epoxy ketones were synthesized by treating corresponding conjugated ketones with an excess of H_2O_2/NaOH in MeOH at 0 °C. High resolution mass spectrometry (HRMS) mass spectra were recorded on a Thermo Scientific Q-Exactive, Accela 1250 pump. NMR (^1H and ^{13}C) spectra were recorded at 400 or 500 MHz (^1H), 100 or 125 MHz {^{13}C, DEPT (distortionless enhancement by polarization transfer)}, 377 MHz (^{19}F), respectively in $CDCl_3$ solutions, if not otherwise specified; chemical shifts (δ) are given in ppm. The ^1H and $^{13}C\{^1H\}$ NMR spectra are referenced to residual solvent signals ($CDCl_3$: δ H = 7.26 ppm, δ C = 77.2 ppm; methanol-d_4: δ H =

3.34 ppm, δ C = 49.5 ppm).

3.4.1 Synthesis of Starting Compounds (Representative Procedure)

(3-Phenyloxiran-2-yl)(4-(prop-2-yn-1-yloxy)phenyl)methanone (10a): In a 50 mL round bottom flask, (*E*)-3-phenyl-1-(4-(prop-2-yn-1-yloxy)phenyl)prop-2-en-1-one (0.8 g, 3.04 mmol) was dissolved in 10 mL MeOH. After cooling the reaction mixture at 0 °C, 1.72 mL H_2O_2 (30% in H_2O) was added dropwise followed by the addition of 3.50 mL of NaOH (10% in H_2O). The reaction mixture was allowed to room temperature and stirred until complete conversion of starting material, which was ensured by TLC. The volatiles were evaporated under vacuo and 20 mL of water was added. The organic compound was extracted in EtOAc (50 mL X 3), the combined extract was dried over Na_2SO_4 and the volatile were evaporated under vacuo. The crude product was purified by column chromatography on silica gel (petroleum ether/EtOAc: 20/1) to yield (3-phenyloxiran-2-yl)(4-(prop-2-yn-1-yloxy)phenyl)methanone (**10a**) (0.22 g, 26%) as a white solid. ^{1}H-NMR (500 MHz, $CDCl_3$): δ = 8.04-8.00 (m, 2H, Ar–H), 7.42-7.35 (m, 5H, Ar–H), 7.05-7.00 (m, 2H, Ar–H), 4.76 (d, *J* = 2.5 Hz, 2H, CH_2), 4.25 (d, *J* = 2.0 Hz, 1H, CH), 4.06 (d, *J* = 1.9 Hz, 1H, CH), 2.56 (t, *J* = 2.5 Hz, 1H, CH). ^{13}C{^{1}H}-NMR (125 MHz, $CDCl_3$): δ = 191.6 (CO), 162.1 (C_q), 135.8 (C_q), 130.8 (2C, CH), 129.4 (C_q), 129.2 (CH), 128.9 (2C, CH), 125.9 (2C, CH), 115.1 (2C, CH), 77.7 (C_q), 76.5 (CH), 61.1 (CH), 59.4 (CH), 56.1 (CH_2).

1-(3-(2,6,6-Trimethylcyclohex-2-en-1-yl)oxiran-2-yl)ethan-1-one (26a): The representative procedure for the synthesis of α,β-epoxy ketones was followed, using α-ionone (1.0 g, 5.20 mmol), 3.12 mL H_2O_2 (30% in H_2O) and 7.28 mL of NaOH (10% in H_2O). Purification by column chromatography on silica gel (petroleum ether/EtOAc: 20/1) yielded **26a** (0.72 g, 66%) as colorless liquid. ^{1}H-NMR (400 MHz, $CDCl_3$): δ = 5.50 (s, 1H, CH), 3.28 (d, *J* = 1.9 Hz, 1H, CH), 2.90 (dd, *J* = 8.6, 2.0 Hz, 1H, CH), 2.05 (s, 3H, CH_3), 2.01-2.00 (m, 2H, CH_2), 1.69 (d, *J* = 1.4 Hz, 3H, CH_3), 1.55-1.47 (m, 1H, CH), 1.37 (d, *J* = 8.6 Hz, 1H, CH), 1.29-1.24 (m, 1H, CH), 1.08 (s, 3H, CH_3), 0.90 (s, 3H, CH_3). ^{13}C{^{1}H}-NMR (100 MHz, $CDCl_3$): δ = 206.2 (CO), 130.5 (C_q), 124.8 (CH), 59.4 (CH), 58.8 (CH), 52.7 (CH), 32.8 (C_q), 31.8 (CH_2), 27.5 (CH_3), 27.1 (CH_3), 26.6 (CH_3), 23.8 (CH_3), 23.1

(CH_2).

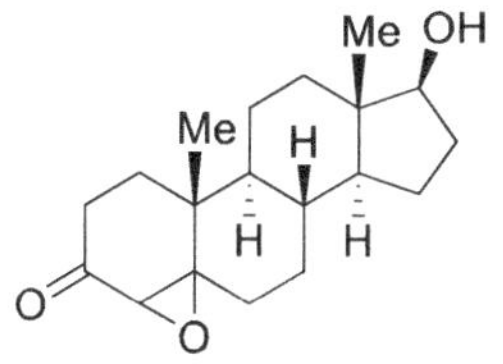

(4a*R*,4b*S*,6a*S*,7*S*,9a*S*,9b*S*)-7-hydroxy-4a,6a-dimethyltetradecahydrocyclopenta[7,8]phenanthrol[1,10a-b]oxiren-2(1a*H*)-one (30a): The representative procedure for the synthesis of α,β-epoxyketones was followed, using Testosterone (1.5 g, 5.20 mmol), 3.12 mL H_2O_2 (30% in H_2O) and 7.28 mL of NaOH (10% in H_2O). Purification by column chromatography on silica gel (petroleum ether/EtOAc: 5/1) yielded **30a** (1.28 g, 81%) as a white solid. **30a** *is obtained as mixture of two isomers (19: 81 ratio) denoted as* **30a'** *and* **30a''**. ^{1}H-NMR (400 MHz, $CDCl_3$): δ = 3.68-3.62 (m, 1H, CH), 3.03 (s, 0.19H, CH; **30a'**), 2.97 (s, 0.81H, CH; **30a''**), 2.32-3.02 (m, 4H, CH), 1.88-1.79 (m, 3H, CH), 1.64-1.54 (m, 5H, CH), 1.50-1.38 (m, 3H, CH), 1.36-1.20 (m, 2H, CH), 1.15 (s, 3H, CH_3), 1.10-1.08 (m, 1H, CH), 1.03-0.95 (m, 2H, CH), 0.77 (s, 0.68H, CH_3; **30a'**), 0.76 (s, 2.49H, CH_3; **30a''**). ^{13}C{^{1}H}-NMR (100 MHz, $CDCl_3$) *For* **30a'**: δ = 207.0 (CO), 81.7 (CH), 70.2 (C_q), 62.9 (CH), 50.8 (CH), 50.3 (CH), 42.9 (C_q), 36.8 (C_q), 36.5 (CH_2), 35.5 (CH), 33.1 (CH_2), 30.4 (CH_2), 29.6 (CH_2), 29.1 (CH_2), 28.5 (CH_2), 23.4 (CH_2), 21.0 (CH_2), 16.6 (CH_3), 11.1 (CH_3). *For* **30a''**: δ = 206.9 (CO), 81.6 (CH), 70.3 (C_q), 62.7 (CH), 50.5 (CH), 46.6 (CH), 43.1 (C_q), 37.3 (C_q), 36.2 (CH_2), 35.1 (CH), 32.6 (CH_2), 30.4 (CH_2), 29.9 (CH_2), 29.8 (CH_2), 26.2 (CH_2), 23.4 (CH_2), 21.2 (CH_2), 19.0 (CH_3), 11.1 (CH_3).

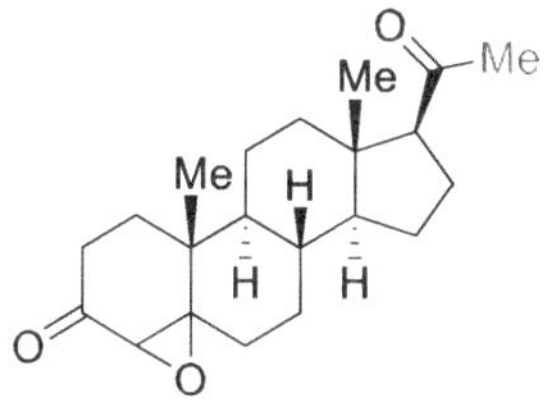

(4a*R*,4b*S*,6a*S*,7*S*,9a*S*,9b*S*)-7-acetyl-4a,6a-dimethyltetradecahydrocyclopenta[7,8]phenanthrol[1,10a-*b*]oxiren-2(1a*H*)-one (32a): The representative procedure for the synthesis of α,β-epoxy ketones was followed, using Progesterone (0.314 g, 1.00 mmol), 0.6 mL H_2O_2 (30% in H_2O) and 1.4 mL of NaOH (10% in H_2O). Purification by column chromatography on silica gel (petroleum ether/EtOAc: 10/1) yielded (4a*R*,4b*S*,6a*S*,7*S*,9a*S*,9b*S*)-7-acetyl-4a,6a-dimethyltetradecahydrocyclopenta [7,8]phenanthrol [1,10a-*b*]oxiren-2(1a*H*)-one (**33a**) (0.238 g, 72%) as a white solid. **32a** *is obtained as mixture of two isomers (22: 78 ratio) denoted as* **32a'** *and* **32a''**. ^{1}H-NMR (500 MHz, $CDCl_3$): δ = 3.03 (s, 0.22H,

CH; **32a'**), 2.97 (s, 0.78H, CH; **32a''**), 2.57-2.49 (m, 1H, CH), 2.42-2.12 (m, 4H, CH), 2.11 (s, 0.77H, CH_3; **32a'**), 2.10 (s, 2.36H, CH_3; **32a''**), 2.07-2.01 (m, 1H, CH), 1.89-1.80 (m, 2H, CH), 1.73-1.64 (m, 3H, CH), 1.60-1.36 (m, 5H, CH), 1.29-1.18 (m, 2H, CH), 1.14 (s, 2.51H, CH_3; **32a''**), 1.11 (s, 0.74H, CH_3; **32a'**), 1.10-1.04 (m, 2H, CH), 0.64 (s, 0.76H, CH_3; **32a'**), 0.63 (s, 2.37H, CH_3; **32a''**). $^{13}C\{^1H\}$-NMR (125 MHz, $CDCl_3$) *For* **32a'**: δ = 209.5 (CO), 207.0 (CO), 70.2 (C_q), 63.8 (CH), 63.0 (CH), 56.0 (CH), 50.7 (CH), 44.2 (C_q), 38.9 (CH_2), 36.9 (C_q), 35.5 (CH), 33.2 (CH_2), 31.7 (CH_3), 30.5 (CH_2), 29.8 (CH_2), 29.3 (CH_2), 29.0 (CH_2), 21.6 (CH_2), 23.0 (CH_2), 16.7 (CH_3), 13.5 (CH_3). *For* **32a''**: δ = 209.4 (CO), 206.9 (CO), 70.4 (C_q), 63.6 (CH), 62.8 (CH), 56.1 (CH), 46.6 (CH), 44.3 (C_q), 38.6 (CH_2), 37.4 (C_q), 35.2 (CH), 32.7 (CH_2), 31.6 (CH_3), 30.5 (CH_2), 29.9 (CH_2), 26.3 (CH_2), 24.5 (CH_2), 23.0 (CH_2), 21.7 (CH_2), 19.1 (CH_3), 13.5 (CH_3).

3.4.2 Representative Procedure for Hydrogenation α,β-Epoxy Ketones and Characterization Data

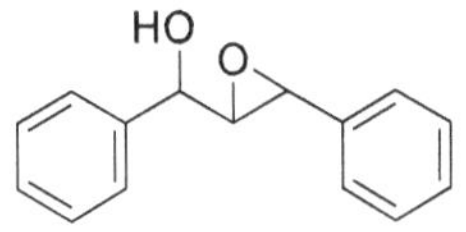

HO O

Synthesis of phenyl(3-phenyloxiran-2-yl)methanol (1): To a dry vial with magnetic bar was introduced **Mn-3** (0.005 g, 0.010 mmol), K_2CO_3 (0.004 g, 0.020 mmol), and phenyl(3-phenyloxiran-2-yl)methanone (**1a**; 0.046 g, 0.205 mmol) inside the glove box. The reaction vial was transferred to an autoclave under argon atmosphere. Then MeOH (1.0 mL) was added and the autoclave was pressurized with H_2 (2 bar) and vented for three times. Finally, the autoclave was pressurized with 2 bar H_2 and stirred (700 rpm) at room temperature (25 °C) for 12 h. The reaction mixture was then concentrated and subjected to column chromatography on silica gel (petroleum ether/EtOAc: 10/1) to yield isomeric mixture of **1** (0.043 g, 93%) as a white solid. **1** *is obtained as mixture of two isomers (56:44 ratio) denoted as* **1'** *and* **1''**.

^1H-NMR (500 MHz, $CDCl_3$): δ = 7.46-7.24 (m, 10H, Ar–H), 5.00 (d, J = 2.8 Hz, 0.56H, CH; **1'**), 4.71 (d, J = 4.3 Hz, 0.44H, CH; **1''**), 4.14 (d, J = 2.0 Hz, 0.56H, CH; **1'**), 4.0 (d, J = 2.0 Hz, 0.44H, CH; **1''**), 3.31-3.30 (d, J = 2.3 Hz, 0.42H, CH; **1''**), 3.30-3.28 (m, 0.57H, CH; **1''**), 2.66 (br s, 0.36H, OH; **1''**), 2.55 (br s, 0.46H, OH; **1'**).

$^{13}C\{^1H\}$-NMR (125 MHz, $CDCl_3$) *For* **1'**: δ = 139.4 (C_q), 136.7 (C_q), 128.9 (2C, CH), 128.7 (2C, CH), 128.5 (CH), 128.5 (CH), 126.7 (2C, CH), 125.9 (2C, CH), 71.4 (CH), 65.1 (CH), 55.2 (CH). *For* **1''**: δ = 140.3 (C_q), 136.5 (C_q), 128.9 (2C, CH), 128.7 (2C, CH), 128.6 (CH), 128.4 (CH), 126.4 (2C, CH), 125.9 (2C, CH), 73.6 (CH), 65.9 (CH), 57.1 (CH).

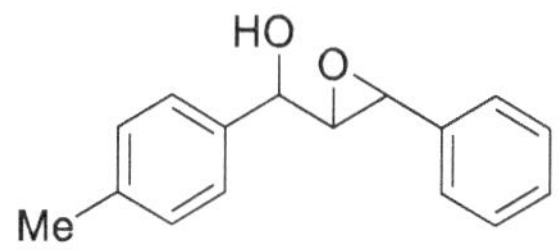

(3-Phenyloxiran-2-yl)(p-tolyl)methanol (2): The representative procedure was followed, using substrate **2a** (0.048 g, 0.201 mmol) and the reaction mixture was stirred at 25 °C for 16 h. Purification by column chromatography on silica gel (petroleum ether/EtOAc: 10/1) yielded an isomeric mixture of **2** (0.047 g, 97%) as a yellow oil. *Compound 2 is obtained as a mixture of three isomers (12:47:41 ratio) denoted as 2', 2'' and 2'''.*

^{1}H-NMR (400 MHz, $CDCl_3$): δ = 7.42-7.23 (m, 7H, Ar–H), 7.18-7.10 (m, 2H, Ar–H), 4.97 (d, J = 2.6 Hz, 0.12, CH; **2'**), 4.95 (d, J = 2.5 Hz, 0.47, CH; **2''**), 4.66 (br s, 0.41H, CH; **2'''**), 4.13 (d, J = 1.9 Hz, 0.48H, CH; **2''**), 4.10 (d, J = 1.9 Hz, 0.11H, CH; **2'**), 3.97 (d, J = 1.9 Hz, 0.41H, CH; **2'''**), 3.28-3.25 (m, 1H, CH), 2.70 (br s, 0.47H, OH; **2''**), 2.60 (br s, 0.45H, OH; **2'''**), 2.33 (s, 1.26H, CH_3; **2'''**), 2.32 (s, 1.42H, CH_3; **2''**), 2.31 (s, 0.35H, CH_3; **2'**).

^{13}C{^{1}H}-NMR (125 MHz, $CDCl_3$) *For* **2''**: δ = 138.2 (C_q), 138.1 (C_q), 137.4 (C_q), 129.5 (2C, CH), 128.6 (2C, CH), 128.4 (CH), 126.7 (2C, CH), 125.9 (2C, CH), 73.4 (CH), 66.0 (CH), 57.0 (CH), 21.3 (CH_3). *For* **2'''**: δ = 136.8 (C_q), 136.6 (C_q), 136.5 (C_q), 129.5 (2C, CH), 128.6 (2C, CH), 128.5 (CH), 126.3 (2C, CH), 125.9 (2C, CH), 71.2 (CH), 65.2 (CH), 65.2 (CH), 21.3 (CH_3). HRMS (ESI): *m/z* Calcd for $C_{16}H_{16}O_2 - H^+$ $[M - H]^+$ 239.1067; Found 239.1065.

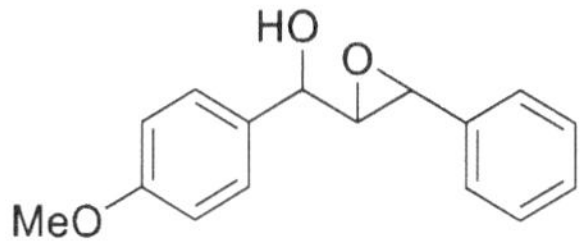

(4-Methoxyphenyl)(3 phenyloxiran-2-yl)methanol (3): The representative procedure was followed, using substrate **3a** (0.051 g, 0.200 mmol), and the reaction mixture was stirred at r.t. for 16 h. Purification by column chromatography on silica gel (petroleum ether/EtOAc: 5/1) yielded an isomeric mixture of **3** (0.041 g, 80%) as a colorless liquid. *Compound 3 is obtained as a mixture of two isomers (56:44 ratio) denoted as* **3'** *and* **3''**.

^{1}H-NMR (500 MHz, $CDCl_3$) **3'**: δ = 7.39-7.26 (m, 7H, Ar–H), 6.92-6.09 (m, 2H, Ar–H), 4.97 (d, J = 2.8 Hz, 0.56H, CH; **3'**), 4.68 (d, J = 4.8 Hz, 0.44H, CH; **3''**), 4.14 (d, J = 2.0 Hz, 0.55H, CH; **3'**), 3.98 (d, J = 2.0 Hz, 0.45H, CH; **3''**), 3.81 (s, 1.33H, CH_3; **3''**), 3.80 (s, 1.76H, CH_3; **3'**), 3.30-3.26 (m, 1H, CH).

^{13}C{^{1}H}-NMR (100 MHz, $CDCl_3$) *For* **3'**: δ = 159.7 (C_q), 136.6 (C_q), 131.5 (C_q), 128.7 (2C, CH), 128.5 (CH), 127.8 (2C, CH), 125.9 (2C, CH), 114.3 (2C, CH), 71.0 (CH), 65.2 (CH), 55.5 (CH_3), 55.2 (CH). *For* **3''**: δ – 159.9 (C_q), 136.8 (C_q), 132.6 (C_q), 128.7 (2C, CH), 128.5 (CH), 128.2 (2C,

CH), 125.9 (2C, CH), 114.3 (2C, CH), 73.2 (CH), 63.0 (CH), 57.0 (CH), 55.5 (CH_3).

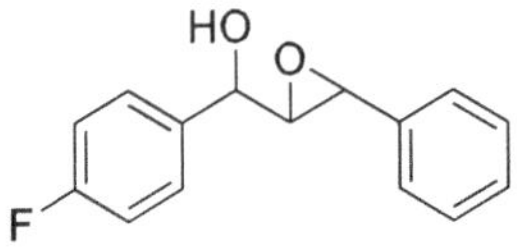

(4-Fluorophenyl)(3-phenyloxiran-2-yl)methanol (4): The representative procedure was followed, using substrate **4a** (0.049 g, 0.202 mmol), and the reaction mixture was stirred at r.t. for 16 h. Purification by column chromatography on silica gel (petroleum ether/EtOAc: 10/1) yielded an isomeric mixture of **4** (0.043 g, 87%) as a colorless liquid. *Compound **4** is obtained as a mixture of two isomers (52:48 ratio) denoted as **4'** and **4''**.*

^{1}H-NMR (400 MHz, $CDCl_3$): δ = 7.44-7.19 (m, 7H, Ar–H), 7.09-6.97 (m, 2H, Ar–H), 4.98 (d, J = 3.0 Hz, 0.52H, CH; **4'**), 4.69 (br s, 0.48H, CH; **4''**), 4.10 (d, J = 2.0 Hz, 0.53H, CH; **4'**), 3.98 (d, J = 2.0 Hz, 0.47H, CH; **4''**), 3.27-3.24 (m, 1H, CH), 2.85 (br s, 0.48H, OH; **4''**), 2.72 (br s, 0.50H, OH; **4'**).

$^{13}C\{^1H\}$-NMR (100 MHz, $CDCl_3$) For **4'**: δ = 162.9 (d, $^1J_{C-F}$ = 246.4 Hz, C_q), 136.5 (C_q), 136.1 (d, $^4J_{C-F}$ = 3.1 Hz, C_q), 128.7 (2C, CH), 128.7 (CH), 128.5 (d, $^2J_{C-F}$ = 8.4 Hz, 2C, CH), 125.9 (2C, CH), 115.9 (d, $^3J_{C-F}$ = 1.5 Hz, 2C, CH), 72.9 (CH), 65.8 (CH), 57.1 (CH). *For* **4''**: δ = 162.8 (d, $^1J_{C-F}$= 247.2 Hz, C_q), 136.3 (C_q), 136.2 (d, $^4J_{C-F}$ = 3.1 Hz, C_q), 128.7 (2C, CH), 128.6 (CH), 128.2 ($^2J_{C-F}$ = 8.4 Hz, 2C, CH), 125.9 (2C, CH), 115.7 (d, $^3J_{C-F}$ = 2.3 Hz, 2C, CH), 70.8 (CH), 65.0 (CH), 55.2 (CH). ^{19}F-NMR (377 MHz, $CDCl_3$): δ = −113.5, −113.7.

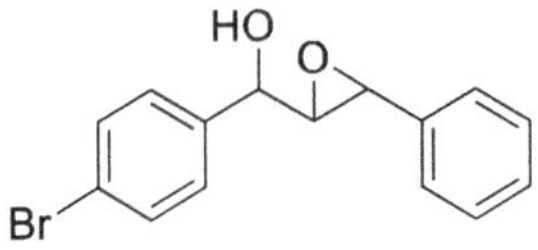

(4-Bromophenyl)(3-phenyloxiran-2-yl)methanol (5): The representative procedure was followed, using substrate **5a** (0.061 g, 0.201 mmol), and the reaction mixture was stirred at r.t. for 16 h. Purification by column chromatography on silica gel (petroleum ether/EtOAc: 10/1) yielded an isomeric mixture of **5** (0.045 g, 73%) as a white solid. *Compound **5** is obtained as mixture of three isomers (14:34:52 ratio) denoted as **5'**, **5''** and **5'''**.*

^{1}H-NMR (400 MHz, $CDCl_3$): δ = 7.50-7.41 (m, 2H, Ar–H), 7.39-7.26 (m, 5H, Ar–H), 7.24-7.10 (m, 2H, Ar–H), 4.97 (d, J = 2.8 Hz, 0.14H, CH; **5'**), 4.93 (d, J = 2.8 Hz, 0.34H, CH; **5''**), 4.67 (d, J = 4.6 Hz, 0.52H, CH; **5'''**), 4.08 (vd, J = 2.8 Hz, 0.15H, CH; **5'**), 4.07 (d, J = 1.8 Hz, 0.34H, CH; **5''**), 3.97 (d, J = 1.9 Hz, 0.52H, CH; **5'''**), 3.25-3.21 (m, 1H, CH), 2.84 (br s, 0.52H, OH; **5'''**), 2.71 (br s, 0.39H, OH; **5''**).

$^{13}C\{^1H\}$-NMR (100 MHz, CDCl3) *For* **5''**: δ = 138.4 (C_q), 136.2 (C_q), 132.0 (2C, CH), 128.7 (2C, CH), 128.6 (CH), 128.1 (2C, CH), 125.9 (2C, CH), 122.3 (C_q), 70.8 (CH), 64.9 (CH), 55.2 (CH). *For* **5'''**: δ = 139.2 (C_q), 136.4 (C_q), 132.0 (2C, CH), 128.7 (2C, CH), 128.6 (CH), 128.3 (2C, CH), 125.9 (2C, CH), 122.4 (C_q), 72.9 (CH), 65.6 (CH), 57.1 (CH). Due to the low concentration of **5'**, we could not interpret $^{13}C\{^1H\}$ for **5'**.

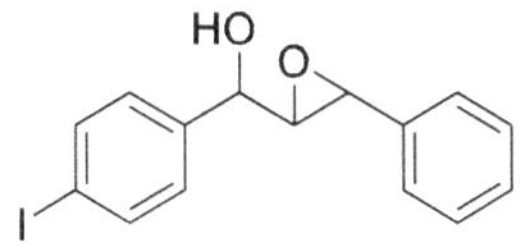

(4-Iodophenyl)(3-phenyloxiran-2-yl)methanol (6): The representative procedure was followed, using substrate **6a** (0.036 g, 0.102 mmol), and the reaction mixture was stirred at r.t. for 16 h. Purification by column chromatography on silica gel (petroleum ether/EtOAc: 10/1) yielded an isomeric mixture of **6** (0.030 g, 84%) as a yellow oil. *Compound **6** is obtained as a mixture of three isomers (10:56:34 ratio) denoted as **6'**, **6''** and **6'''**.*

^1H-NMR (400 MHz, CDCl3): δ = 7.73-7.63 (m, 2H, Ar–H), 7.43-7.29 (m, 3H, Ar–H), 7.26-6.99 (m, 4H, Ar–H), 5.00 (d, J = 2.8 Hz, 0.10H, CH; **6'**), 4.96 (d, J = 2.9 Hz, 0.56H, CH; **6''**), 4.68 (d, J = 4.6 Hz, 0.34H, CH; **6'''**), 4.08 (d, J = 2.0 Hz, 0.64H, CH; **6'** and **6''**), 3.99 (d, J = 2.1 Hz, 0.36H, CH; **6'''**), 3.26-3.23 (m, 1H, CH).

$^{13}C\{^1H\}$-NMR (100 MHz, CDCl3) *For* **6''**: δ = 139.9 (C_q), 138.0 (2C, CH), 136.4 (C_q), 128.8 (2C, CH), 128.6 (CH), 128.5 (2C, CH), 125.9 (2C, CH), 94.1 (C_q), 73.0 (CH), 65.6 (CH), 57.1 (CH). *For* **6'''**: δ = 139.1 (C_q), 137.9 (2C, CH), 136.2 (C_q), 128.7 (2C, CH), 128.6 (CH), 128.3 (2C, CH), 125.9 (2C, CH), 94.0 (C_q), 70.9 (CH), 64.8 (CH), 55.1 (CH). Due to low concentration of **6'**, we could not interpret $^{13}C\{^1H\}$-NMR for **6'**.

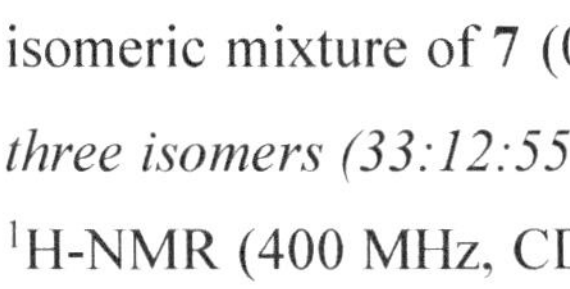

(3-Phenyloxiran-2-yl)(4-(trifluoromethyl)phenyl)methanol (7): The representative procedure was followed, using substrate **7a** (0.059 g, 0.202 mmol), and the reaction mixture was stirred at r.t. for 16 h. Purification by column chromatography on silica gel (petroleum ether/EtOAc: 10/1) yielded an isomeric mixture of **7** (0.044 g, 74%) as a colorless liquid. *Compound 7 is obtained as a mixture of three isomers (33:12:55 ratio) denoted as 7', 7'' and 7'''.*

^1H-NMR (400 MHz, CDCl3): δ = 7.66-7.55 (m, 4H, Ar–H), 7.44-7.36 (m, 1H, Ar–H), 7.34-7.30 (m, 2H, Ar–H), 7.27-7.22 (m, 2H, Ar–H), 5.07 (d, J = 2.9 Hz, 0.33H, CH; 7'), 5.03 (d, J = 2.9 Hz, 0.12H,

CH; **7''**), 4.80 (d, J = 4.6 Hz, 0.55H, CH; **7'''**), 4.20 (d, J = 1.8 Hz, 0.12H, CH; **7''**), 4.10 (d, J = 2.1 Hz, 0.33H, CH; **7'**), 4.03 (d, J = 2.1 Hz, 0.55H, CH; **7'''**), 3.30-3.29 (m, 0.32H, CH; **7'**), 3.29-3.27 (m, 0.56H, CH; **7'''**), 3.26-3.25 (m, 0.13H, CH; **7''**).

$^{13}C\{^1H\}$-NMR (100 MHz, $CDCl_3$) *For* **7'**: δ = 143.4 (C_q), 136.2 (C_q), 130.5 (C_q), 128.8 (2C, CH), 128.7 (CH), 126.9 (2C, CH), 125.9 (2C, CH), 125.8 (2C, CH), 124.2 (q, $^1J_{C-F}$ = 272.4 Hz, C_q), 70.8 (CH), 64.8 (CH), 55.2 (CH). *For* **7''**: δ = 144.1 (C_q), 136.2 (C_q), 130.8 (C_q), 128.8 (2C, CH), 128.8 (CH), 126.7 (2C, CH), 125.9 (2C, CH), 125.8 (2C, CH), 124.2 (q, $^1J_{C-F}$ = 272.4 Hz, C_q), 72.9 (CH), 65.6 (CH), 57.2 (CH). ^{19}F-NMR (377 MHz, $CDCl_3$): δ = −62.5, −62.6. Due to the low concentration of **7''**, we could not interpret $^{13}C\{^1H\}$-NMR for **7''**.

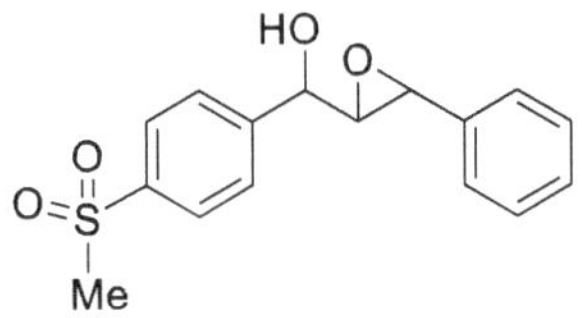

(4-(Methylsulfonyl)phenyl)(3-phenyloxiran-2-yl)methanol (8): The representative procedure was followed, using substrate **8a** (0.061 g, 0.202 mmol) and the reaction mixture was stirred at r.t. for 16 h. Purification by column chromatography on silica gel (petroleum ether/EtOAc: 2/1) yielded an isomeric mixture of **8** (0.049 g, 80%) as a white solid. *Compound **8** is obtained as a mixture of three isomers (35:13:52 ratio) denoted as **8'**, **8''** and **8'''**.*

^1H-NMR (400 MHz, $CDCl_3$): δ = 7.96-7.22 (m, 9H, Ar–H), 5.09 (d, J = 3.3 Hz, 0.35H, CH; **8'**), 5.03 (d, J = 2.9 Hz, 0.13H, CH; **8''**), 4.86 (d, J = 4.8 Hz, 0.52H, CH; **8'''**), 4.23 (d, J = 1.9 Hz, 0.13H, CH; **8''**), 4.09 (d, J = 2.0 Hz, 0.35H, CH; **8'**), 4.05 (d, J = 2.1 Hz, 0.52H, CH; **8'''**), 3.30-3.29 (m, 0.35H, CH; **8'**), 3.28-3.26 (m, 0.52H, CH; **8'''**), 3.24-4.23 (m, 0.14H, CH; **8''**), 3.04 (s, 1.11H, CH_3; **8'**), 3.04 (s, 1.55H, CH_3; **8'''**), 3.02 (s, 0.37H, CH_3; **8''**), 1.68 (br s, 1H, OH).

$^{13}C\{^1H\}$-NMR (125 MHz, $CDCl_3$) *For* **8'**: δ = 146.0 (C_q), 140.2 (C_q), 135.9 (C_q), 128.8 (2C, CH), 128.7 (CH), 127.8 (2C, CH), 127.3 (2C, CH), 125.8 (2C, CH), 71.0 (CH), 64.7 (CH), 55.4 (CH), 44.6 (CH_3). *For* **8'''**: δ = 146.5 (C_q), 140.3 (C_q), 136.1 (C_q), 128.8 (2C, CH), 128.7 (CH), 127.9 (2C, CH), 127.4 (2C, CH), 125.9 (2C, CH), 72.7 (CH), 65.4 (CH), 57.1 (CH), 44.6 (CH_3). HRMS (ESI): *m/z* Calcd for $C_{16}H_{16}O_4S + H^+$ $[M + H]^+$ 305.0848; Found 305.0851. Due to the lower concentration of **8''**, we could not interpret $^{13}C\{^1H\}$-NMR for **8''**.

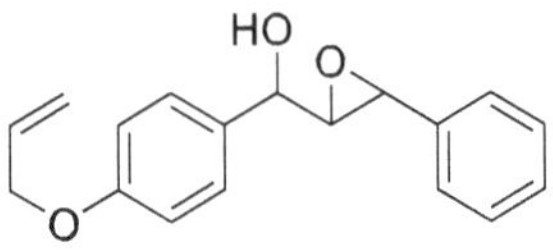

(4-(Allyloxy)phenyl)(3-phenyloxiran-2-yl)methanol (9): The representative procedure was followed, using substrate **9a** (0.057 g, 0.203 mmol), and the reaction mixture was stirred at r.t. for 16 h. Purification by column chromatography on silica gel (petroleum ether/EtOAc: 10/1) yielded an isomeric mixture of **9** (0.029 g, 51%) as a yellow oil. *Compound **9** is obtained as a mixture of two isomers (60:40 ratio) denoted as **9'** and **9''**.*

^{1}H-NMR (500 MHz, $CDCl_3$): δ = 7.38-7.24 (m, 7H, Ar–H), 6.94-6.89 (m, 2H, Ar–H), 6.09-5.99 (m, 1H, CH), 5.43-5.37 (m, 1H, CH), 5.30-5.26 (m, 1H, CH), 4.96 (d, J = 2.8 Hz, 0.60H, CH; **9'**), 4.67 (d, J = 4.8 Hz, 0.40H, CH; **9''**), 4.54-4.51 (m, 2H, CH_2), 4.13 (d, J = 2.0 Hz, 0.60H, CH; **9'**), 3.98 (d, J = 2.0 Hz, 0.40H, CH; **9''**), 3.29-3.26 (m, 1H, CH). 3.29-3.26 (m, 1H, CH).

$^{13}C\{^1H\}$-NMR (125 MHz, $CDCl_3$) *For* **9'**: δ = 158.7 (C_q), 136.6 (C_q), 133.3 (CH), 132.7 (C_q), 128.7 (2C, CH), 128.5 (CH), 127.7 (2C, CH), 125.9 (2C, CH), 117.9 (CH_2), 115.1 (2C, CH), 73.1 (CH), 69.0 (CH_2), 65.9 (CH), 57.0 (CH_2). *For* **9''**: δ = 158.9 (C_q), 136.8 (C_q), 133.3 (CH), 131.7 (C_q), 128.7 (2C, CH), 128.5 (CH), 128.1 (2C, CH), 125.9 (2C, CH), 117.9 (CH_2), 115.1 (2C, CH), 71.0 (CH), 69.0 (CH_2), 65.1 (CH), 55.2 (CH).

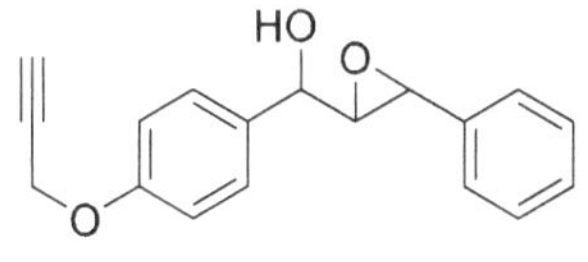

HO O O

(3-Phenyloxiran-2-yl)(4-(prop-2-yn-1-yloxy)phenyl)methanol (10): The representative procedure was followed, using substrate **10a** (0.057 g, 0.205 mmol) and the reaction mixture was stirred at r.t. for 16 h. Purification by column chromatography on silica gel (petroleum ether/EtOAc: 5/1) yielded an isomeric mixture of **10** (0.039 g, 68%) as a colorless liquid. *Compound **10** is obtained as a mixture of two isomers (58:42 ratio) denoted as **10'** and **10''**.*

^{1}H-NMR (400 MHz, $CDCl_3$): δ = 7.40-7.24 (m, 7H, Ar–H), 7.00-6.96 (m, 2H, Ar–H), 4.96 (d, J = 2.8 Hz, 0.58H, CH; **10'**), 4.69 (s, 0.42H, CH; **10''**), 4.68-4.67 (m, 2H, CH_2), 4.13 (d, J = 2.0 Hz, 0.58H, CH; **10''**), 3.98 (d, J = 2.0 Hz, 0.42H, CH; **10''**), 3.29-3.28 (m, 0.42H, CH; **10''**), 3.27-3.26 (m, 0.59H, CH; **10'**), 2.53-2.50 (m, 1H, CH), 2.45 (br s, 1H, OH).

$^{13}C\{^1H\}$-NMR (100 MHz, $CDCl_3$) *For* **10'**: δ = 157.8 (C_q), 136.7 (C_q), 132.4 (C_q), 128.7 (2C, CH), 128.5 (CH), 128.2 (2C, CH), 125.9 (2C, CH), 115.3 (2C, CH), 77.4 (C_q), 75.8 (CH), 70.9 (CH), 65.1 (CH), 56.0 (CH_2), 55.2 (CH). *For* **10''**: δ = 157.7 (C_q), 136.5 (C_q), 133.5 (C_q), 128.7 (2C, CH), 128.6 (CH), 127.8 (2C, CH), 125.9 (2C, CH), 115.3 (2C, CH), 78.6 (CH), 77.4 (C_q), 73.1 (CH), 65.9 (CH), 57.0 (CH), 56.0 (CH_2).

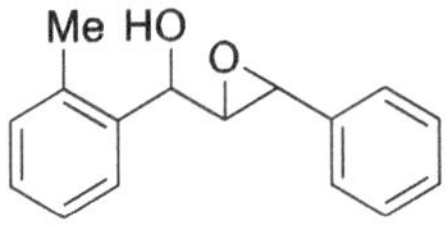

(3-Phenyloxiran-2-yl)(o-tolyl)methanol (11): The representative procedure was followed, using substrate **11a** (0.048 g, 0.201 mmol), and the reaction mixture was stirred at r.t. for 16 h. Purification by column chromatography on silica gel (petroleum ether/EtOAc: 10/1) yielded an isomeric mixture of **11** (0.045 g, 93%) as a colorless liquid. *Compound **11** is obtained as a mixture of three isomers (52:26:22 ratio) denoted as **11'**, **11''** and **11'''**.*

^{1}H-NMR (500 MHz, $CDCl_3$): δ = 7.57-7.07 (m, 9H, Ar–H), 5.26 (s, 0.52H, CH; **11'**), 5.03 (s, 0.26H, CH; **11''**), 4.97 (s, 0.22H, CH; **11'''**), 4.21 (s, 0.26H, CH; **11''**), 4.12 (s, 0.52H, CH; **11'**), 4.01 (s, 0.22H, CH; **11'''**), 3.31 (s, 0.26H, CH; **11''**), 3.27 (s, 0.52H, CH; **11'**), 3.14 (s, 0.22H, CH; **11'''**), 2.66 (br s, 0.22H, OH; **11'''**), 2.56 (br s, 0.78H, OH; **11'** and **11''**), 2.40 (s, 2.32H, CH_3; **11'** and **11''**), 2.20 (s, 0.68H, CH_3; **11'''**).

$^{13}C\{^1H\}$-NMR (125 MHz, $CDCl_3$) *For* **11'**: δ = 137.4 (C_q), 136.8 (C_q), 135.8 (C_q), 130.7 (CH), 128.7 (2C, CH), 128.5 (CH), 128.3 (CH), 126.6 (CH), 126.3 (CH), 125.9 (2C, CH), 67.7 (CH), 64.5 (CH), 55.2 (CH), 19.3 (CH_3). *For* **11''**: δ = 138.5 (C_q), 136.6 (C_q), 135.6 (C_q), 130.8 (CH), 128.7 (2C, CH), 128.6 (CH), 128.3 (CH), 126.6 (CH), 126.4 (CH), 125.9 (2C, CH), 69.7 (CH), 65.2 (CH), 57.0 (CH), 19.5 (CH_3). HRMS (ESI): *m/z* Calcd for $C_{16}H_{16}O_2$ + Na $[M + Na]^+$ 263.1043; Found 263.1038. Due to the low concentration of **11'''**, we could not interpret $^{13}C\{^1H\}$-NMR for **11'''**.

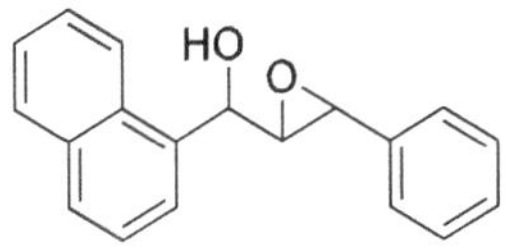

Naphthalen-1-yl(3-phenyloxiran-2-yl)methanol (12): The representative procedure was followed, using substrate **12a** (0.055 g, 0.20 mmol) and the reaction mixture was stirred at r.t. for 16 h. Purification by column chromatography on silica gel (petroleum ether/EtOAc: 10/1) yielded isomeric mixture of **12** (0.052 g, 94%) as yellow oil. *Compound **12** is obtained as a mixture of two isomers (76:24 ratio) denoted as 12' and **12''**.*

^{1}H-NMR (500 MHz, $CDCl_3$): δ = 8.10-8.05 (m, 1H, Ar–H), 7.81-7.71 (m, 2H, Ar–H), 7.66-7.58 (m, 1H, Ar–H), 7.46-7.36 (m, 3H, Ar–H), 7.21-7.16 (m, 5H, Ar–H), 5.73 (s, 0.76H, CH; **12'**), 5.45 (s, 0.24H, CH; **12''**), 4.10 (s, 0.74H, CH; **12'**), 4.03 (s, 0.24H, CH; **12''**), 3.44 (s, 0.24H, CH; **12''**), 3.41 (s, 0.77H, CH; **12'**), 2.65 (br s, 0.72H, OH).

$^{13}C\{^1H\}$-NMR (125 MHz, $CDCl_3$) *For* **12'**: δ = 136.6 (C_q), 136.2 (C_q), 134.1 (C_q), 131.0 (C_q), 129.1 (CH), 129.0 (CH), 128.7 (2C, CH), 128.5 (CH), 126.6 (CH), 125.9 (CH), 125.9 (2C, CH), 125.7

(CH), 124.1 (CH), 123.1 (CH), 67.8 (CH), 64.7 (CH), 55.5 (CH). *For* **12''**: δ = 136.6 (C_q), 135.0 (C_q), 133.9 (C_q), 131.1 (C_q), 129.1 (CH), 129.0 (CH), 128.7 (2C, CH), 128.5 (CH), 126.6 (CH), 125.9 (CH), 125.9 (2C, CH), 125.7 (CH), 124.1 (CH), 123.1 (CH), 67.8 (CH), 64.7 (CH), 55.5 (CH).

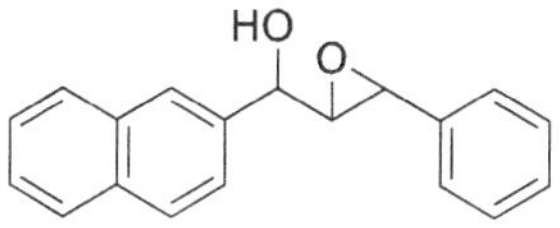

Naphthalen-2-yl(3-phenyloxiran-2-yl)methanol (13): The representative procedure was followed, using substrate **13a** (0.055 g, 0.200 mmol) and the reaction mixture was stirred at r.t. for 16 h. Purification by column chromatography on silica gel (petroleum ether/EtOAc: 10/1) yielded an isomeric mixture of **13** (0.047 g, 85%) as a yellow solid. *Compound **13** is obtained as a mixture of three isomers (40:14:46 ratio) denoted as **13'**, **13''** and **13'''**.*

^{1}H-NMR (500 MHz, $CDCl_3$): δ = 7.89-7.75 (m, 4H, Ar–H), 7.54-7.43 (m, 3H, Ar–H), 7.37-7.22 (m, 5H, Ar–H), 5.12 (d, J = 2.9 Hz, 0.40H, CH; **13'**), 5.01 (d, J = 2.9 Hz, 0.14H, CH; **13''**), 4.85 (d, J = 4.6 Hz, 0.46H, CH; **13'''**), 4.28 (d, J = 1.9 Hz, 0.14H, CH; **13''**), 4.16 (d, J = 2.0 Hz, 0.40H, CH; **13'**), 4.03 (d, J = 2.0 Hz, 0.47H, CH; **13'''**), 3.37-3.34 (m, 1H, CH), 2.98 (br s, 0.29H, OH; **13'''**), 2.84 (br s, 0.19H, OH; **13'**).

^{13}C{^{1}H}-NMR (125 MHz, $CDCl_3$) *For* **13'**: δ = 136.9 (C_q), 136.4 (C_q), 133.4 (C_q), 133.3 (C_q), 128.7 (CH), 128.6 (2C, CH), 128.5 (CH), 128.2 (CH), 127.9 (CH), 126.4 (CH), 126.3 (CH), 125.9 (2C, CH), 125.4 (CH), 124.3 (CH), 71.6 (CH), 65.2 (CH), 55.4 (CH). *For* **13'''**: δ = 137.7 (C_q), 136.6 (C_q), 133.4 (C_q), 133.4 (C_q), 128.9 (CH), 128.7 (2C, CH), 128.5 (CH), 128.2 (CH), 127.9 (CH), 126.5 (CH), 126.3 (CH), 125.9 (2C, CH), 125.8 (CH), 124.3 (CH), 73.6 (CH), 65.9 (CH), 57.1 (CH).

HRMS (ESI): *m/z* Calcd for $C_{19}H_{16}O_2 + Na^+$ $[M + Na]^+$ 299.1043; Found 299.1045. Due to the low concentration of **13''**, we could not interpret ^{13}C{^{1}H}-NMR for **13''**.

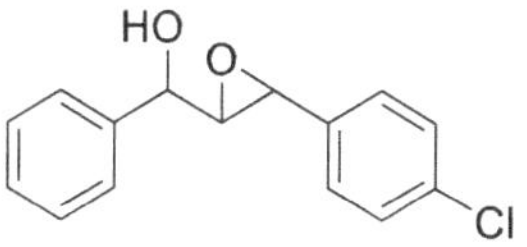

(3-(4-Chlorophenyl)oxiran-2-yl)(phenyl)methanol (14): The representative procedure was followed, using substrate **14a** (0.052 g, 0.201 mmol) and the reaction mixture was stirred at r.t. for 16 h. Purification by column chromatography on silica gel (petroleum ether/EtOAc: 10/1) yielded an isomeric mixture of **14** (0.050 g, 95%) as a yellow oil. *Compound **14** is obtained as a mixture of two isomers (55:45 ratio) denoted as **14'** and **14''**.*

^{1}H-NMR (400 MHz, $CDCl_3$): δ = 7.44-7.26 (m, 7H, Ar–H), 7.17 (d, J = 8.5 Hz, 2H, Ar–H), 4.97 (d,

$J = 2.9$ Hz, 0.55H, CH; **14'**), 4.70 (d, $J = 4.6$ Hz, 0.45H, CH; **14''**), 4.10 (d, $J = 2.0$ Hz, 0.55H, CH; **14'**), 3.97 (d, $J = 2.0$ Hz, 0.45H, CH; **14''**), 3.25-3.22 (m, 1H, CH), 2.63 (br s, 1H, OH).

$^{13}C\{^1H\}$-NMR (100 MHz, $CDCl_3$) *For* **14'**: δ = 140.2 (C_q), 135.3 (C_q), 134.3 (C_q), 129.0 (2C, CH), 128.9 (2C, CH), 128.6 (CH), 127.2 (2C, CH), 126.7 (2C, CH), 71.3 (CH), 65.2 (CH), 56.3 (CH). *For* **14''**: δ = 139.3 (C_q), 135.1 (C_q), 134.4 (C_q), 128.9 (2C, CH), 128.9 (CH), 128.5 (CH), 127.2 (2C, CH), 126.4 (CH), 73.3 (CH), 65.9 (CH), 54.6 (CH). HRMS (ESI): *m/z* Calcd for $C_{15}H_{13}ClO_2 - H^+$ $[M - H]^+$ 259.0520; Found 259.0516.

Phenyl(3-(4-(trifluoromethyl)phenyl)oxiran-2-yl)methanol (15): The representative procedure was followed, using substrate **15a** (0.059 g, 0.202 mmol) and the reaction mixture was stirred at r.t. for 16 h. Purification by column chromatography on silica gel (petroleum ether/EtOAc: 10/1) yielded isomeric mixture of **15** (0.054 g, 91%) as yellow solid. *Compound **15** is obtained as mixture of three isomers (21:30:49 ratio) denoted as **15'**, **15''** and **15'''**.*

^1H-NMR (500 MHz, $CDCl_3$): δ = 7.63-7.53 (m, 2H, Ar–H), 7.45-7.21 (m, 7H, Ar–H), 5.03 (d, $J = 2.0$ Hz, 0.21H, CH; **15'**), 4.99 (d, $J = 2.4$ Hz, 0.30H, CH; **15''**), 4.74 (s, 0.49H, CH; **15'''**), 4.19 (d, $J = 1.6$ Hz, 0.30H, CH; **15''**), 4.09 (d, $J = 1.9$ Hz, 0.22H, CH; **15'**), 4.05 (d, $J = 1.8$ Hz, 0.49H, CH; **15'''**), 3.28-3.27 (m, 0.49H, CH; **15'''**), 3.26 (d, $J = 2.1$ Hz, 0.22H, CH; **15'**), 3.24 (t, $J = 2.4$ Hz, 0.30H, CH; **15''**), 2.82 (br s, 0.20H, OH; **15'**), 2.76 (br s, 0.45H, OH; **15'''**), 2.66 (br s, 0.27H, OH; **15''**).

$^{13}C\{^1H\}$-NMR (100 MHz, $CDCl_3$) For **15''**: δ = 141.0 (C_q), 139.2 (C_q), 130.9 (d, $^2J_{C-F}$ = 32.8 Hz, C_q), 129.0 (2C, CH), 128.7 (CH), 126.6 (2C, CH), 126.1 (2C, CH), 125.7 (q, $^3J_{C-F}$ = 3.8 Hz, 2C, CH), 122.7 (q, $^1J_{C-F}$ = 272.4 Hz), 71.3 (CH), 65.6 (CH), 54.5 (CH).

For **15'''**: δ = 140.7 (C_q), 140.1 (C_q), 130.9 (d, $^2J_{C-F}$ = 32.8 Hz, C_q), 129.0 (2C, CH), 128.7 (CH), 126.4 (2C, CH), 126.1 (2C, CH), 125.7 (q, $^3J_{C-F}$ = 3.8 Hz, 2C, CH), 122.7 (q, $^1J_{C-F}$ = 272.4 Hz), 73.3 (CH), 66.2 (CH), 56.2 (CH). HRMS (ESI): *m/z* Calcd for $C_{16}H_{13}F_3O_2 - H^+$ $[M - H]^+$ 293.0788; Found 293.1767. ^{19}F-NMR (377 MHz, $CDCl_3$): δ = −62.6, −62.7. Due to the low concentration of **15'**, we could not interpret $^{13}C\{^1H\}$-NMR for **15'**.

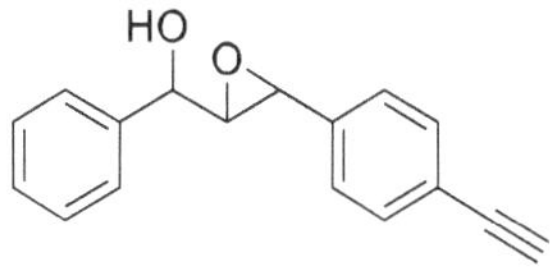

(3-(4-ethynylphenyl)oxiran-2-yl)(phenyl)methanol (16): The representative procedure was followed, using substrate **16a** (0.050 g, 0.201 mmol) and the reaction mixture was stirred at r.t. for 16 h. Purification by column chromatography on silica gel (petroleum ether/EtOAc: 10/1) yielded an isomeric mixture of **16** (0.037 g, 74%) as a colorless liquid. *Compound **16** is obtained as a mixture of two isomers (56:44 ratio) denoted as **16'** and **16''**.*

^{1}H-NMR (400 MHz, $CDCl_3$): δ = 7.50-7.19 (m, 9H, Ar–H), 5.00 (d, J = 2.9 Hz, 0.56H, CH; **16'**), 4.73 (d, J = 4.6 Hz, 0.44H, CH; **16''**), 4.12-4.08 (m, 0.56H, CH; **16'**), 4.00 (d, J = 2.0 Hz, 0.44H, CH; **16''**), 3.28-3.24 (m, 1H, CH), 3.08 (s, 0.43H, CH; **16''**), 3.07 (s, 0.56H, CH; **16'**), 2.56 (br s, 1H, OH).

^{13}C{^{1}H}-NMR (125 MHz, $CDCl_3$) *For* **16'**: δ = 140.2 (C_q), 139.3 (C_q), 137.5 (C_q), 132.5 (2C, CH), 129.0 (2C, CH), 128.6 (CH), 126.7 (2C, CH), 125.8 (2C, CH), 122.3 (C_q), 83.4 (d, J = 2.3 Hz, CH), 73.3 (CH), 66.0 (CH), 56.5 (CH). *For* **16''**: δ = 140.1 (C_q), 137.3 (C_q), 136.4 (C_q), 132.4 (2C, CH), 128.9 (2C, CH), 128.5 (CH), 126.4 (2C, CH), 125.8 (2C, CH), 122.2 (C_q), 78.0 (d, J = 5.3 Hz, CH), 71.3 (CH), 65.3 (CH), 54.8 (CH).

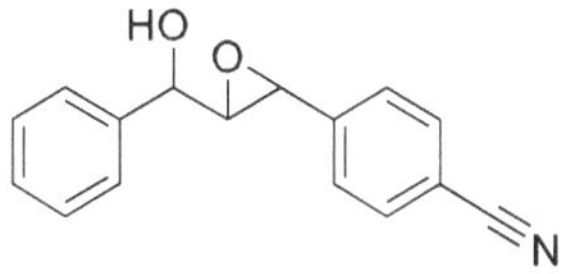

4-(3-(Hydroxy(phenyl)methyl)oxiran-2-yl)benzonitrile (17): The representative procedure was followed, using substrate **17a** (0.050 g, 0.20 mmol) and the reaction mixture was stirred at r.t. for 16 h. Purification by column chromatography on silica gel (petroleum ether/EtOAc: 5/1) yielded an isomeric mixture of **17** (0.037 g, 73%) as a yellow oil. *Compound 17 is obtained as a mixture of three isomers (16:28:56 ratio) denoted as 17', 17'' and 17'''.*

^{1}H-NMR (400 MHz, $CDCl_3$): δ = 7.66-7.54 (m, 2H, Ar–H), 7.45-7.20 (m, 7H, Ar–H), 5.00 (d, J = 3.3 Hz, 0.16H, CH; **17'**), 4.99 (d, J = 2.9 Hz, 0.28H, CH; **17''**), 4.77 (d, J = 4.4 Hz, 0.56H, CH; **17'''**), 4.17 (d, J = 1.8 Hz, 0.29H, CH; **17''**), 4.05 (d, J = 1.9 Hz, 0.71H, CH; **17'** and **17'''**), 3.27-3.23 (m, 0.72H, CH; **17'** and **17'''**), 3.23-3.21 (m, 0.28H, CH; **17''**).

^{13}C{^{1}H}-NMR (100 MHz, $CDCl_3$) *For* **17''**: δ = 142.4 (C_q), 139.1 (C_q), 132.5 (2C, CH), 129.0 (2C, CH), 128.7 (2C, CH), 127.1 (CH), 126.4 (2C, CH), 118.7 (C_q), 112.1 (C_q), 71.3 (CH), 65.7 (CH), 54.4 (CH). *For* **17'''**: δ = 142.2 (C_q), 139.9 (C_q), 132.5 (2C, CH), 129.0 (2C, CH), 128.7 (2C, CH), 126.6 (CH), 126.5 (2C, CH), 118.7 (C_q), 112.2 (C_q), 73.1 (CH), 66.3 (CH), 55.9 (CH). HRMS (ESI):

m/z Calcd for $C_{16}H_{13}NO_2 + H^+$ $[M + H]^+$ 252.1025; Found 252.1014. Due to the low concentration of **17'**, we could not interpret $^{13}C\{^1H\}$-NMR for **17'**.

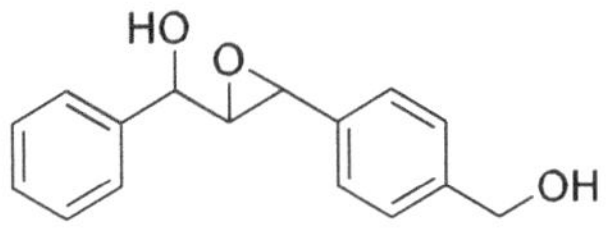

(3-(4-(Hydroxymethyl)phenyl)oxiran-2-yl)(phenyl)methanol (18): The representative procedure was followed, using substrate **18a** (0.051 g, 0.201 mmol) and the reaction mixture was stirred at r.t. for 16 h. Purification by column chromatography on silica gel (petroleum ether/EtOAc: 5/1) yielded an isomeric mixture of **18** (0.050 g, 97%) as yellow oil. *Compound **18** is obtained as a mixture of two isomers (69:31 ratio) denoted as **18'** and **18''**.*

^1H-NMR (400 MHz, $CDCl_3$): δ = 7.44-7.21 (m, 9H, Ar–H), 4.97 (d, J = 2.9 Hz, 0.69H, CH; **18'**), 4.70 (d, J = 4.9 Hz, 0.31H, CH; **18''**), 4.63 (s, 1.14H, CH_2; **18'**), 4.63 (s, 0.88H, CH_2; **18''**), 4.12 (d, J = 2.0 Hz, 0.69H, CH; **18'**), 3.98 (d, J = 3.1 Hz, 0.30H, CH; **18''**), 3.28-3.26 (m, 1H, CH), 2.80 (br s, 0.31H, OH; **18''**), 2.69 (br s, 0.68H, OH; **18'**).

$^{13}C\{^1H\}$-NMR (125 MHz, $CDCl_3$) *For* **18'**: δ = 141.2 (C_q), 139.4 (C_q), 136.1 (C_q), 128.9 (2C, CH), 128.6 (CH), 127.3 (2C, CH), 126.7 (2C, CH), 126.1 (2C, CH), 71.4 (CH), 65.2 (CH), 65.0 (CH_2), 55.1 (CH). *For* **18''**: δ = 141.3 (C_q), 140.3 (C_q), 135.9 (C_q), 128.9 (2C, CH), 128.7 (CH), 127.4 (2C, CH), 126.9 (2C, CH), 126.4 (2C, CH), 73.5 (CH), 65.9 (CH), 65.0 (CH_2), 56.9 (CH).

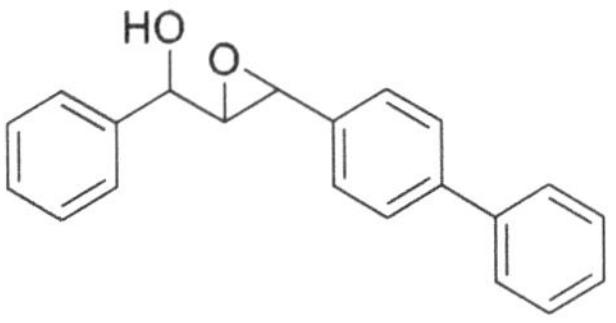

(3-([1,1'-Biphenyl]-4-yl)oxiran-2-yl)(phenyl)methanol (19): The representative procedure was followed, using substrate **19a** (0.061 g, 0.203 mmol) and the reaction mixture was stirred at r.t. for 16 h. Purification by column chromatography on silica gel (petroleum ether/EtOAc: 10/1) yielded an isomeric mixture of **19** (0.054 g, 88%) as a white solid. *Compound **19** is obtained as a mixture of three isomers (15:37:48 ratio) denoted as **19'**, **19''** and **19'''**.*

^1H-NMR (400 MHz, $CDCl_3$): δ = 7.59-7.49 (m, 4H, Ar–H), 7.46-7.25 (m, 10H, Ar–H), 5.01 (d, J = 2.8 Hz, 0.15H, CH; **19'**), 4.99 (d, J = 2.5 Hz, 0.37H, CH; **19''**), 4.71 (d, J = 3.5 Hz, 0.48H, CH; **19'''**), 4.17 (d, J = 1.6 Hz, 0.38H, CH; **19''**), 4.15 (d, J = 1.8 Hz, 0.15H, CH; **19'**), 4.03 (d, J = 1.8 Hz, 0.46H; **19'''**), 3.34-3.30 (m, 1H, CH), 2.86 (br s, 0.40H, OH; **19'''**), 2.72 (br s, 0.44H, OH; **19'** and **19''**).

$^{13}C\{^1H\}$-NMR (100 MHz, $CDCl_3$) *For* **19''**: δ = 141.6 (C_q), 140.7 (C_q), 139.4 (C_q), 135.5 (C_q), 129.0

(2C, CH), 128.9 (2C, CH), 128.4 (CH), 127.6 (CH), 127.4 (2C, CH), 127.2 (2C, CH), 126.7 (2C, CH), 126.4 (2C, CH), 73.6 (CH), 66.0 (CH), 55.1 (CH). *For* **19'''**: δ = 141.5 (C_q), 140.7 (C_q), 140.3 (C_q), 135.7 (C_q), 129.0 (2C, CH), 128.9 (2C, CH), 128.5 (CH), 127.6 (CH), 127.4 (2C, CH), 127.2 (2C, CH), 126.7 (2C, CH), 126.4 (2C, CH), 71.4 (CH), 65.2 (CH), 56.9 (CH). Due to the low concentration of **19'**, we could not interpret $^{13}C\{^1H\}$-NMR for **19'**.

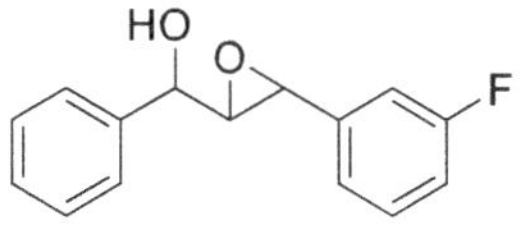

(3-(3-Fluorophenyl)oxiran-2-yl)(phenyl)methanol (20): The representative procedure was followed, using substrate **20a** (0.049 g, 0.202 mmol) and the reaction mixture was stirred at r.t. for 16 h. Purification by column chromatography on silica gel (petroleum ether/EtOAc: 10/1) yielded an isomeric mixture of **20** (0.043 g, 87%) as a colorless liquid. *Compound **20** is obtained as a mixture of two isomers (54:46 ratio) denoted as **20'** and **20''**.*

^1H-NMR (400 MHz, $CDCl_3$): δ = 7.45-6.92 (m, 9H, Ar–H), 4.97 (d, J = 2.3 Hz, 0.54H, CH; **20'**), 4.70 (d, J = 3.6 Hz, 0.46H, CH; **20''**), 4.12-4.09 (m, 0.54H, CH; **20'**), 3.99 (d, J = 1.9 Hz, 0.46H, CH; **20''**), 3.27-3.22 (m, 1H, CH), 2.76 (br s, 0.54H, OH; **20'**), 2.65 (br s, 0.33H, OH; **20''**).

$^{13}C\{^1H\}$-NMR (100 MHz, $CDCl_3$) *For* **20'**: δ = 163.2 (d, $^1J_{C-F}$ = 246.3 Hz, C_q), 140.2 (C_q), 139.3 (d, $^3J_{C-F}$ = 10.2 Hz, C_q), 130.3 (d, $^3J_{C-F}$ = 8.7 Hz, CH), 129.0 (2C, CH), 128.5 (CH), 126.4 (2C, CH), 121.7 (CH), 115.5 (d, $^2J_{C-F}$ = 21.1 Hz, CH), 112.8 (CH), 73.3 (CH), 66.0 (CH), 56.3 (d, $^5J_{C-F}$ = 2.2 Hz, CH). *For* **20''**: δ = 163.2 (d, $^1J_{C-F}$ = 246.3 Hz, C_q), 139.5 (d, $^3J_{C-F}$ = 7.3 Hz, C_q), 139.3 (C_q), 130.3 (d, $^3J_{C-F}$ = 8.7 Hz, CH), 128.9 (2C, CH), 128.6 (CH), 126.7 (2C, CH), 121.6 (CH), 115.4 (d, $^2J_{C-F}$ = 21.8 Hz, CH), 112.6 (CH), 71.3 (CH), 65.3 (CH), 54.6 (d, $^5J_{C-F}$ = 1.5 Hz, CH). HRMS (ESI): *m/z* Calcd for $C_{15}H_{13}FO_2 + H^+$ $[M + H]^+$ 245.0978; Found 245.0880. ^{19}F-NMR (377 MHz, $CDCl_3$): δ = −112.7, −112.9.

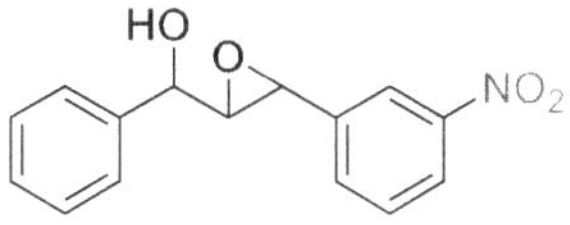

(3-(3-Nitrophenyl)oxiran-2-yl)(phenyl)methanol (21): The representative procedure was followed, using substrate **21a** (0.054 g, 0.201 mmol) and the reaction mixture was stirred at r.t. for 16 h. Purification by column chromatography on silica gel (petroleum ether/EtOAc: 10/1) yielded an isomeric mixture of **21** (0.046 g, 84%) as a colorless liquid. *Compound **21** is obtained as a mixture of three isomers (16:30:54 ratio) denoted as **21'**, **21''** and **21'''**.*

^1H-NMR (400 MHz, $CDCl_3$): δ = 8.34-7.76 (m, 2H, Ar–H), 7.59-7.22 (m, 7H, Ar–H), 5.08 (d, J =

3.3 Hz, 0.16H, CH; **21'**), 5.01 (d, J = 2.9 Hz, 0.30H, CH; **21''**), 4.78 (d, J = 4.4 Hz, 0.54H, CH; **21'''**), 4.23 (d, J = 2.0 Hz, 0.29H, CH; **21''**), 4.11 (d, J = 2.0 Hz, 0.70H, CH; **21'** and **21'''**), 3.22-3.28 (m, 1H, CH).

$^{13}C\{^1H\}$-NMR (125 MHz, $CDCl_3$) *For* **21''**: δ = 141.7 (C_q), 139.3 (C_q), 136.0 (C_q), 131.8 (CH), 129.7 (CH), 129.0 (2C, CH), 128.8 (CH), 126.6 (2C, CH), 123.3 (CH), 120.9 (CH), 71.3 (CH), 65.6 (CH), 54.1 (CH). *For* **21'''**: δ = 148.6 (C_q), 139.3 (C_q), 139.1 (C_q), 131.8 (CH), 129.8 (CH), 129.0 (2C, CH), 128.7 (CH), 126.4 (2C, CH), 123.4 (CH), 120.9 (CH), 73.1 (CH), 66.2 (CH), 55.7 (CH). Due to the low concentration of **21'**, we could not interpret $^{13}C\{^1H\}$-NMR for **21'**.

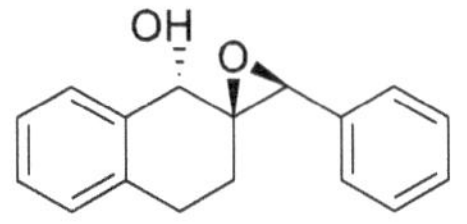

3'-Phenyl-3,4-dihydro-1*H*-spiro[naphthalene-2,2'-oxiran]-1-ol (22): The representative procedure was followed, using substrate **22a** (0.100 g, 0.40 mmol) and the reaction mixture was stirred at r.t. for 16 h. Purification by column chromatography on silica gel (petroleum ether/EtOAc: 10/1) yielded **22'** (0.022 g, 22%) and mixture of **22''** and **22'''** (0.042 g, 41%) as white solid.

^1H-NMR (400 MHz, $CDCl_3$) for **22'**: δ = 7.57-7.55 (m, 1H, Ar–H), 7.39-7.28 (m, 5H, Ar–H), 7.27-7.21 (m, 2H, Ar–H), 7.11 (d, J = 6.9 Hz, 1H, Ar–H), 4.94 (s, 1H, CH), 4.38 (s, 1H, CH), 2.96-2.98 (m, 1H, CH), 2.69-2.62 (m, 1H, CH), 2.47 (br s, 1H, OH), 1.99-1.92 (m, 1H, CH), 1.69-1.63 (m, 1H, CH). $^{13}C\{^1H\}$-NMR (100 MHz, $CDCl_3$): δ = 137.8 (C_q), 136.6 (C_q), 135.3 (C_q), 129.0 (CH), 128.5 (CH), 128.3 (2C, CH), 128.1 (CH), 127.9 (CH), 126.7 (CH), 126.6 (2C, CH), 69.9 (CH), 66.4 (C_q), 61.8 (CH), 27.4 (CH_2), 22.4 (CH_2). $^{13}C\{^1H\}$-NMR (125 MHz, $CDCl_3$) for **22'**: δ = 137.8 (C_q), 136.6 (C_q), 135.3 (C_q), 129.0 (CH), 128.5 (CH), 128.3 (2C, CH), 128.1 (CH), 127.9 (CH), 126.7(CH), 126.6 (2C, CH), 69.9 (CH), 66.4 (C_q), 61.8 (CH), 27.4 (CH_2), 22.4 (CH_2).

^1H-NMR (500 MHz, $CDCl_3$) for **22''** and **22'''**: δ = 7.57-7.17 (m, 8H, Ar–H), 7.08-7.04 (m, 1H, Ar–H), 4.98 (s, 0.73H, CH; **22''**), 4.89 (s, 0.27H, CH; **22'''**), 4.44 (s, 0.27H, CH; **22'''**), 4.23 (s, 0.73H, CH; **22''**), 2.87-2.83 (m, 0.73H, CH; **22''**), 2.81 (br s, 0.73H, OH; **22''**), 2.78-2.75 (m, 0.27H, CH; **22'''**), 2.64-2.61 (m, 0.47H, CH), 2.55-2.50 (m, 0.73H, CH; **22''**), 2.14-2.10 (m, 0.73H, CH; **22''**), 1.93-1.87 (m, 0.27H, CH; **22'''**), 1.77-1.74 (m, 0.27H, CH; **22'''**), 1.66-1.60 (m, 0.73H, CH; **22''**).

$^{13}C\{^1H\}$-NMR (125 MHz, $CDCl_3$): *For* **22''** δ = 139.5 (C_q), 137.5 (C_q), 132.0 (C_q), 129.8 (CH), 128.9 (CH), 128.7 (2C, CH), 128.5 (CH), 128.5 (CH), 127.4 (2C, CH), 126.4 (CH), 74.3 (CH), 66.9 (C_q), 56.0 (CH), 25.6 (CH_2), 22.8 (CH_2). *For* **22'''**: δ = 137.2 (C_q), 136.3 (C_q), 135.5 (C_q), 128.5 (CH), 128.2 (2C, CH), 127.9 (CH), 127.7 (CH), 127.7 (CH), 126.6 (CH), 126.6 (2C, CH), 70.7 (CH),

66.8 (C_q), 60.0 (CH), 27.6 (CH_2), 23.3 (CH_2). HRMS (ESI): *m/z* Calcd for $C_{17}H_{16}O_2 + H^+$ $[M + H]^+$ 253.1229; Found 253.0527.

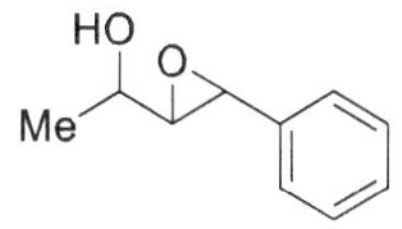

1-(3-Phenyloxiran-2-yl)ethan-1-ol (24): The representative procedure was followed, using substrate **24a** (0.033 g, 0.203 mmol) and the reaction mixture was stirred at r.t. for 16 h. Purification by column chromatography on silica gel (petroleum ether/EtOAc: 5/1) yielded an isomeric mixture of **24** (0.028 g, 84%) as a colorless liquid. *Compound **24** is obtained as a mixture of two isomers (46:54 ratio) denoted as **24'** and **24''**.*

^{1}H-NMR (500 MHz, $CDCl_3$): δ = 7.36-7.26 (m, 5H, Ar–H), 4.10-3.95 (m, 0.46H, CH; **24'**), 3.96 (d, J = 2.1 Hz, 0.46H, CH; **24'**), 3.86 (d, J = 2.1 Hz, 0.54H, CH; **24''**), 3.84 (br s, 0.54H, CH; **24''**), 3.08 (dd, J = 2.8, 2.3 Hz, 0.46H, CH; **24'**), 3.04 (dd, J = 2.6, 2.1 Hz, 0.54H, CH; **24''**), 2.29 (br s, 1H, OH), 1.33 (d, J = 6.6 Hz, 1.62H, CH_3; **24''**), 1.31 (d, J = 6.6 Hz, 1.38H, CH_3; **24'**).

$^{13}C\{^1H\}$-NMR (125 MHz, $CDCl_3$) *For* **24'**: δ = 136.9 (C_q), 128.7 (2C, CH), 128.4 (CH), 125.9 (2C, CH), 66.5 (CH), 65.0 (CH), 54.9 (CH), 18.9 (CH_3). *For* **24''**: δ = 137.1 (C_q), 128.7 (2C, CH), 128.5 (CH), 125.9 (2C, CH), 67.4 (CH), 65.8 (CH), 56.7 (CH), 20.1 (CH_3). HRMS (ESI): *m/z* Calcd for $C_{10}H_{12}O_2 + H^+$ $[M + H]^+$ 165.0910; Found 165.0912.

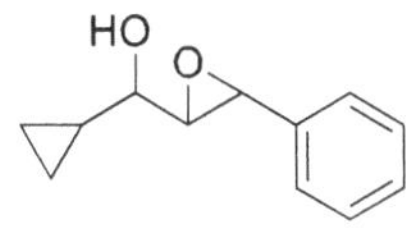

Cyclopropyl(3-phenyloxiran-2-yl)methanol (25): The representative procedure was followed using substrate **25a** (0.038 g, 0.202 mmol) and the reaction mixture was stirred at r.t. for 16 h. Purification by column chromatography on silica gel (petroleum ether/EtOAc: 10/1) yielded an isomeric mixture of **25** (0.033 g, 86%) as a colorless liquid. *Compound **25** is obtained as a mixture of two isomers (35:65 ratio) denoted as **25'** and **25''**.*

^{1}H-NMR (400 MHz, $CDCl_3$): δ = 7.36-7.26 (m, 5H, Ar–H), 4.03 (s, 0.35H, CH; **25'**), 3.89 (s, 0.65H, CH; **25''**), 3.28 (d, J = 8.5 Hz, 0.35H, CH; **25'**), 3.23-3.19 (m, 1H, CH), 3.09-3.07 (m, 0.65H, CH; **25''**), 2.10 (br s, 0.50H, OH), 1.13-0.88 (m, 1H, CH), 0.64-0.60 (m, 2H, CH_2), 0.45-0.36 (m, 2H, CH_2).

$^{13}C\{^1H\}$-NMR (100 MHz, $CDCl_3$) *For* **25'**: δ = 137.2 (C_q), 128.7 (2C, CH), 128.4 (CH), 125.9 (2C, CH), 73.4 (CH), 64.5 (CH), 54.8 (CH), 13.3 (CH), 2.9 (2C, CH_2). *For* **25''**: δ = 137.0 (C_q), 128.7 (2C, CH), 128.5 (CH), 125.9 (2C, CH), 74.8 (CH), 65.1 (CH), 56.4 (CH), 15.0 (CH), 2.1 (2C, CH_2).

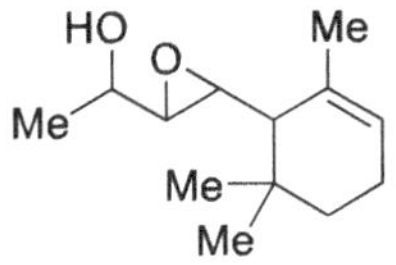

1-(3-(2,6,6-Trimethylcyclohex-2-en-1-yl)oxiran-2-yl)ethan-1-ol (26): The representative procedure was followed using substrate **26a** (0.084 g, 0.403 mmol) and the reaction mixture was stirred at r.t. for 16 h. Purification by column chromatography on silica gel (petroleum ether/EtOAc: 10/1) yielded an isomeric mixture of **26** (0.076 g, 90%) as a colorless liquid. *Compound **26** is obtained as a mixture of two isomers, denoted as **26'** and **26''**.*

^{1}H-NMR (400 MHz, $CDCl_3$) **26'**: δ = 5.46 (s, 1H, CH), 3.95-3.89 (m, 1H, CH), 2.88-2.74 (m, 2H, CH_2), 2.18 (br s, 1H, OH), 1.99-1.98 (m, 2H, CH_2), 1.70-1.67 (m, 3H, CH_3), 1.57-1.47 (m, 1H, CH), 1.33-1.28 (m, 1H, CH), 1.25 (d, J = 6.5 Hz, 3H, CH_3), 1.22-1.19 (m, 1H, CH), 1.05 (s, 3H, CH_3), 0.88 (s, 3H, CH_3).

Due to low concentration of **26''**, we could not interpret ^{1}H-NMR for **26''**.

^{13}C{^{1}H}-NMR (100 MHz, $CDCl_3$) *For* **26'**: δ = 131.2 (C_q), 124.0 (CH), 65.6 (CH), 60.5 (CH), 56.0 (CH), 52.4 (CH), 32.7 (C_q), 31.5 (CH_2), 27.9 (CH_3), 26.9 (CH_3), 24.0 (CH_3), 23.2 (CH_2), 19.1 (CH_3). *For* **26''**: δ= 131.1 (C_q), 124.1 (CH), 67.9 (CH), 61.6 (CH), 67.6 (CH), 52.4 (CH), 32.7 (C_q), 31.7 (CH_2), 27.7 (CH_3), 27.0 (CH_3), 23.9 (CH_3), 23.2 (CH_2), 20.0 (CH_3). HRMS (ESI): m/z Calcd for $C_{13}H_{22}O_2 + H^+$ $[M + H]^+$ 211.1693; Found 211.1692.

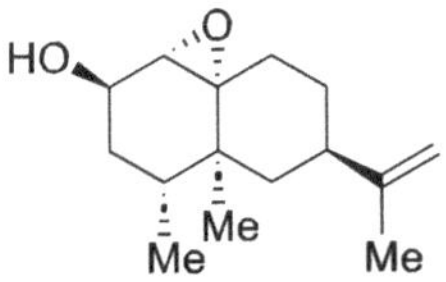

(4*R*,4a*S*,6*R*)-4,4a-Dimethyl-6-(prop-1-en-2-yl)octahydro-3*H*-naphtho[1,8a-*b*]oxiren-2-ol (27): The representative procedure was followed, using substrate **27a** (0.047 g, 0.201 mmol) and the reaction mixture was stirred at r.t. for 16 h. Purification by column chromatography on silica gel (petroleum ether/EtOAc: 5/1) yielded **27** (0.033 g, 69%) as colorless liquid.

^{1}H-NMR (400 MHz, $CDCl_3$): δ = 4.71 (d, J = 8.0 Hz, 2H, CH_2), 4.08 (t, J = 8.6 Hz, 1H, CH), 2.92 (s, 1H, OH), 2.37-2.29 (m, 1H, CH), 2.16 (td, J = 13.7, 4.3 Hz, 1H, CH), 1.72 (s, 3H, CH_3), 1.69-1.50 (m, 6H, CH_2), 1.22-1.19 (m, 2H, CH_2), 1.12-1.07 (m, 1H, CH), 1.03 (s, 3H, CH_3), 0.74 (d, J = 6.9 Hz, 3H, CH_3). ^{13}C{^{1}H}-NMR (125 MHz, $CDCl_3$): δ = 150.1 (C_q), 109.1 (CH_2), 67.1 (C_q), 65.9 (CH), 64.4 (CH), 40.5 (CH), 39.8 (CH_2), 36.1 (C_q), 35.6 (CH_2), 31.0 (CH), 30.4 (CH_2), 28.7 (CH_2), 20.9 (CH_3), 15.3 (CH_3), 14.9 (CH_3).

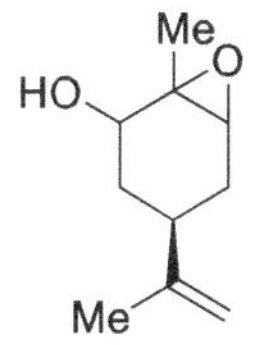

(4*S*)-1-Methyl-4-(prop-1-en-2-yl)-7-oxabicyclo[4.1.0]heptan-2-ol (28): The representative procedure was followed, using substrate **28a** (0.036 g, 0.217 mmol) and H_2 (5 bar) and the reaction mixture was stirred at r.t. for 16 h. Purification by column chromatography on silica gel (petroleum ether/EtOAc: 10/1) yielded isomeric mixture of **28** (0.023 g, 63%) as a colorless liquid. *Compound **28** is obtained as mixture of two isomers (16:84), denoted as **28'** and **28''**.*

^{1}H-NMR (400 MHz, $CDCl_3$) for **28'** and **28''**: δ = 4.79 (m, 2H, CH_2), 4.16-4.15 (m, 0.16H, CH; **28'**), 3.93-3.87 (m, 0.84H, CH; **28''**), 3.29-3.13 (m, 0.84H, CH; **28'**), 2.21-2.14 (m, 2H, CH_2), 1.70 (s, 3H, CH_3), 1.67-1.61 (m, 2H, CH_2), 1.42 (s, 2.17H, CH_3; **28''**), 1.39 (s, 0.89H, CH_3; **28'**), 1.38-1.30 (m, 1H, CH).

^{13}C{^{1}H}-NMR (100 MHz, $CDCl_3$) *For* **28''**: δ= 148.1 (C_q), 109.7 (CH_2), 77.4 (C_q), 68.2 (CH), 63.4 (CH), 36.0 (CH_2), 31.8 (CH), 30.3 (CH_2), 21.7 (CH_3), 21.2 (CH_3).

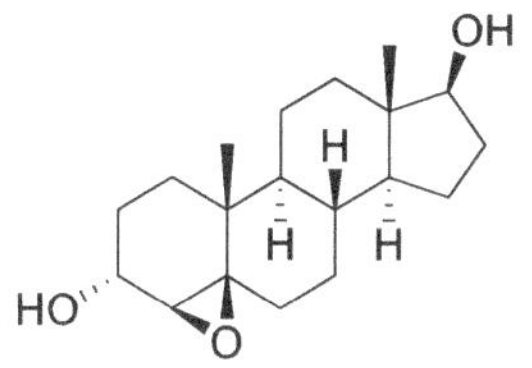

(1a*S*,2*R*,4a*R*,4b*S*,6a*S*,7*S*,9a*S*,9b*S*,11a*R*)-4a,6a-Dimethylhexadecahydrocyclopenta[7,8]phenanthrol[1,10a-*b*]oxirene-2,7-diol (30): The representative procedure was followed, using substrate **30a** (0.061 g, 0.20 mmol) and the reaction mixture was stirred at r.t. for 16 h. Purification by column chromatography on silica gel (petroleum ether/EtOAc: 1/1) yielded **30** (0.037 g, 60%) as a white solid. ^{1}H-NMR (400 MHz, CD_3OD): δ = 4.59 (br s, 1H, OH), 3.81 (dd, J = 9.6, 7.5 Hz, 1H, CH), 3.60 (t, J = 8.6 Hz, 1H, CH), 2.82 (s, 1H, CH), 2.14-2.10 (m, 1H, CH), 2.04-1.98 (m, 1H, CH), 1.88-1.80 (m, 2H, CH_2), 1.70-1.65 (m, 2H, CH_2), 1.58-1.53 (m, 2H, CH_2), 1.47-1.27 (m, 7H, CH_2, CH), 1.16-1.06 (m, 4H, CH_2), 1.02 (s, 3H, CH_3), 0.77 (s, 3H, CH_3). ^{13}C{^{1}H}-NMR (100 MHz, CD_3OD): δ = 83.3 (CH), 69.1 (C_q), 68.3 (CH), 68.1 (CH), 52.9 (C_q), 48.3 (CH), 44.9 (C_q), 38.7 (CH_2), 38.4 (CH), 37.4 (CH), 32.8 (CH_2), 31.7 (CH_2), 31.5 (CH_2), 28.1 (CH_2), 27.2 (CH_2), 25.2 (CH_2), 22.7 (CH_2), 20.0 (CH_3), 12.4 (CH_3). HRMS (ESI): *m/z* Calcd for $C_{19}H_{30}O_3 + H^+$ $[M + H]^+$ 307.2268; Found 307.2262.

(1a*S*,2*R*,4a*R*,4b*S*,6a*S*,7*S*,9a*S*,9b*S*,11a*R*)-2-hydroxy-4a,6a-dimethylhexadecahydrocyclopenta[7,8]phenanthro[1,10a-*b*]oxiren-7-yl acetate (31): The representative procedure was followed, using substrate **31a** (0.139 g, 0.401 mmol), 1.0 mL of MeOH:DCM (4:1) and the reaction mixture was stirred at r.t. for 16 h. Purification by column chromatography on silica gel (petroleum ether/EtOAc: 1/1) yielded **31** (0.081 g, 58%) as a white solid. ^{1}H-NMR (500 MHz, $CDCl_3$): δ = 4.58 (dd, J = 8.9 8.1 Hz, 1H, CH), 3.95 (dd, J = 9.5, 7.4 Hz, 1H, CH), 2.84 (s, 1H, CH), 2.22-2.12 (m, 1H, CH), 2.03 (s, 3H, CH_3), 1.79-1.61 (m, 5H, CH), 1.54-1.46 (m, 3H, CH), 1.38-1.23 (m, 6H, CH), 1.19-1.05 (m, 5H, CH), 1.00 (s, 3H, CH_3), 0.79 (s, 3H, CH_3). ^{13}C{^{1}H}-NMR (125 MHz, $CDCl_3$): δ = 171.4 (CO), 82.9 (CH), 66.9 (C_q), 66.9 (CH), 66.0 (CH), 50.6 (CH), 46.3 (CH), 42.8 (C_q), 36.9 (CH_2), 36.6 (C_q), 35.1 (CH), 30.9 (CH_2), 29.9 (CH_2), 27.7 (CH_2), 26.2 (CH_2), 25.9 (CH_2), 23.8 (CH_2), 21.4 (CH_3), 20.8 (CH_2), 18.9 (CH_3), 12.2 (CH_3). HRMS (ESI): m/z Calcd for $C_{21}H_{32}O_4 + H^+$ $[M + H]^+$ 349.2373; Found 349.2371.

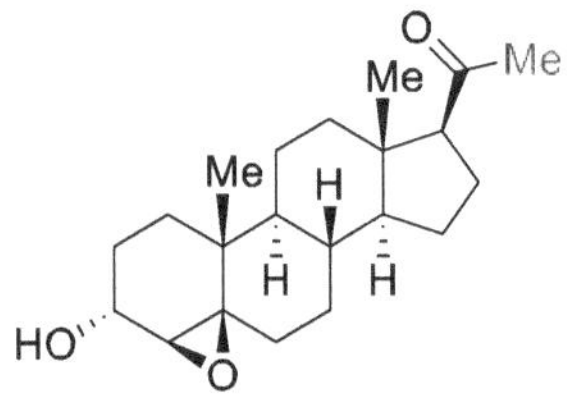

1-((4a*R*,4b*S*,6a*S*,7*S*,9a*S*,9b*S*)-2-hydroxy-4a,6a-dimethylhexadecahydrocyclopenta[7,8]phenanthro[1,10a-*b*]oxiren-7-yl)ethan-1-one (32): The representative procedure was followed, using substrate **32a** (0.133 g, 0.402 mmol), 2.0 mL of MeOH:DCM (4:1) as solvent. The reaction mixture was stirred at r.t. for 16 h. Purification by column chromatography on silica gel (petroleum ether/EtOAc: 5/1) yielded **32** (0.073 g, 55%) as a white solid. ^{1}H-NMR (400 MHz, $CDCl_3$): δ = 3.97-3.93 (m, 1H, CH), 2.83 (s, 1H, CH), 2.51 (t, J = 9.0 Hz, 1H, CH), 1.44 (br s, 1H, OH), 2.18-2.13 (m, 1H, CH), 2.10 (s, 3H, CH_3), 2.07-1.99 (m, 2H, CH), 1.78-1.61 (m, 4H, CH), 1.53-1.34 (m, 5H, CH), 1.30-1.18 (m, 4H, CH), 1.09-1.02 (m, 3H, CH), 0.97 (s, 3H, CH_3), 0.61 (s, 3H, CH_3). ^{13}C{^{1}H}-NMR (100 MHz, $CDCl_3$): δ = 209.8 (CO), 66.9 (C_q), 66.7 (CH), 66.0 (CH), 63.8 (CH), 56.4 (CH), 46.1 (CH), 44.3 (C_q), 38.9 (CH_2), 36.5 (C_q), 35.2 (CH), 31.6 (CH_3), 30.9 (CH_2), 30.3 (CH_2), 26.2 (CH_2), 25.7 (CH_2), 24.6 (CH_2), 23.0 (CH_2), 21.3 (CH_2), 18.8 (CH_3), 13.5 (CH_3). HRMS (ESI): m/z Calcd for $C_{21}H_{32}O_3 + H^+$ $[M + H]^+$ 333.2424; Found

333.2420.

3.4.3 Derivatization of *α,β*-Epoxy Alcohols

Nucleophilic Addition to Epoxide

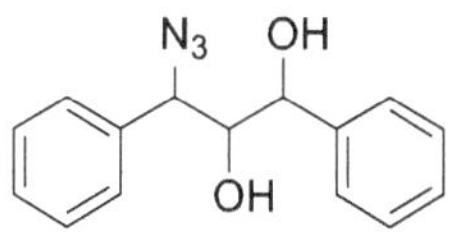

Synthesis of 3-azido-1,3-diphenylpropane-1,2-diol (33): To an oven dried round bottom flask was introduced phenyl(3-phenyloxiran-2-yl)methanol (**1**; 0.1 g, 0.44 mmol) and dissolved in 5 mL of dry DMF. To this NaN_3 (0.057 g, 0.88 mmol) and NH_4Cl (0.88 mmol, 0.047 g) was added and stirred for 10 minutes. Then the reaction flask was heated at 100 °C for 16 h. After completion of reaction, the crude mixture was subjected to the column chromatography (silica gel, petroleum ether/EtOAc: 10/1) to yield an isomeric mixture of 3-azido-1,3-diphenylpropane-1,2-diol (**33**; 0.075 g, 63%) as a colorless liquid. *Compound 33 is obtained as a mixture of two isomers (50:50 ratio) denoted as 33′ and 33″.*

^{1}H-NMR (500 MHz, $CDCl_3$): δ = 7.38-7.23 (m, 10H, Ar–H), 4.72-4.71 (d, J = 3.3 Hz, 0.49H, CH; **33'**), 4.54 (d, J = 5.9 Hz, 0.50H, CH; **33''**), 4.50 (d, J = 7.0 Hz, 0.49H, CH; **33'**), 4.36 (d, J = 7.0 Hz, 0.51H, CH; **33''**), 3.98 (vt, J = 6.5 Hz, 0.52H, CH; 33'), 3.81 (dd, J = 6.9, 3.4 Hz, 0.48H, CH; **33''**), 2.76 (br s, 1H, OH), 2.24 (br s, 1H, OH).

^{13}C{^{1}H}-NMR (125 MHz, $CDCl_3$) *For* **33'**: δ = 140.9 (C_q), 136.1 (C_q), 129.0 (2C, CH), 128.8 (2C, CH), 128.7 (2C, CH), 128.4 (CH), 128.1 (CH), 127.6 (2C, CH), 77.2 (CH), 74.5 (CH), 66.9 (CH). *For* **33''**: δ = 139.9 (C_q), 135.9 (C_q), 128.9 (2C, CH), 128.9 (2C, CH), 128.7 (2C, CH), 128.5 (CH), 128.0 (CH), 126.5 (2C, CH), 76.1 (CH), 72.5 (CH), 66.8 (CH).

Protection of OH group

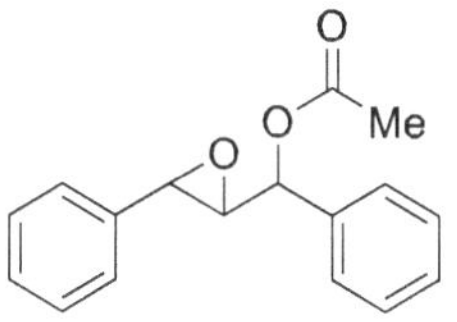

Synthesis of phenyl(3-phenyloxiran-2-yl)methyl acetate (34): To an oven dried round bottom flask was introduced phenyl(3-phenyloxiran-2-yl)methanol (**1**; 0.1 g, 0.44 mmol) and dissolved in 5 mL of dry CH_2Cl_2. To this reaction mixture, DMAP (0.011 g, 0.088 mmol) and Et_3N (0.53 mmol, 0.074 mL) was added at 0 °C and stirred for 10 minutes followed by dropwise addition of acetic anhydride (0.53 mmol, 0.050 mL) and the stirring was continued for 1 h. After completion of reaction, the volatiles were evaporated under vacuo and 10 mL of water was added. The organic compound was extracted in EtOAc and the crude product was subjected to the column chromatography on silica gel (petroleum ether/EtOAc: 40/1) to yield two isomers **34'** (0.048 g, 41%) and **34''** (0.041 g, 35%) as colorless liquids.

For **34'**: ^{1}H-NMR (400 MHz, $CDCl_3$): δ = 7.45-7.24 (m, 10H, Ar–H), 5.92 (d, J = 4.4 Hz, 1H, CH), 3.90 (d, J = 1.8 Hz, 1H, CH), 3.28 (dd, J = 4.3, 1.9 Hz, 1H, CH), 2.12 (s, 3H, CH_3). ^{13}C{^{1}H}-NMR (100 MHz, $CDCl_3$): δ = 170.0 (CO), 136.5 (C_q), 136.3 (C_q), 128.9 (CH), 128.8 (2C, CH), 128.7 (2C, CH), 128.5 (CH), 127.5 (2C, CH), 125.8 (2C, CH), 74.0 (CH), 62.8 (CH), 56.8 (CH), 21.2 (CH_3).

For **34''**: ^{1}H-NMR (500 MHz, $CDCl_3$): δ = 7.42-7.26 (m, 8H, Ar–H), 7.24-7.20 (m, 2H, Ar–H), 5.75 (d, J = 6.1 Hz, 1H, CH), 3.85 (d, J = 1.9 Hz, 1H, CH), 3.86 (dd, J = 6.2, 2.0 Hz, 1H, CH), 2.16 (s, 3H, CH_3). ^{13}C{^{1}H}-NMR (125 MHz, $CDCl_3$): δ = 170.1 (CO), 136.7 (C_q), 136.2 (C_q), 128.9 (2C, CH), 128.8 (CH), 128.7 (2C, CH), 128.6 (CH), 127.0 (2C, CH), 125.8 (2C, CH), 75.8 (CH), 63.5 (CH), 57.0 (CH), 21.3 (CH_3). HRMS (ESI): m/z Calcd for $C_{17}H_{16}O_3 + H^+$ $[M + H]^+$ 269.1172; Found 269.1169.

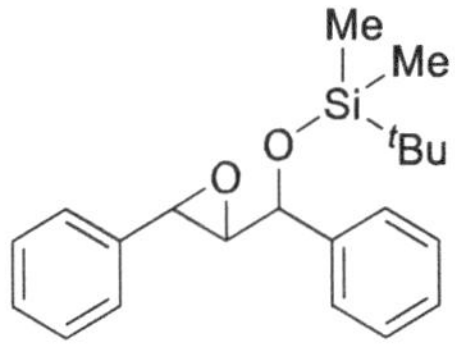

Synthesis of *tert*-butyldimethyl(phenyl(3-phenyloxiran-2-yl)methoxy)silane (35): To an oven dried 25 mL round bottom flask was introduced phenyl(3-phenyloxiran-2-yl)methanol (1; 0.072 g, 0.318 mmol) and dissolved in 5 mL of dry DMF. To this imidazole (0.11 g, 1.6 mmol) and tert-butyldimethylsilyl chloride (0.8 mmol, 0.12 g) was added at 0 °C and stirred for 16 h. After completion of reaction, the crude mixture was subjected to the column chromatography on silica gel (petroleum ether/EtOAc: 100/1) to yield isomeric mixture of *tert*-butyldimethyl(phenyl(3-phenyloxiran-2-yl)methoxy)silane (**35**; 0.085 g, 78%) as colorless liquid. *Compound 35 is obtained as mixture of two isomers (50:50 ratio) denoted as **35'** and **35''**.*

^{1}H-NMR (400 MHz, $CDCl_3$): δ = 7.40-7.39 (m, 2H, Ar–H), 7.34-7.17 (m, 8H, Ar–H), 4.81 (d, J = 3.5 Hz, 0.50H, CH; **35'**), 4.60 (d, J = 5.9 Hz, 0.50H, CH; 35''), 4.01 (vd, 0.50H, CH; **35'**), 3.85 (d, J = 1.6 Hz, 0.50H, CH; **35''**), 3.16 (dd, J = 5.9, 1.9 Hz, 0.50H, CH; **35'**), 3.08 (dd, J = 3.5, 1.9 Hz, 0.50H, CH; **35''**), 0.94 (s, 4.5H, CH_3; **35'**), 0.15 (s, 1.5H, CH_3; **35'**), 0.11 (s, 1.5H, CH_3; **35''**), 0.90 (s, 4.5H, CH_3; **35''**), 0.02 (s, 1.5H, CH_3; **35'**), 0.00 (s, 1.5H, CH_3; **35''**).

^{13}C{^{1}H}-NMR (100 MHz, $CDCl_3$) *For* **35'**: δ = 141.4 (C_q), 137.5 (C_q), 128.6 (2C, CH), 128.5 (2C, CH), 128.3 (CH), 127.9 (CH), 126.4 (2C, CH), 125.8 (2C, CH), 76.0 (CH), 66.9 (CH), 57.1 (CH),

26.0 (CH_3), 18.5 (C_q), -4.5 (CH_3), -4.7 (3C, CH_3). *For* **35''**: δ = 141.0 (C_q), 137.0 (C_q), 128.6 (2C, CH), 128.6 (2C, CH), 128.1 (CH), 127.9 (CH), 126.3 (2C, CH), 125.8 (2C, CH), 73.5 (CH), 66.2 (CH), 55.9 (CH), 26.0 (CH_3), 18.5 (C_q), -4.6 (CH_3), -4.7 (3C, CH_3). HRMS (ESI): *m/z* Calcd for $C_{21}H_{28}O_2Si + H^+$ $[M + H]^+$ 341.1937; Found 341.2631.

3.4.4 Mechanistic Experiments

Procedure for Deuterium Labelling Experiment. A standard catalytic hydrogenation reaction was performed using CD_3OD as a solvent. After completion of reaction time, the reaction mixture was evaporated under vacuo. Purification by column chromatography provided the hydrogenated product **1** without any deuterium incorporation.

(1a) + H_2 (2 bar) —[**Mn-3** (5 mol%), K_2CO_3 (10 mol%); CD_3OD, RT, 16 h]→ (1) D/HO, H (0% D incorporation)

Experiment to Understand Non-innocent Behaviour of Ligand *NH*. The representative procedure was followed, using phenyl(3-phenyloxiran-2-yl)methanone (**1a**; 0.046 g, 0.205 mmol), **Mn-3^{Me}** (0.0054 g, 0.010 mmol), K_2CO_3 (0.0028 g, 0.020 mmol), and the reaction mixture was stirred at r.t for 16 h. The reaction mixture was analysed by GC that ensured no formation of hydrogenated product.

Attempted Hydrogenation using Dearomatized Species. The representative procedure was followed, using phenyl(3-phenyloxiran-2-yl)methanone (**1a**; 0.046 g, 0.205 mmol), **Mn-3'** (0.0043 g, 0.010 mmol) and the reaction mixture was stirred at r.t. for 16 h. Purification by column chromatography on silica gel (petroleum ether/EtOAc: 1/1) yielded **1** (0.044 g, 95%) as a white solid.

3.4.5 X-ray Structural Data

X-ray intensity data measurement of compounds **30-32** were carried out on a Bruker D8 VENTURE Kappa Duo PHOTON II CPAD diffractometer equipped with Incoatech multilayer mirrors optics. The intensity measurements were carried out with Cu and Mo micro-focus sealed tube diffraction sources (CuKα= 1.54178 Å for **31** and **32** and MoKα= 0.71073 Å for **30**) at 100(2) K temperature. The X-ray generator was operated at 50 kV and 1.1 A for Cu and 50 kV and 1.4 mA for Mo radiations. A preliminary set of cell constants and an orientation matrix were calculated from two sets of 40 frames for Cu radiation and three sets of 36 frames for Mo radiation. Data were collected with ω scan width of 0.5° at different settings of φ and 2θ with a frame time of 20-30 secs (depending on the diffraction power of the crystals) keeping the sample-to-detector distance fixed at 5.00 cm. The unit-cell measurements, data collection, integration, scaling, and absorption corrections were performed by using Bruker APEX3 software.[46] The diffraction images were integrated by using Bruker SAINT Programs. The data were corrected for Lorentz-polarization and absorption effects by using the multi-scan method by using SADABS with the transmission coefficients. Using APEX3 program suite, the structure was solved with the SHELXT 2014/5 (Sheldrick, 2014) structure solution program,[47] using direct methods. The model was refined with a version of ShelXL-2018/3 (Sheldrick, 2018),[48] using Least Squares minimization based on F^2. All the H-atoms (except –OH bound H atoms) were placed in a geometrically idealized positions and constrained to ride on their parent atoms by using the HFIX command in SHELX-TL. The –OH bound H atoms were located in the difference Fourier map and refined isotropically. The crystallographic details for all the compounds are summarized in Table 3.2.

Table 3.2. Crystallography details of compounds **30**, **31** and **32**.

Crystal Data	**30**	**31**	**32**
Formula	$C_{19}H_{30}O_3$, CH_3OH	$C_{21}H_{32}O_4$	$C_{21}H_{32}O_3$
M_r	338.47	348.46	332.46
Crystal Size, (mm)	0.12×0.09×0.07	0.11×0.10×0.08	0.16×0.11×0.05
Temp. (K)	100(2)	100(2)	100(2)
Wavelength (Å)	MoKα: 0.71073	CuKα: 1.54178	CuKα: 1.54178
Crystal Syst.	orthorhombic	monoclinic	monoclinic
Space Group	$P2_12_12_1$	$P2_1$	$P2_1$
a/Å	7.4467(2)	7.3919(5)	7.9965(7)
b/Å	11.3626(3)	11.4351(8)	11.2314(9)
c/Å	21.7245(5)	10.8839(7)	10.0681(8)
$\alpha/^0$	90	90	90
$\beta/^0$	90	102.920(2)	93.820(4)
$\gamma/^0$	90	90	90
V/Å^3	1838.19(8)	896.69(10)	902.23(13)
Z	4	2	2
D_{calc}/g cm^{-3}	1.223	1.291	1.224
μ/mm^{-1}	0.083	0.697	0.625
F(000)	744	380	364
Ab. Correct.	multi-scan	multi-scan	multi-scan
T_{min}/T_{max}	0.5697/0.7457	0.6289/0.7536	0.6477/0.7536
$2\theta_{max}$	56.6	144.6	149
Total reflns.	29701	15982	40077
Uni. reflns.	4531	3410	3527
Obs. reflns.	4398	3388	3183
h, k, l (min, max)	(-9, 9), (-9, 0), (-28, 28)	(-9, 9), (-14, 14), (-13, 13)	(-9, 9), (-13, 13), (-12, 12)
R_{int}	0.0345	0.0432	0.0745
R_{sig}	0.0574	0.0327	0.0315
No. of para.	230	233	224
No. of restraints	0	1	1
$R1$ [I> $2\sigma(I)$]	0.0309	0.0299	0.0388
$wR2$[I> $2\sigma(I)$]	0.0810	0.0784	0.0855
$R1$ [all data]	0.0320	0.0301	0.0468
$wR2$ [all data]	0.0820	0.0785	0.0909
goodness-of-fit	1.050	1.074	1.119
$\Delta\rho_{max}$, $\Delta\rho_{min}$(eÅ^{-3})	+0.296, -0.168	+0.234, -0.145	+0.158, -0.173
CCDC No.	2330236	2330235	2330237

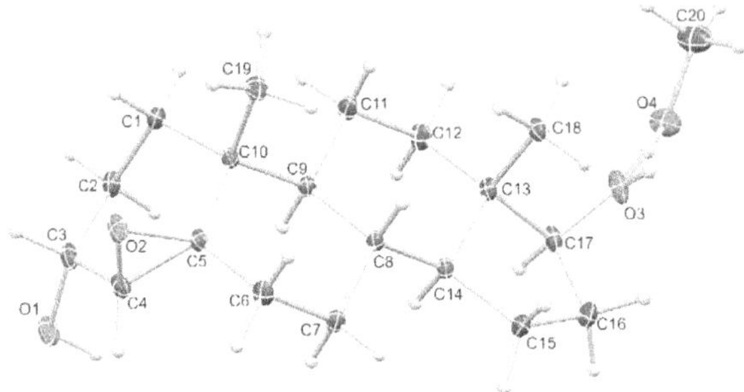

Figure 3.2. ORTEP of compound **30** showing the atom-numbering scheme. Displacement ellipsoids are drawn at the 50% probability level and H atoms are shown as small spheres with arbitrary radii. H-atoms are shown as small spheres with arbitrary radii. The dotted line (cyan) indicate the H-bonding interactions engaging methanol molecule and hydroxy Oxygen of **30**.

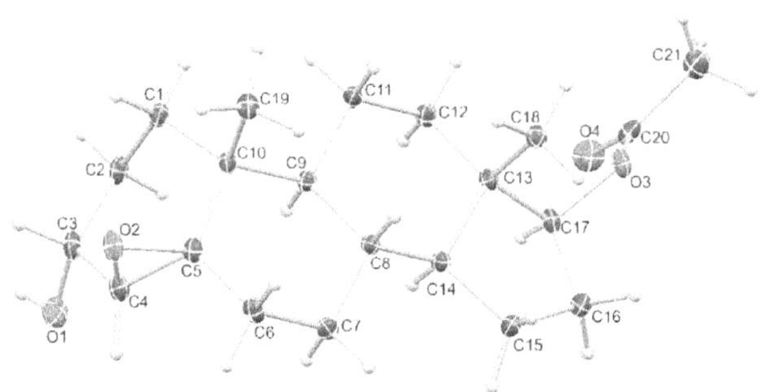

Figure 3.3. ORTEP of compound **31** showing the atom-numbering scheme. Displacement ellipsoids are drawn at the 50% probability level and H atoms are shown as small spheres with arbitrary radii. H-atoms are shown as small spheres with arbitrary radii.

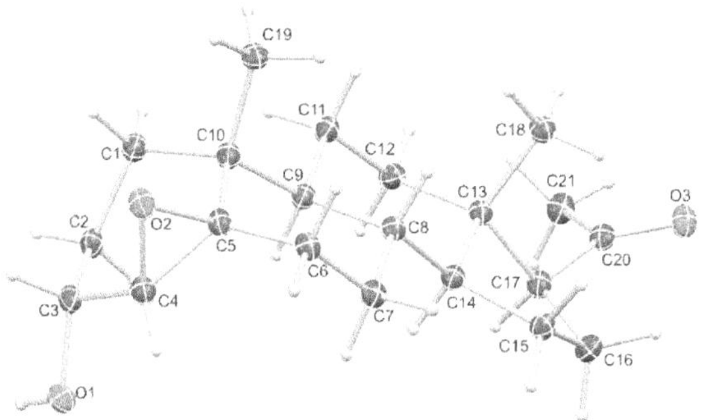

Figure 3.4. ORTEP of compound **32** showing the atom-numbering scheme. Displacement ellipsoids are drawn at the 30% probability level and H atoms are shown as small spheres with arbitrary radii. H-atoms are shown as small spheres with arbitrary radii.

3.4.6 ^{1}H and ^{13}C{^{1}H} NMR Spectra of Selected Hydrogenated Compounds

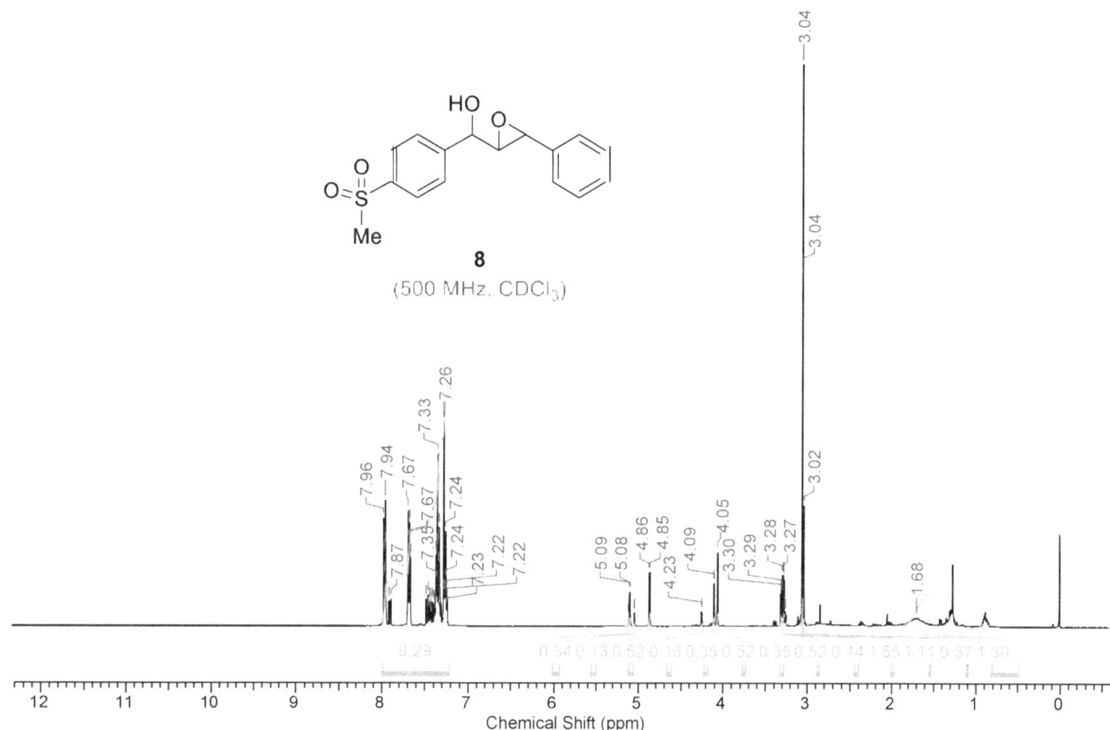

^{1}H-NMR spectrum of **8**

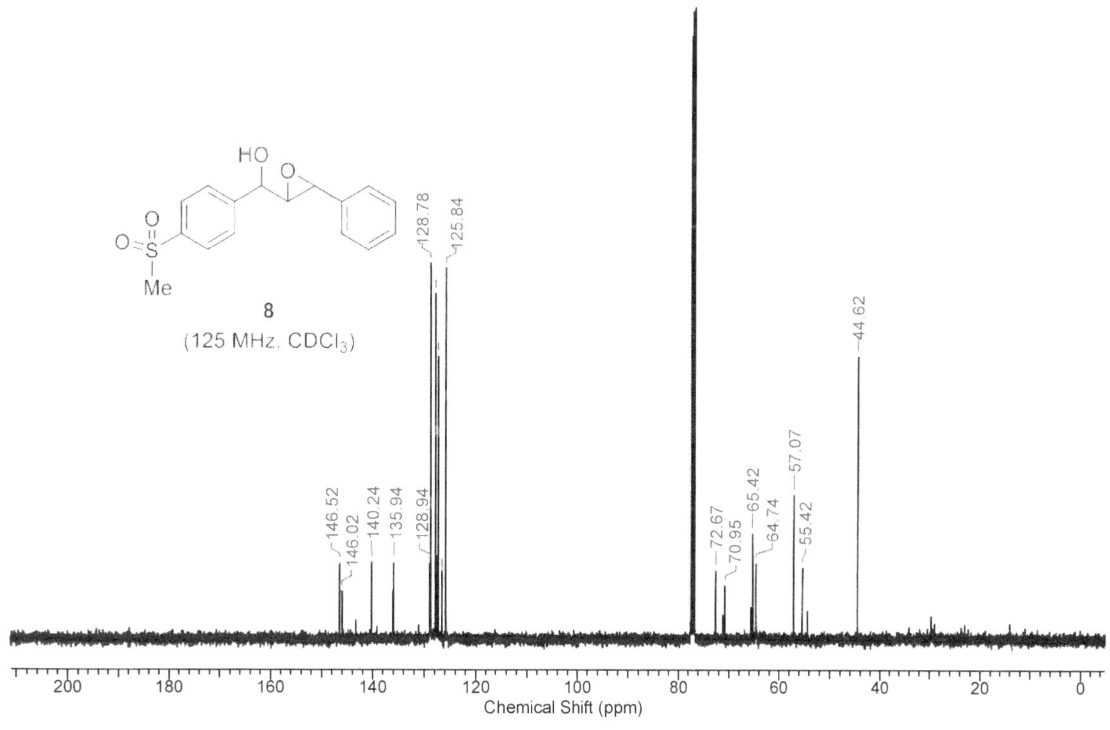

^{13}C{^{1}H}-NMR spectrum of **8**

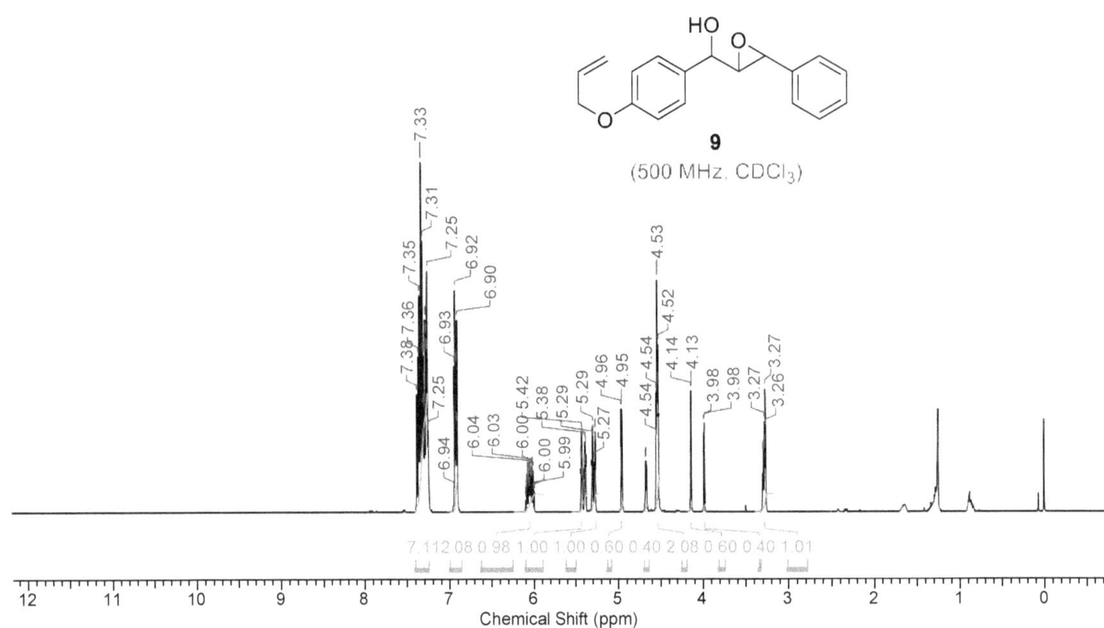

^{1}H-NMR spectrum of **9**

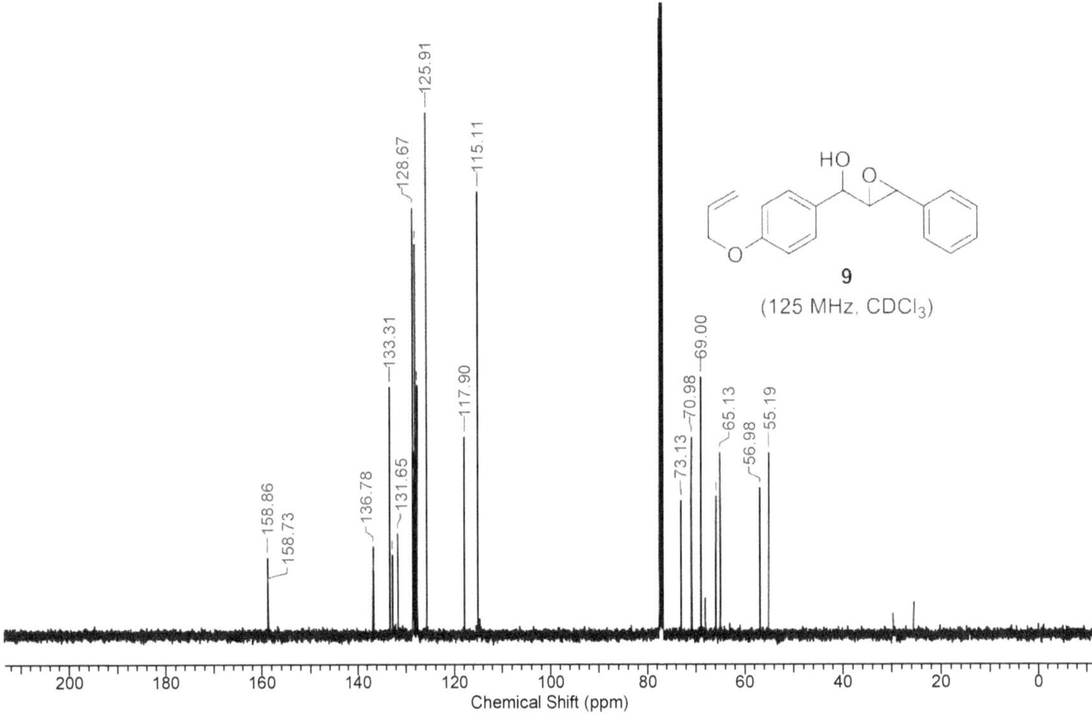

$^{13}C\{^{1}H\}$-NMR spectrum of **9**

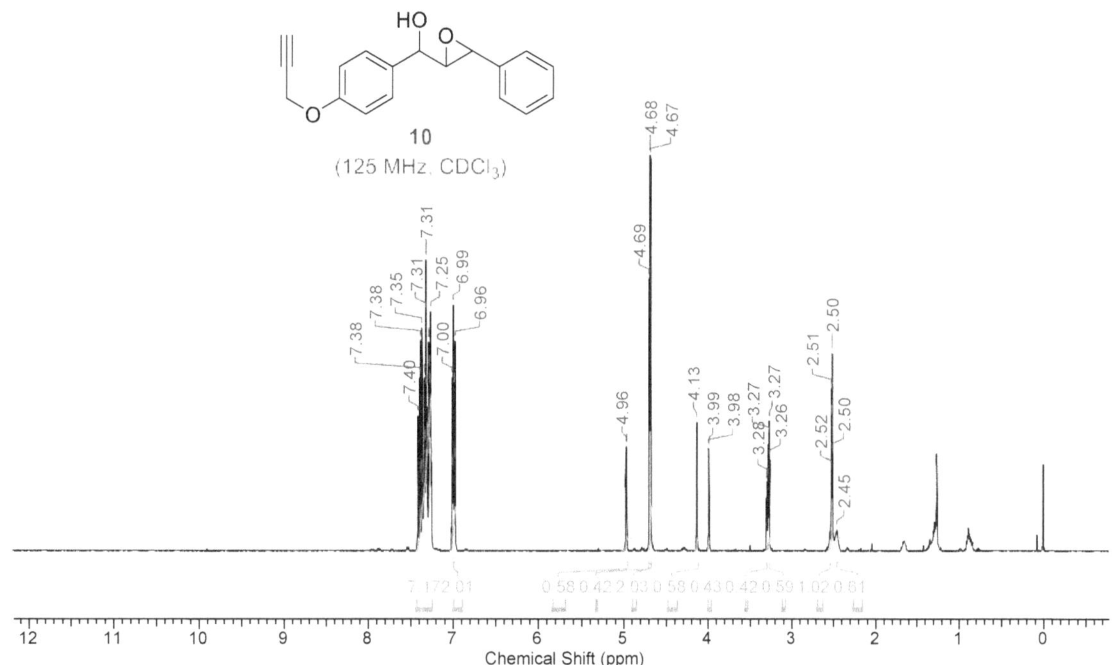

^{1}H-NMR spectrum of **10**

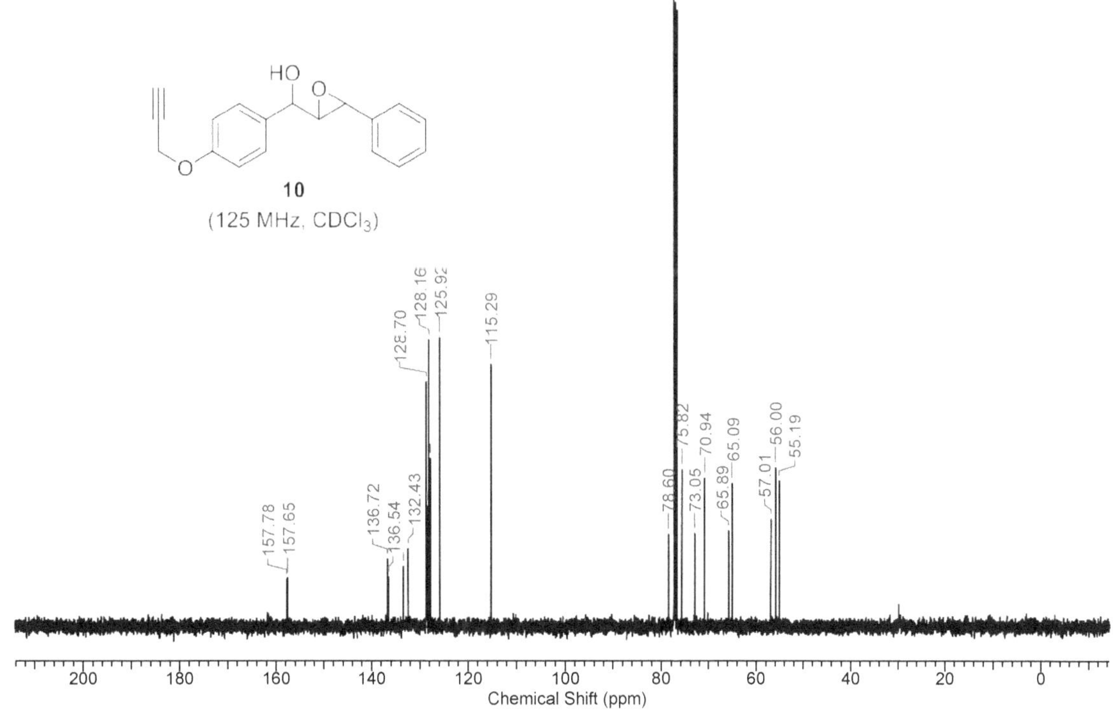

^{13}C{^{1}H}-NMR spectrum of **10**

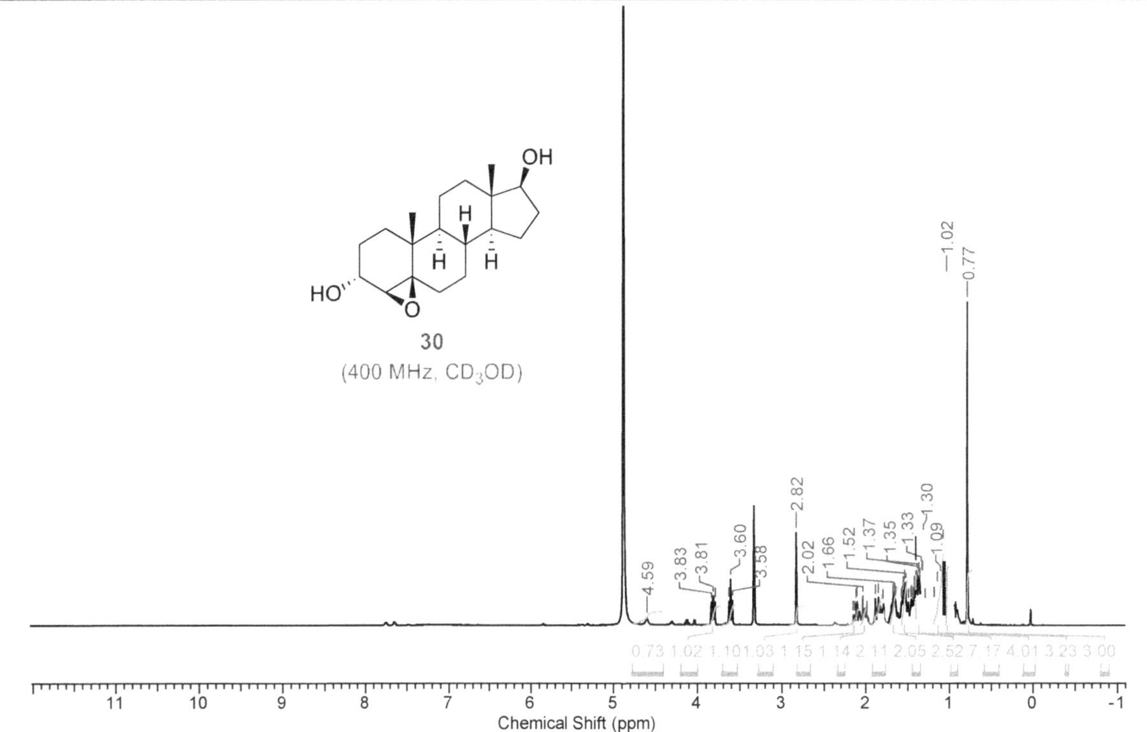

^{1}H-NMR spectrum of **30**

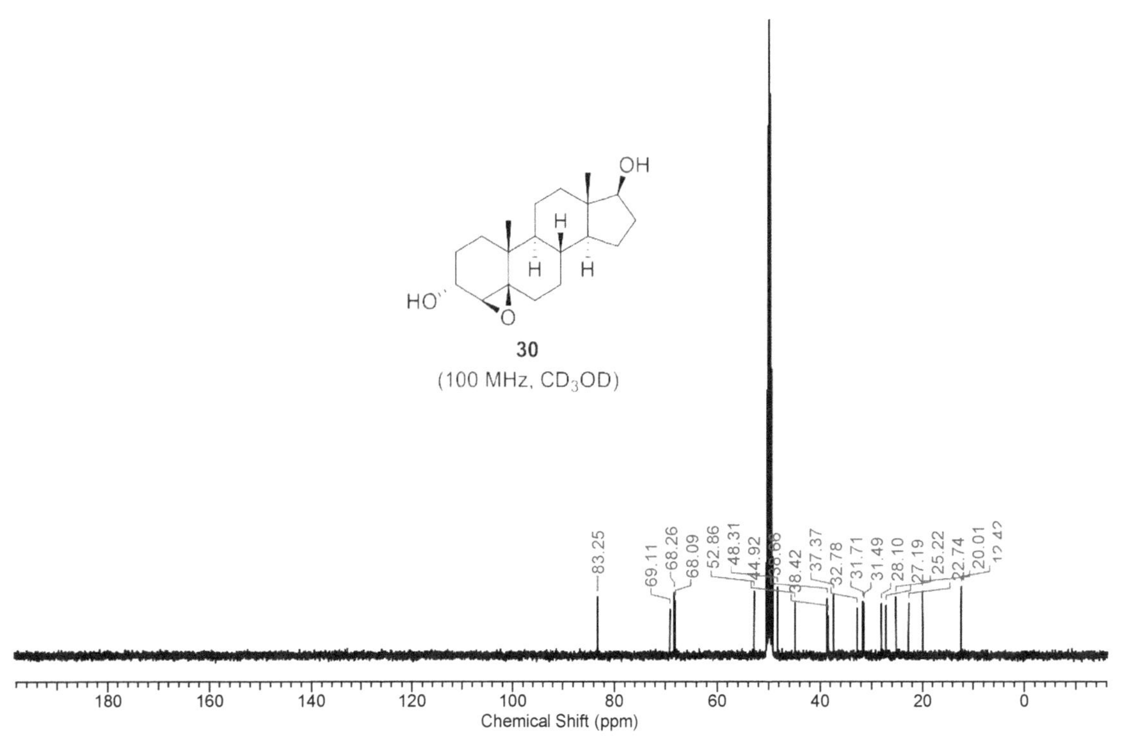

^{13}C{^{1}H}-NMR spectrum of **30**

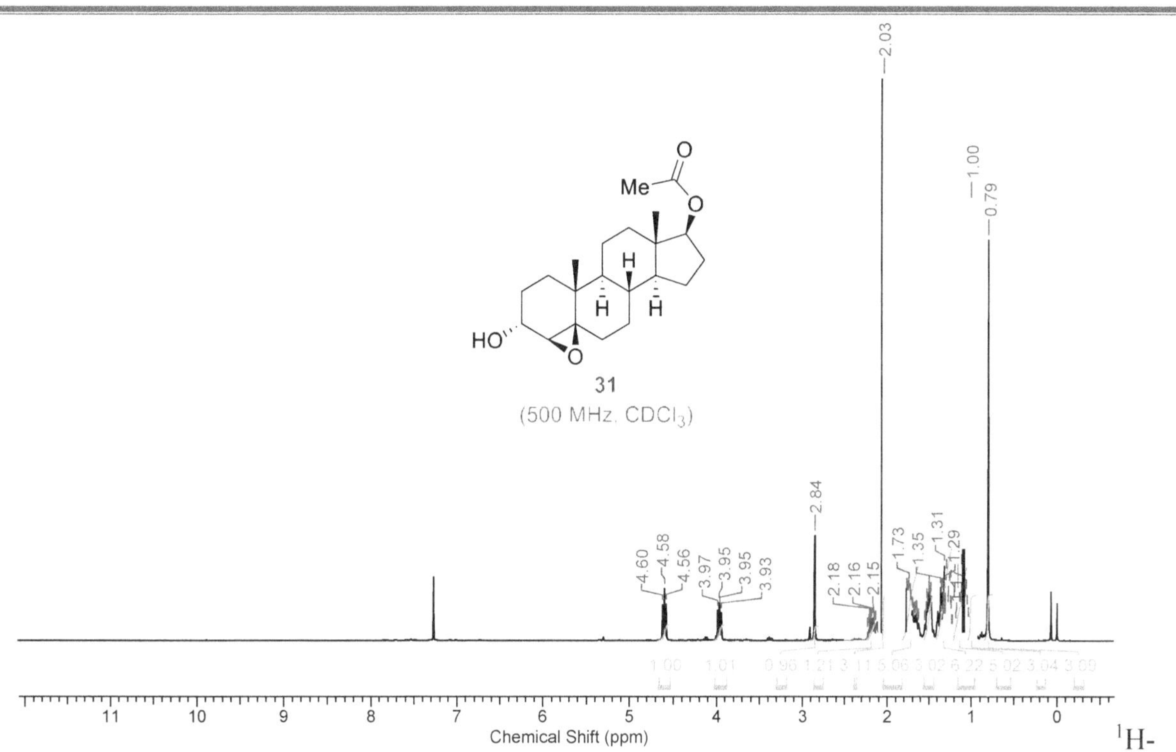

^{1}H-NMR spectrum of **31**

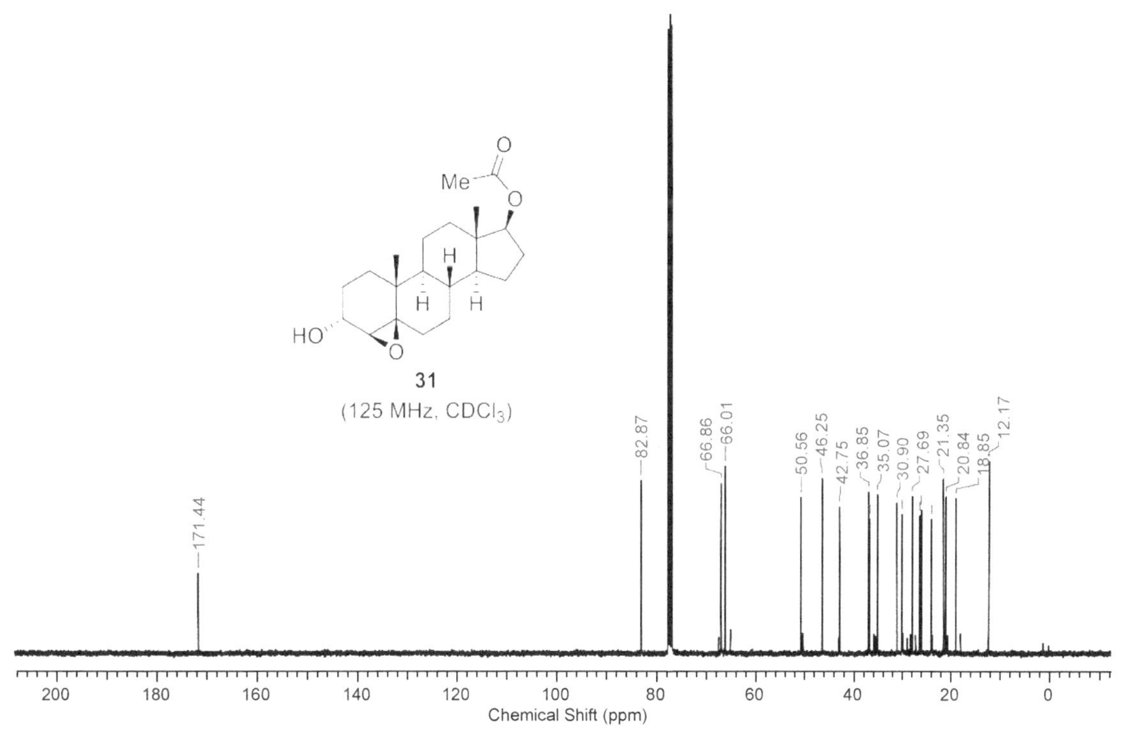

$^{13}C\{^1H\}$-NMR spectrum of **31**

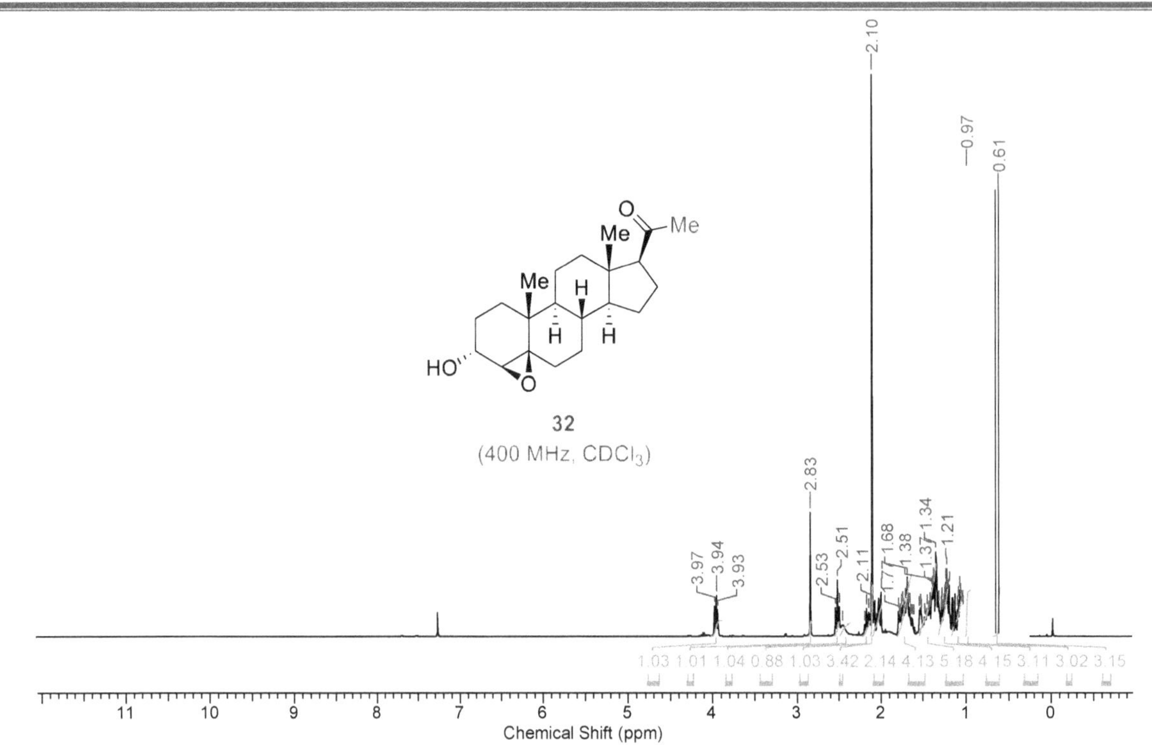

$^{13}C\{^{1}H\}$-NMR spectrum of **32**

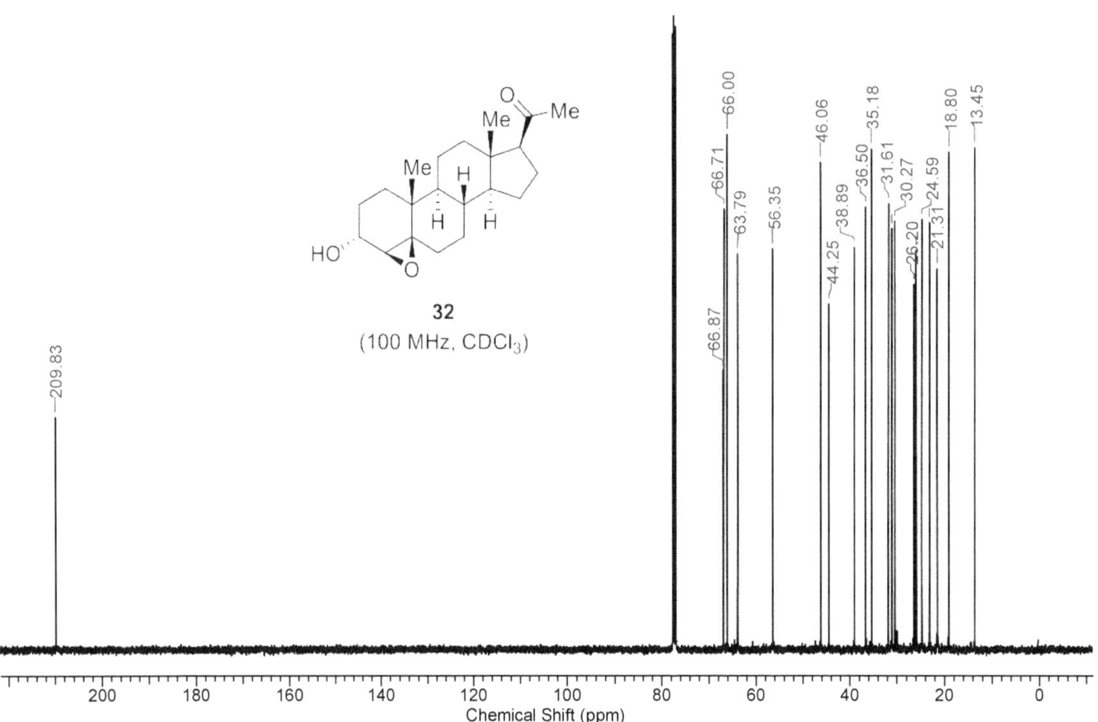

$^{13}C\{^{1}H\}$-NMR spectrum of **32**

3.5 REFERENCES

(1) Hanson, R. M., *Chem. Rev.* **1991**, *91*, 437-475.

(2) Gomes, A. R.; Varela, C. L.; Tavares-da-Silva, E. J.; Roleira, F., *Eur. J. Med. Chem.* **2020**, *201*, 112327.

(3) Cui, H.; Shen, Y.; Chen, Y.; Wang, R.; Wei, H.; Fu, P.; Lei, X.; Wang, H.; Bi, R.; Zhang, Y., *J. Am. Chem. Soc.***2022**, *144*, 8938-8944.

(4) Hatakeyama, S.; Sakurai, K.; Numata, H.; Ochi, N.; Takano, S., *J. Am. Chem. Soc.* **1988**, *110*, 5201-5203.

(5) Katsuki, T.; Sharpless, K. B., *J. Am. Chem. Soc.* **1980**, *102*, 5974-5976.

(6) Howe, G. P.; Wang, S.; Procter, G., *Tetrahedron Lett.* **1987**, *28*, 2629-2632.

(7) Nakata, T.; Tanaka, T.; Oishi, T., *Tetrahedron Lett.* **1981**, *22*, 4723-4726.

(8) Kawakami, T.; Shibata, I.; Baba, A.; Matsuda, H., *J. Org. Chem.* **1993**, *58*, 7608-7609.

(9) Hojo, M.; Fujii, A.; Murakami, C.; Aihara, H.; Hosomi, A., *Tetrahedron Lett.* **1995**, *36*, 571-574.

(10) Forkel, N. V.; Henderson, D. A.; Fuchter, M., *Tetrahedron Lett.* **2014**, *55*, 5511-5514.

(11) Chakraborty, S.; Bhattacharya, P.; Dai, H.; Guan, H., *Acc. Chem. Res.* **2015**, *48*, 1995-2003.

(12) Zell, T.; Milstein, D., *Acc. Chem. Res.* **2015**, *48*, 1979-1994.

(13) Filonenko, G. A.; van Putten, R.; Hensen, E. J. M.; Pidko, E.A., *Chem. Soc. Rev.* **2018**, *47*, 1459- 1483.

(14) Zell, T.; Langer, R., *ChemCatChem* **2018**, *10*, 1930-1940.

(15) Liu, W.; Sahoo, B.; Junge, K.; Beller, M., *Acc. Chem. Res.* **2018**, *51*, 1858- 1869.

(16) Alig, L.; Fritz, M.; Schneider, S., *Chem. Rev.* **2019**, *119*, 2681-2751.

(17) Fleischer, S.; Zhou, S.; Junge, K.; Beller, M., *Angew. Chem., Int. Ed.* **2013**, *52*, 5120-5124.

(18) Wienhöfer, G.; Westerhaus, F. A.; Junge, K.; Ludwig, R.; Beller, M., *Chem. Eur. J.* **2013**, *19*, 7701-7707.

(19) Gorgas, N.; Stöger, B.; Veiros, L. F.; Kirchner, K., *ACS Catal.* **2016**, *6*, 2664-2672.

(20) Shevlin, M.; Friedfeld, M. R.;Sheng, H.; Pierson, N. A.; Hoyt, J. M.; Campeau, L.-C.; Chirik, P. J., *J. Am. Chem. Soc.* **2016**, *138*, 3562-3569.

(21) Glatz, M.;Stöger, B.; Himmelbauer, D.; Veiros, L. F.; Kirchner, K., *ACS Catal.* **2018**, *8*, 4009-4016.

(22) Zimmermann, B. M.; Kobosil, S. C. K.; Teichert, J. F., *Chem. Commun.* **2019**, *55*, 2293-2296.

(23) Guan, J.; Chen, J.; Luo, Y.; Guo, L.; Zhang, W., *Angew. Chem. Int. Ed.* **2023**, *62*, e202306380.

(24) Blaser, H.-U.; Malan, C.; Pugin, B.; Spindler, F.; Steiner, H.; Studer, M., *Adv. Synth. Catal.* **2003**,*345*, 103-151.

(25) Sharma, D. M.; Punji, B., *Chem. Asian J.* **2020**, *15*, 690-708.

(26) Druais, V.; Meyer, C.; Cossy, J., *Org. Lett.* **2012**, *14*, 516-519.

(27) Zhao, Z.; Bagdi, P. R.; Yang, S.; Liu, J.; Xu, W.; Fang, X., *Org. Lett.* **2019**, *21*, 5491-5494.

(28) Widegren, M. B.; Harkness, G. J.; Slawin, A. M. Z.; Cordes, D. B.; Clarke, M. L., *Angew. Chem. Int. Ed.* **2017**, *56*, 5825-5828.

(29) Mukherjee, A.; Milstein, D., *ACS Catal.* **2018**, *8*,11435-11469.

(30) Buhaibeh, R.; Filippov, O. A.; Bruneau-Voisine, A.; Willot, J.; Duhayon, C.; Valyaev, D. A.; Lugan, N.; Canac, Y.; Sortais, J.-B., *Angew. Chem., Int. Ed.* **2019**, *58*, 6727-6731.

(31) Ling, F.; Hou,H.; Chen, J.; Nian, S.; Yi, X.; Wang, Z.; Song, D.; Zhong, W., *Org. Lett.* **2019**, *21*, 3937-3941.

(32) Yang, W.; Chernyshov, I. Y.; van Schendel, R. K. A.; Weber, M.; Müller, C.; Filonenko, G. A.; Pidko, E. A., *Nat. Commun.* **2021**, *12*, 12.

(33) Maji, B.; Barman, M. K., *Synthesis* **2017**, *49*, 3377-3393.

(34) Garbe, M.; Junge,K.; Beller, M., *Eur. J. Org. Chem.* **2017**, *2017*, 4344-4362.

(35) Kallmeier, F.; Kempe, R., *Angew. Chem. Int. Ed.* **2018**, *57*, 46-60.

(36) Gorgas, N.;Kirchner, K., *Acc. Chem. Res.* **2018**, *51*, 1558-1569.

(37) Wang, Y.; Wang, M.; Li, Y.; Liu, Q., *Chem* **2021**, *7*, 1180-1223.

(38) Freitag, F.; Irrgang, T.; Kempe, R., *J. Am. Chem. Soc.* **2019**, *141*, 11677-11685.

(39) Weber, S.; Stöger, B.; Veiros, L. F.; Kirchner, K., *ACS Catal.* **2019**, *9*, 9715-9720.

(40) Zhang, L.; Wang, Z.; Han, Z.; Ding, K., *Angew. Chem. Int. Ed.* **2020**, *59*, 15565-15569.

(41) Wang, Z.; Zhao, X.; Huang, A.; Yang, Z.; Cheng, Y.; Chen, J.; Ling, F.; Zhong, W., *Tetrahedron Lett.* **2021**, *82*, 153389.

(42) Weber, S.; Brünig, J.; Veiros, L. F.; Kirchner, K., *Organometallics* **2021**, *40*, 1388-1394.

(43) Rahaman, S. M. W.; Pandey, D. K.; Rivada-Wheelaghan, O.; Dubey, A.; Fayzullin, R. R.; Khusnutdinova, J. R., *ChemCatChem* **2020**, *12*, 5912-5918.

(44) Gonçalves, T. P.; Huang, K.-W., *J. Am. Chem. Soc.* **2017**, *139*, 13442-13449.

(45) Shimbayashi, T.; Fujita, K.-i., *Catalysts* **2020**, *10*, 635.

(46) Bruker, *APEX3, SAINT and SADABS.* Bruker AXS Inc., Madison, Wisconsin, USA. **2016**.

(47) Sheldrick, G. M., A Short History of SHELX. *Acta Crystallogr.* **2008**, *A64*, 112-122.

(48) Sheldrick, G. M., Crystal Structure Refinement with SHELXL. *Acta Crystallogr.* **2015**, *C71*, 3-8.

Chapter 4

Manganese-Catalyzed Chemoselective Direct Hydrogenation of α-Ketoamides

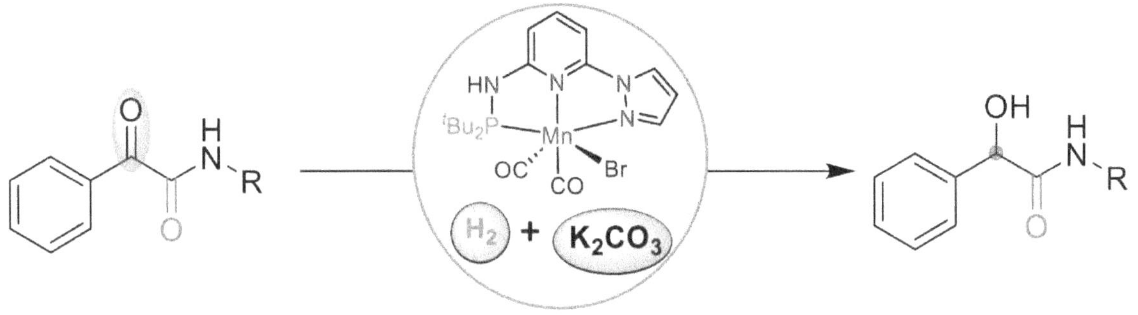

This chapter has been partly adapted from the publication "Manganese-Catalyzed Chemoselective Direct Hydrogenation of α,β-Epoxy Ketones and α-Ketoamides at Room Temperature" **Shabade, A. B.**; Singh, R. K.; Gonnade, R. G.; and Punji, B. *Adv. Synth. Catal.*, **2024**, doi.org/10.1002/adsc.202400267.

4.1 INTRODUCTION

Chemoselective hydrogenation is one of the key processes in the synthesis of organic compounds. This method minimizes the requirement of additional steps for synthesizing a particular moiety by eliminating protection and deprotection reactions in the complete process. Transition metal-catalyzed hydrogenation has been developed using noble metal catalysts such as rhodium, ruthenium, palladium, and iridium.[1] However, the high costs and the lower abundance of these metals are the major concerns related to the developed protocols. Hence there is a need to develop a sustainable protocol using base metal catalysts towards chemoselective reduction of various unsaturated moieties. Recently, iron, cobalt, and manganese have been explored for the hydrogenation of ketone, imine, and amide.[2–7] However, it is quite difficult to selectively hydrogenate carbonyl groups when there are other reducible functionalities present, especially using base metal catalysts.[8,9] In particular, the chemoselective hydrogenation of α-ketoamides to α-hydroxy amides are very important reaction. The α-hydroxy amides present in structural backbones in a wide range of organic compounds are used in different organic transformations. Additionally, they are essential structural motifs in a variety of drugs and biologically active compounds. (Scheme 4.1a).[13–15] The chemoselective hydrogenation of α-ketoamides is well developed using silanes, formate, and DMF as a hydrogen source under either transition metal catalysis, organocatalysis, or metal-free conditions (Scheme 4.1b).[16–21] These protocols are associated with limitations that require high temperatures and result in the production of stoichiometric waste. Hence the use of molecular hydrogen an environmentally friendly hydrogen source, under 3d metal-catalyzed hydrogenation is the current interest of the research community.

Manganese is the third most prevalent element in the crust of the Earth (with a concentration of 950 mg kg^{-1}), which has piqued the interest of scientists.[22] Notably, manganese is less toxic compared to the noble metals. It is commonly found in living organisms and plays crucial roles in metabolism and antioxidant mechanisms.[23,24] Recently, manganese has been well explored towards independent reduction of polar double bonds. However, the chemoselective hydrogenation of unsaturated bonds catalyzed by manganese using molecular hydrogen is rare, and previous approaches have often required high temperatures and hydrogen pressures, presenting a major drawback (Scheme 4c).[3,6] Unfortunately, the chemoselective hydrogenation of α-ketoamides using industrially compatible molecular hydrogen and sustainable, cost-effective 3d metal catalysts has not been precedented. Acknowledging the significance of α-hydroxy amides in organic synthesis, we introduce an effective method for selectively hydrogenating α-ketoamides to yield valuable α-

hydroxy amides at room temperature (25°C) using a Mn(I) catalyst and moderate H_2 pressure (Scheme 4.1d).

Scheme 4.1 a) Bioactive Compounds Containing α-Hydroxy Amides b) Organo-Catalyzed, Metal-Catalyzed, or Metal-Free Chemoselective Hydrogenation of α-Ketoamides, c) Manganese-Catalyzed Direct Hydrogenation of Ketone, Imine, and Amide, d) Present Work: Mn-Catalyzed Chemoselective Direct Hydrogenation of α-Ketoamides.

4.2 RESULTS AND DISCUSSION

4.2.1 Optimization of Reaction Conditions

Initially, we started optimization for the chemoselective hydrogenation of α-ketoamide by using *N*-benzyl-2-oxo-2-phenylacetamide (**1a**) as a model substrate. We were interested in implementing the recently developed manganese catalysts using molecular hydrogen for this process. Firstly, we screened the manganese catalysts (**Mn-1** to **Mn-3**; Table 1) using 20 bar H_2 at room temperature. Among these catalysts, **Mn-3** in combination with catalytic KO^{t}Bu exhibited superior conversion (Table 4.1, entries 1-3). Hence, we continued further optimization with **Mn-3**. Reducing the hydrogen pressure from 20 bar to 2 bar did not affect the overall conversion of **1a** (entries 4 and 5). At this stage, we screened other mild bases including K_3PO_4, K_2CO_3, and KOAc, using 2 bar H_2. Among these bases, K_2CO_3 was suitable to achieve a 95% isolated yield of **1** (entries 6-8). A reduction in the yield of **1** was seen when the hydrogen pressure was lowered from 2 to 1 bar (entry 9). Lowering the catalyst concentration from 5 mol% to 3 mol% resulted in lower yield of **1** (entry 10). The reaction using other alcoholic solvents including ethanol and isopropanol resulted in only 55% and 39% of **1** respectively (entries 11, 12). However, the reaction did not proceed under non-alcoholic solvents, such as 1,4-dioxane or 2MeTHF as a solvent (entry 13). The in-situ method using the equimolar ratio of 5 mol% $Mn(CO)_5Br$ and **L3** was not an optimal condition, resulting in only 62% of **1** (entry 14). Again, at this stage, the **Mn-1** and **Mn-2** were compared, which showed inferior yield for **1** than that observed with **Mn-3** (entries 15, 16). Other controlled reactions suggested the necessity of 5 mol% of **Mn-3**, and 10 mol% K_2CO_3, in 1.0 mL MeOH, using 2 bar H_2 at room temperature for 16 h reaction time as an optimal condition (Entries 12-18).

Table 4.1 Optimization of Reaction Conditions. [a]

(1a) → [Mn] (5 mol%), base (10 mol%), H_2 (bar), MeOH, rt, 16 h → (1)

Entry	[Mn]	Base	H_2 (bar)	Yields [b]
1	**Mn-1**	KO^{t}Bu	20	82 (79)
2	**Mn-2**	KO^{t}Bu	20	95 (93)
3	**Mn-3**	KO^{t}Bu	20	98 (95)
4	**Mn-3**	KO^{t}Bu	5	98
5	**Mn-3**	KO^{t}Bu	2	98
6	**Mn-3**	K_3PO_4	2	97
7	**Mn-3**	**K_2CO_3**	**2**	**97 (95)**
8	**Mn-3**	KOAc	2	19
9	**Mn-3**	K_2CO_3	1	24
10	**Mn-3** (3 mol%)	K_2CO_3	2	92 (88)
11[c]	**Mn-3**	K_2CO_3	2	55
12[d]	**Mn-3**	K_2CO_3	2	39

13 [e]	**Mn-3**	K_2CO_3	2	--
14	$Mn(CO)_5Br$ + **L3**	K_2CO_3	2	62
15	**Mn-1**	K_2CO_3	2	04
16	**Mn-2**	K_2CO_3	2	62
17	$Mn(CO)_5Br$	K_2CO_3	2	--
18	--	K_2CO_3	2	--

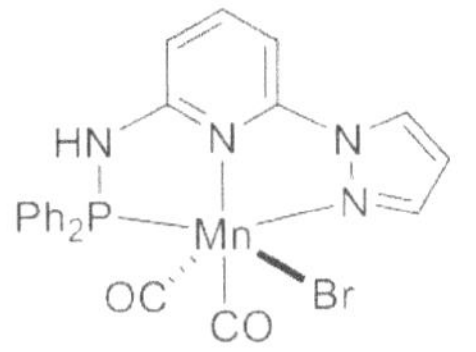

(Mn-1) 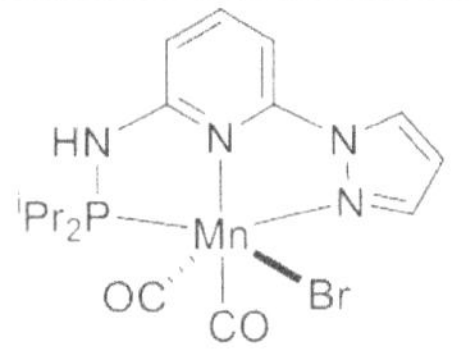 **(Mn-2)** 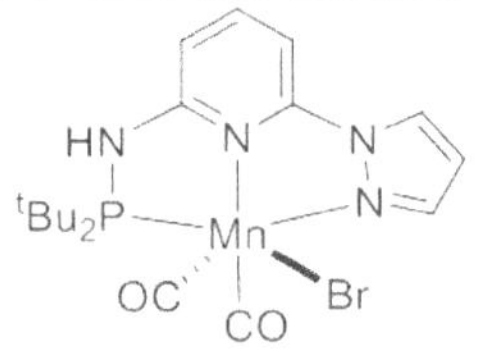 **(Mn-3)** 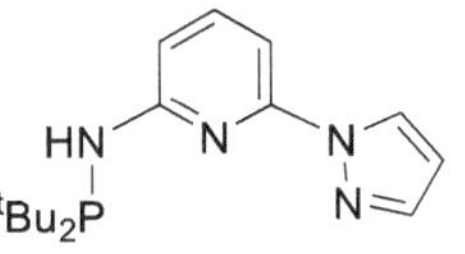 L3

[a] Reaction Conditions: **1a** (0.048 g, 0.20 mmol), base (0.02 mmol), [**Mn**] catalyst (0.01 mmol, 5 mol%), solvent (1.0 mL). [b] ^{1}H-NMR yield using CH_2Br_2 as standard. Isolated yields are given in parentheses. [c] Using EtOH as solvent. [d] Using ^{i}PrOH as solvent. [e] Using 1,4-dioxane or 2MeTHF as solvent. All three manganese complexes contain a mixture of two geometrical isomers. RT = room temperature (~ 25 °C).

4.2.2 Substrate Scope for Substituted α-Ketoamides

Next, we explored the generality and limitations of chemoselective hydrogenation of *N*-substituted α-ketoamides using optimized reaction conditions (Scheme 4.2). Notably, we have used a mixture of methanol and dichloromethane ($MeOH:CH_2Cl_2$ in 4:1) to address the solubility issue of other derivatives of α-ketoamides. The α-ketoamides containing electronically distinct substituents on the amide nitrogen including alkyl and aryl substitution were tested under standard reaction conditions. The α-ketoamide derived from substituted benzylamines resulted in the formation of **2** and **3** in 59% and 91% respectively. These examples show the susceptibility of alcoholic OH and heteroaromatic moieties, under standard conditions. In addition to alkylamines, ketoamides derived from aromatic amines were suitable for hydrogenation using our protocol (**4** and **5**). The sensitive substitutions like halo groups remained untouched to give corresponding α-hydroxy amides in good yields (**6-9**). This method was very selective for the C=O bond reduction of α-ketoamide without affecting reducible functional groups including acetyl and azobenzene remained unaffected (**10** and **11**). These examples highlight the importance of the current protocol for the chemoselective hydrogenation of α-ketoamides to α-hydroxy amides. Highly coordinative groups such as thiomethyl, trifluoromethoxy, and *N*-heterocyclic moieties, which could potentially poison the catalyst, are suitable for this transformation (**13-17**). Unfortunately, tertiary ketoamides were not suitable under the optimized conditions due to the competitive esterification of the amide group in methanol (**18a** and **19a**).

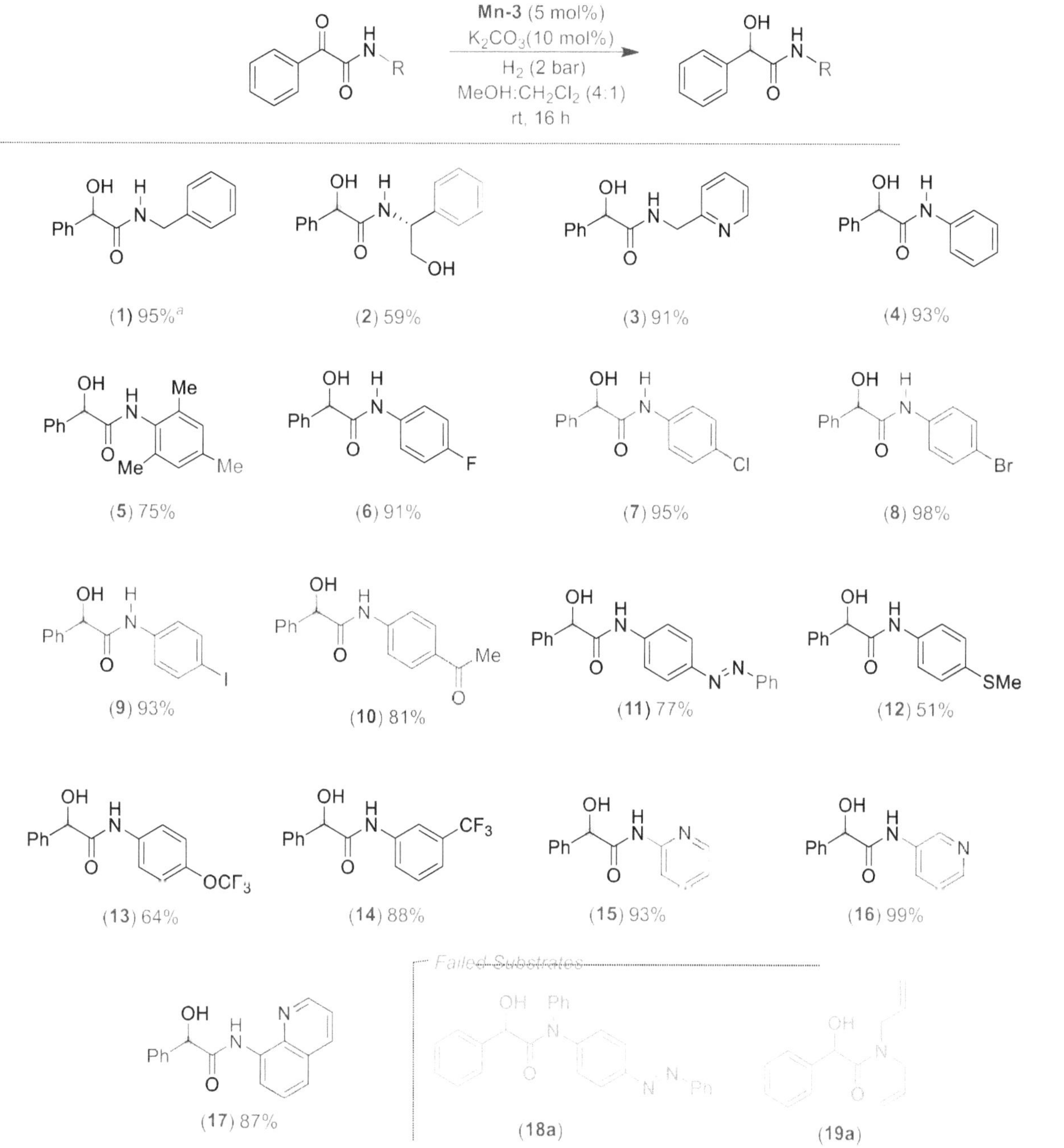

Scheme 4.2 Scope for Chemoselective Hydrogenation of Substituted α-Ketoamides. Conditions: α-ketoamide (0.20 mmol), H_2 (2 bar), **Mn-3** (0.005 g, 0.01 mmol), K_2CO_3 (0.0028 g, 0.02 mmol) using 1 mL of MeOH:CH_2Cl_2 (4:1). Due to the solubility problem of substrates in MeOH, a mixture of MeOH and CH_2Cl_2 was used. [a] MeOH was used as a solvent. Yields are of isolated compounds.

4.2.3 Mechanistic Aspects

To get into mechanistic insight we performed different controlled experiments. The attempted hydrogenation using the **Mn-3^{Me}** complex did not yield the expected product (Scheme 4.3a). Additionally, the hydrogenation of **1a** using the dearomatized complex **Mn-3'** as a catalyst resulted in 94% of **1**. These two experiments suggest there is the necessity of *N*H group in the ligand backbone and this reaction is proceeding *via* the metal-ligand cooperation pathway for H_2 activation. Further to know the hydrogen source of the reaction, we performed a standard reaction using CD_3OD as methanol is the solvent. However, we did not observe any deuterium incorporation in product **1** (not even at hydroxy). To verify the role of MeOH, we performed reactions using aprotic solvents, including 2MeTHF and dioxane, where **1a** remained unreacted. This suggests, that although no deuterium incorporation was detected at oxygen (which might be due to the proton exchange) probably methanol provides proton for this reaction and molecular hydrogen serves hydride.

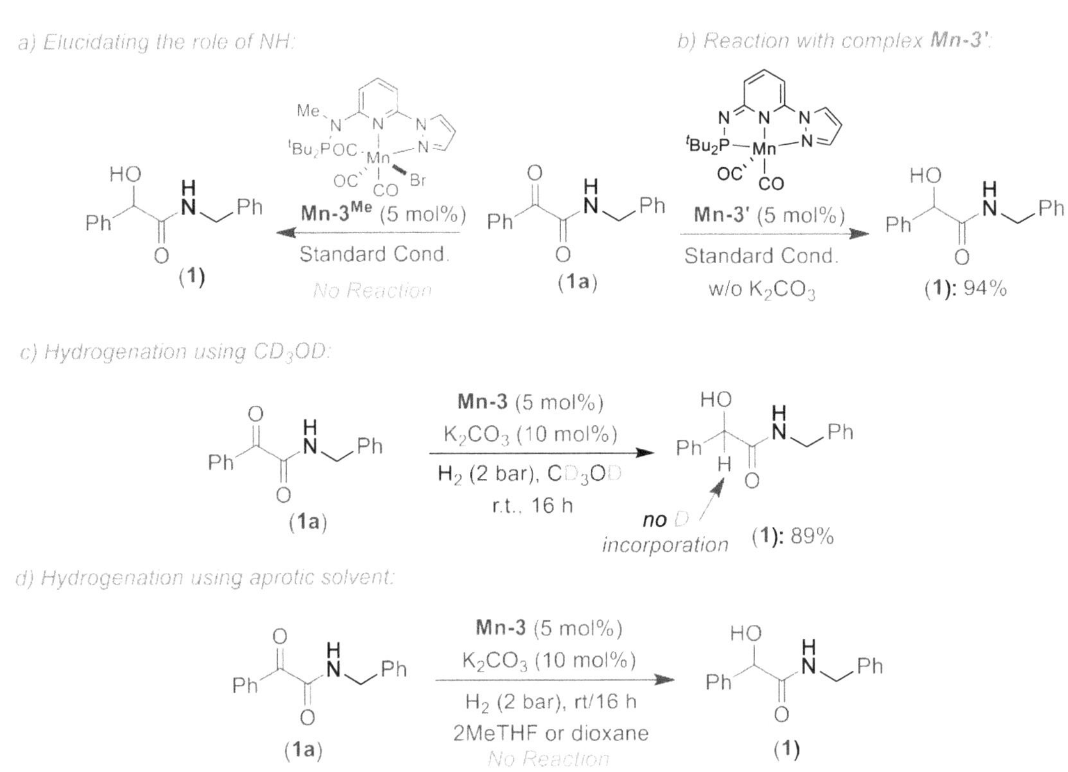

Scheme 4.3 Mechanistic Experiments.

4.2.4 Plausible Catalytic Cycle

For chemoselective hydrogenation of ketoamides under manganese catalysis, we propose a plausible catalytic cycle based on previous literature (Figure 4.1). The cycle starts from the reaction of K_2CO_3 with precatalyst **Mn-3** to form the dearomatized active intermediate **Mn-3'**. This active intermediate upon H_2 activation (via **B**) results in the formation of manganese hydride species **C**. Then the hydride would be transferred to the carbonyl group of **1a** to form alkoxy intermediate **D**. Followed by the protonation with methanol, the final product will be delivered along with the formation of methoxy manganese intermediate **E**, which upon deprotonation of *NH* group in the ligand by methoxy group regenerates the active intermediate **Mn-3'**.

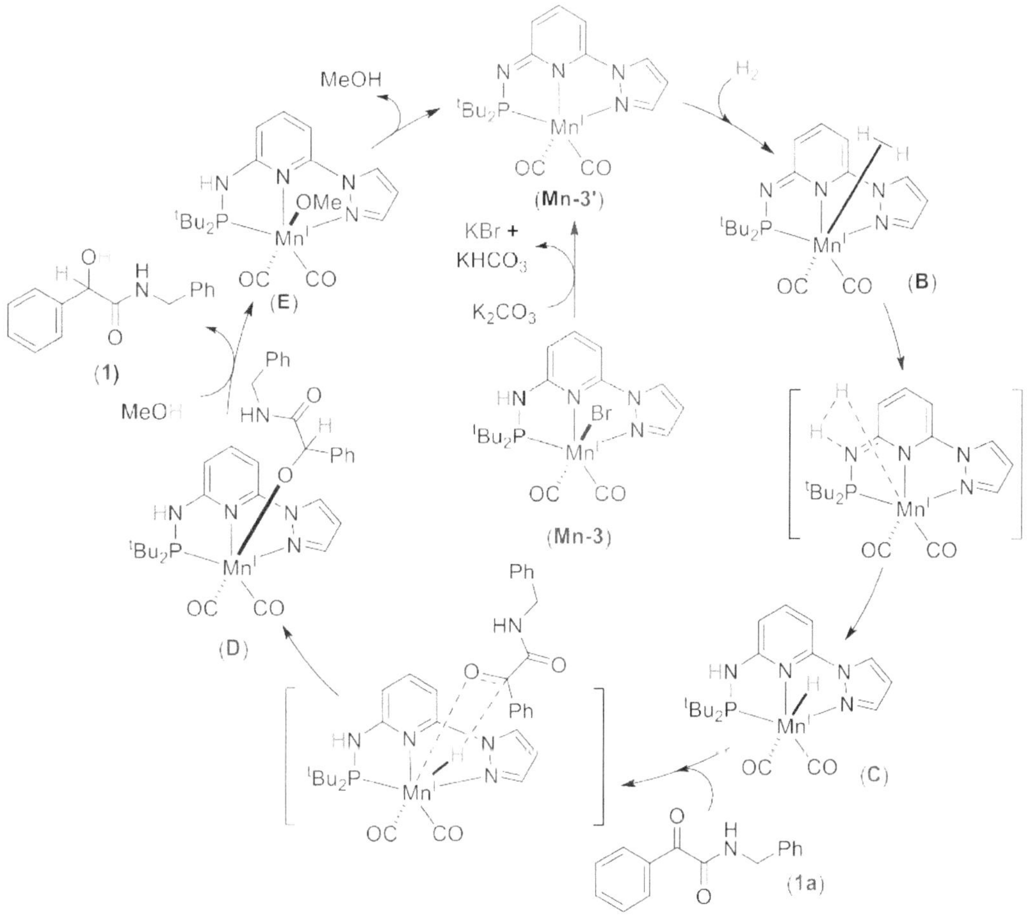

Figure 4.1 The catalytic cycle of Mn(I) catalyzed direct hydrogenation of ketoamides.

4.3 CONCLUSION

In this chapter, we have developed a streamlined and chemoselective process for hydrogenating α-ketoamides to produce α-hydroxy amides with excellent selectivity. This method utilizes an affordable manganese catalyst and an environmentally friendly source of hydrogen, improving efficiency and sustainability. This method represents a greener process for the synthesis of valuable α-hydroxy amides which offers advantages over conventional approaches hazardous hydrogen sources, and harsh conditions. Various substituted α-ketoamides were successfully converted to α-hydroxy amides with the retention of crucial and delicate functionalities essential for synthesis. These functionalities include halides, hydroxy, alkoxy, thiomethyl, acetyl, and azo groups. Initial mechanistic findings suggest this reaction proceeds via the H_2 activation by metal-ligand cooperation and protonation by MeOH.

4.4 EXPERIMENTAL SECTION

All the manipulations were conducted under an argon atmosphere either in a glove box or using standard Schlenk techniques in pre-dried glassware. The catalytic reactions were performed in oven-dried glass vials with magnetic bars by placing them in the pressure reactor. Solvents were dried over Na/benzophenone or Mg and distilled prior to use. Liquid reagents were flushed with argon prior to use. All other chemicals were obtained from commercial sources and were used without further purification. High-resolution mass spectrometry (HRMS) mass spectra were recorded on a Thermo Scientific Q-Exactive, Accela 1250 pump. NMR (^{1}H and ^{13}C) spectra were recorded at 400 or 500 MHz (^{1}H), 100 or 125 MHz {^{13}C, DEPT (distortionless enhancement by polarization transfer)}, 377 MHz (^{19}F), respectively in $CDCl_3$ solutions, if not otherwise specified; chemical shifts (δ) are given in ppm. The ^{1}H and ^{13}C{^{1}H} NMR spectra are referenced to residual solvent signals ($CDCl_3$: δ H = 7.26 ppm, δ C = 77.2 ppm; methanol-d_4: δ H = 3.34 ppm, δ C = 49.5 ppm).

4.4.1 Representative Procedure for Catalytic Hydrogenation and Characterization Data

OH H N O

***N*-Benzyl-2-hydroxy-2-phenylacetamide (1)**: To a dry vial with a magnetic bar was introduced **Mn-3** (0.005 g, 0.010 mmol), K_2CO_3 (0.0028 g, 0.0202 mmol), and *N*-benzyl-2-oxo-2-phenylacetamide (**1a**; 0.048 g, 0.201 mmol), inside the glove box. The reaction vial was transferred to an autoclave under an argon atmosphere. Then, MeOH (1.0 mL) was added and the autoclave was pressurized with H_2 (2 bar) and vented three times. Finally, the autoclave was pressurized with 2 bar H_2 and stirred (700 rpm) at room temperature (~ 25 °C) for 16 h. The reaction mixture was then concentrated and subjected to column chromatography on silica gel (petroleum ether/EtOAc: 2/1) to yield **1** (0.046 g, 95%) as a white solid. ^{1}H-NMR (400 MHz, $CDCl_3$): δ = 7.40-7.26 (m, 8H, Ar–H), 7.22–7.11 (m, 2H, Ar–H), 5.05 (s, 1H, CH), 4.44–4.41 (m, 2H, CH_2), 3.73 (br s, 1H, OH). ^{13}C{^{1}H}-NMR (100 MHz, $CDCl_3$) δ = 172.6 (CO), 139.6 (C_q), 137.7 (C_q), 128.7 (2C, CH), 128.5 (2C, CH), 127.6 (2C,

CH), 126.8 (2C, CH), 127.9 (CH), 127.6 (CH), 74.1 (CH), 43.3 (CH_2).

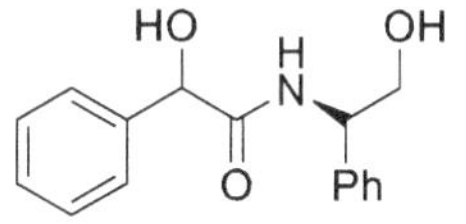

2-Hydroxy-*N*-((*S*)-2-hydroxy-1-phenylethyl)-2-phenylacetamide (2): The representative procedure was followed, using substrate **2a** (0.054 g, 0.201 mmol), 1.0 mL of MeOH:CH_2Cl_2 (4:1) and the reaction mixture was stirred at 50 °C for 16 h. Purification by column chromatography on silica gel (petroleum ether/EtOAc: 2/1) yielded **2** (0.032 g, 59%) as a colorless liquid. *Compound **2** is obtained as a mixture of two isomers (57:45 ratio) denoted as **2'** and **2''**.* ^{1}H-NMR (400 MHz, CD_3OD): δ = 7.46-7.31 (m, 10H, Ar–H), 5.36 (s, 1H, CH; **2''**), 5.36 (s, 1H, CH; **2'**), 4.69-4.57 (m, 2H, CH_2), 4.44-4.36 (m, 1H, CH). ^{13}C{^{1}H}-NMR (100 MHz, CD_3OD) For **2'**: δ = 174.1 (CO), 140.7 (C_q), 135.4 (C_q), 131.5 (CH), 131.2 (2C, CH), 130.5 (2C, CH), 130.4 (CH), 129.3 (2C, CH), 128.7 (2C, CH), 75.2 (CH), 67.0 (CH_2), 55.9 (CH). For **2''**: δ = 174.3 (CO), 140.8 (C_q), 135.5 (C_q), 131.6 (CH), 131.3 (2C, CH), 130.5 (2C, CH), 130.5 (CH), 129.3 (2C, CH), 128.9 (2C, CH), 75.2 (CH), 67.1 (CH_2), 55.8 (CH).

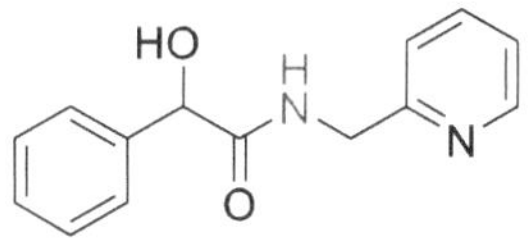

2-Hydroxy-2-phenyl-*N*-(pyridin-2-ylmethyl)acetamide (3): The representative procedure was followed, using substrate **3a** (0.049 g, 0.204 mmol), 1.0 mL of MeOH:CH_2Cl_2 (4:1) and the reaction mixture was stirred at r.t. for 16 h. Purification by column chromatography on silica gel (petroleum ether/EtOAc: 2/1) yielded **3** (0.045 g, 91%) as a yellow solid. ^{1}H-NMR (400 MHz, DMSO-d_6): δ = 7.50 (d, J = 6.9 Hz, 2H, Ar–H), 7.38-7.35 (m, 2H, Ar–H), 7.32-7.24 (m, 5H, Ar–H), 5.10 (s, 1H, CH), 4.44 (s, 2H, CH_2). ^{13}C{^{1}H}-NMR (100 MHz, DMSO-d_6) δ = 176.4 (CO), 142.6 (C_q), 140.8 (C_q), 130.3 (2C, CH), 130.2 (2C, CH), 129.9 (CH), 129.2 (CH), 129.0 (CH), 128.8 (CH), 128.8 (CH), 76.4 (CH), 44.5 (CH_2).

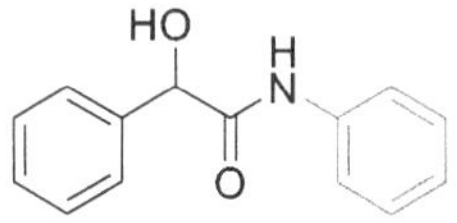

2-Hydroxy-*N*,2-diphenylacetamide (4): The representative procedure was followed, using substrate **4a** (0.046 g, 0.204 mmol), 1.0 mL of MeOH:CH_2Cl_2 (4:1) and the reaction mixture was stirred at r.t. for 16 h. Purification by column chromatography on silica gel (petroleum ether/EtOAc: 5/1) yielded

4 (0.043 g, 93%) as a white solid. ^{1}H-NMR (400 MHz, $CDCl_3$): δ = 8.75 (br s, 0.33H, *N*H), 7.52 (d, J = 8.1 Hz, 2H, Ar–H), 7.46 (d, J = 7.0 Hz, 2H, Ar–H), 7.36-7.27 (m, 5H, Ar–H), 7.10 (t, J = 7.4 Hz, 1H, Ar–H), 5.10 (s, 1H, CH). ^{13}C{^{1}H}-NMR (100 MHz, $CDCl_3$) δ = 171.0 (CO), 139.5 (C_q), 137.2 (C_q), 129.1 (2C, CH), 128.8 (2C, CH), 128.6 (CH), 126.8 (2C, CH), 124.7 (CH), 119.9 (2C, CH), 74.5 (CH).

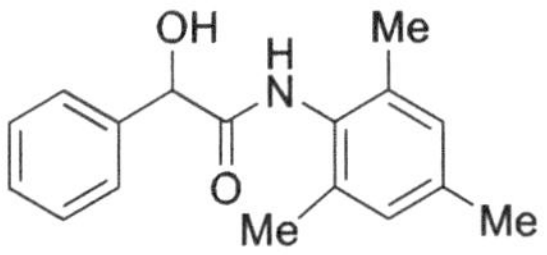

2-Hydroxy-*N*-mesityl-2-phenylacetamide (5): The representative procedure was followed, using substrate **5a** (0.054 g, 0.202 mmol), 1.0 mL of MeOH:CH_2Cl_2 (4:1) and the reaction mixture was stirred at r.t. for 16 h. Purification by column chromatography on silica gel (petroleum ether/EtOAc: 5/1) yielded **5** (0.041 g, 75%) as a white solid. ^{1}H-NMR (400 MHz, $CDCl_3$): δ = 7.48 (dd, J = 8.0, 1.0 Hz, 2 H), 7.42–7.34 (m, 3 H), 6.82 (s, 2 H), 5.12 (s, 1 H), 2.22 (s, 3 H), 2.0 (s, 6 H). ^{13}C{^{1}H}-NMR (100 MHz, $CDCl_3$) δ = 171.1 (CO), 139.6 (C_q), 137.3 (C_q), 135.1 (C_q), 130.3 (2C, CH), 129.0 (2C, CH), 128.9 (CH), 128.8 (C_q), 126.8 (2C, CH), 74.4 (CH), 21.0 (CH_3), 18.1 (CH_3).

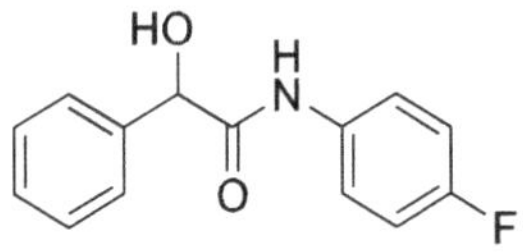

***N*-(4-Fluorophenyl)-2-hydroxy-2-phenylacetamide (6)**: The representative procedure was followed, using substrate **6a** (0.049 g, 0.202 mmol), 1.0 mL of MeOH:CH_2Cl_2 (4:1) and the reaction mixture was stirred at r.t. for 16 h. Purification by column chromatography on silica gel (petroleum ether/EtOAc: 1/1) yielded **6** (0.045 g, 91%) as a white solid. ^{1}H-NMR (400 MHz, $CDCl_3$): δ = 7.45-7.55 (m, 4H), 7.44-7.33 (m, 3H), 7.01 (t, J = 8.6 Hz, 2H), 5.20 (s, 1H). ^{13}C{^{1}H}-NMR (100 MHz, $CDCl_3$): δ = 169.8 (CO), 159.6 (d, $^{1}J_{C\text{–}F}$ = 243.0 Hz, C_q), 138.9 (C_q), 133.1 (C_q), 129.04 (2C, CH), 129.0 (CH), 126.8 (2C, CH), 121.6 (d, $^{3}J_{C\text{–}F}$ = 8.0 Hz, 2C, CH), 115.7 (d, $^{2}J_{C\text{–}F}$ = 22.0 Hz, 2C, CH), 74.7 (CH).

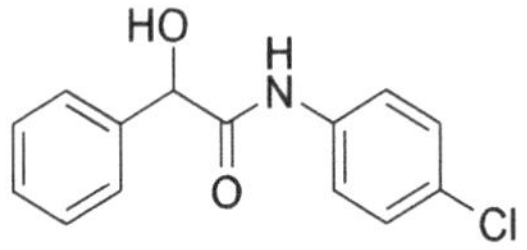

***N*-(4-Chlorophenyl)-2-hydroxy-2-phenylacetamide (7)**: The representative procedure was followed, using substrate **7a** (0.052 g, 0.200 mmol), 1.0 mL of MeOH:CH_2Cl_2 (4:1) and the reaction mixture was stirred at r.t. for 16 h. Purification by column chromatography on silica gel (petroleum ether/EtOAc: 2/1) yielded **7** (0.050 g, 95%) as a white solid. ^{1}H-NMR (400 MHz, CD_3OD): δ = 10.82 (br s, 1H, *N*H), 8.37 (d, *J* = 8.9 Hz, 2H, Ar–H), 8.26 (d, *J* = 7.1 Hz, 2H, Ar–H), 8.17-8.08 (m, 5H, Ar–H), 5.91 (s, 1H, CH). ^{13}C{^{1}H}-NMR (100 MHz, CD_3OD): δ = 183.3 (CO), 151.3 (C_q), 147.9 (C_q), 140.1 (2C, CH), 139.9 (2C, CH), 139.6 (CH), 139.5 (C_q), 137.9 (2C, CH), 133.2 (2C, CH), 85.2 (CH).

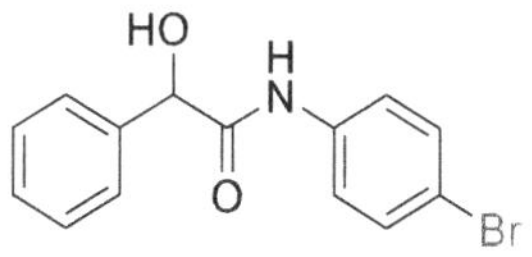

HO H N O Br

***N*-(4-Bromophenyl)-2-hydroxy-2-phenylacetamide (8)**: The representative procedure was followed, using substrate **8a** (0.061 g, 0.201 mmol), 1.0 mL of MeOH:CH_2Cl_2 (4:1) and the reaction mixture was stirred at r.t. for 16 h. Purification by column chromatography on silica gel (petroleum ether/EtOAc: 2/1) yielded **8** (0.060 g, 98%) as a white solid. ^{1}H-NMR (500 MHz, $CDCl_3$): δ = 8.83 (br s, *N*H), 7.44-7.41 (m, 4H, Ar–H), 7.38-7.25 (m, 5H, Ar–H), 5.39 (br s, OH), 5.09 (s, 1H, CH). ^{13}C{^{1}H}-NMR (125 MHz, $CDCl_3$) δ = 171.1 (CO), 139.4 (C_q), 136.4 (C_q), 132.1 (2C, CH), 128.8 (2C, CH), 128.6 (CH), 126.8 (2C, CH), 121.4 (2C, CH), 117.2 (C_q), 74.4 (CH).

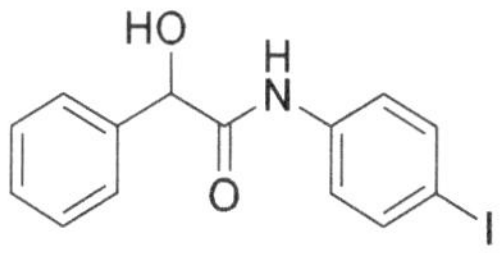

HO H N O I

2-Hydroxy-*N*-(4-iodophenyl)-2-phenylacetamide (9): The representative procedure was followed, using substrate **9a** (0.071 g, 0.202 mmol), 1.0 mL of MeOH:CH_2Cl_2 (4:1) and the reaction mixture was stirred at r.t. for 16 h. Purification by column chromatography on silica gel (petroleum ether/EtOAc: 2/1) yielded **9** (0.066 g, 93%) as a white solid. ^{1}H-NMR (400 MHz, $CDCl_3$): δ = 7.62 (d, *J* = 8.8 Hz, 2H, Ar–H), 7.48-7.47 (m, 2H, Ar–H), 7.42-7.37 (m, 3H, Ar–H), 7.33 (d, *J* = 8.8 Hz, 2H, Ar–H), 5.20 (s, 1H, CH). ^{13}C{^{1}H}-NMR (100 MHz, $CDCl_3$) δ = 170.0 (CO), 138.9 (C_q), 138.2 (2C, CH), 137.1 (C_q), 129.3 (2C, CH), 129.0 (CH), 127.0 (2C, CH), 127.8 (2C, CH), 77.4 (C_q), 75.0 (CH).

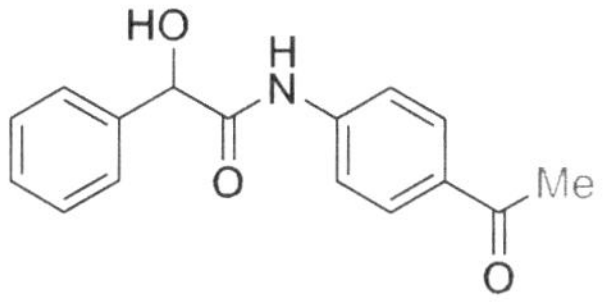

HO H N O Me O

***N*-(4-Acetylphenyl)-2-hydroxy-2-phenylacetamide (10)**: The representative procedure was followed, using substrate **10a** (0.054 g, 0.202 mmol), 1.0 mL of MeOH:CH_2Cl_2 (4:1) and the reaction mixture was stirred at r.t. for 16 h. Purification by column chromatography on silica gel (petroleum ether/EtOAc: 10/1) yielded **10** (0.044 g, 81%) as a white solid. ^{1}H-NMR (400 MHz, CD_3OD): δ = 8.01-7.97 (m, 2H, Ar–H), 7.82-7.79 (m, 2H, Ar–H), 7.58-7.56 (m, 2H, Ar–H), 7.42-7.37 (m, 2H, Ar–H), 7.36-7.32 (m, 1H, Ar–H), 5.21 (s, 1H, CH), 2.59 (s, 3H, CH_3). ^{13}C{^{1}H}-NMR (100 MHz, CD_3OD) δ = 200.3 (CO), 174.9 (CO), 144.8 (C_q), 142.2 (C_q), 135.0 (C_q), 131.5 (2C, CH), 130.4 (2C, CH), 130.1 (CH), 128.8 (2C, CH), 121.5 (2C, CH), 76.7 (CH), 27.3 (CH_3). HRMS (ESI): *m/z* Calcd for $C_{16}H_{15}NO_3 + H^+$ $[M + H]^+$ 270.1125; Found 270.1120.

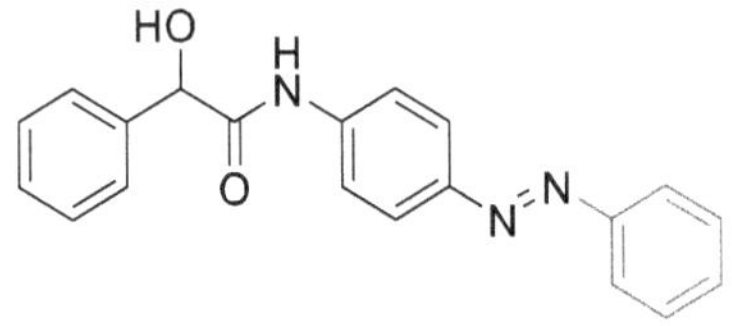

(*E*)-2-Hydroxy-2-phenyl-*N*-(4-(phenyldiazenyl)phenyl)acetamide (11): The representative procedure was followed, using substrate **11a** (0.066 g, 0.200 mmol), 1.0 mL of MeOH:CH_2Cl_2 (4:1) and the reaction mixture was stirred at r.t. for 16 h. Purification by column chromatography on silica gel (petroleum ether/acetone: 30/1) yielded **11** (0.051 g, 77%) as an orange solid. ^{1}H-NMR (400 MHz, CD_3OD): δ = 9.65 (br s, 1H, *N*H), 7.86 (m, 5H, Ar–H), 7.65-7.29 (m, 9H, Ar–H), 5.20 (s, 1H, CH). ^{13}C{^{1}H}-NMR (100 MHz, CD_3OD) δ = 181.6 (CO), 162.5 (C_q), 158.6 (C_q), 150.9 (C_q), 150.3 (C_q), 140.8 (CH), 139.1 (2C, CH), 138.4 (2C, CH), 138.0 (CH), 136.8 (2C, CH), 133.8 (2C, CH), 132.6 (2C, CH), 129.9 (2C, CH), 83.3 (CH).

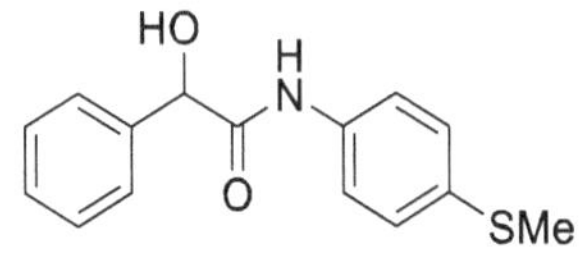

2-Hydroxy-*N*-(4-(methylthio)phenyl)-2-phenylacetamide (12): The representative procedure was followed, using substrate **12a** (0.055 g, 0.203 mmol), 1.0 mL of MeOH:CH_2Cl_2 (4:1) and the reaction mixture was stirred at r.t. for 16 h. Purification by column chromatography on silica gel (petroleum ether/EtOAc: 2/1) yielded **12** (0.028 g, 51%) as a white solid. ^{1}H-NMR (400 MHz, $CDCl_3$): δ = 7.12-7.08 (m, 4H, Ar–H), 6.99-6.90 (m, 3H, Ar–H), 6.83 (d, J = 8.8 Hz, 2H, Ar–H), 4.75 (s, 1H, CH), 2.06 (s, 3H, CH_3). ^{13}C{^{1}H}-NMR (100 MHz, $CDCl_3$) δ = 171.1 (CO), 139.6 (C_q), 134.9 (C_q), 133.9 (C_q), 128.7 (2C, CH), 128.5 (CH), 128.0 (2C, CH), 126.8 (2C, CH), 120.5 (2C, CH), 74.4 (CH), 16.6 (CH_3).

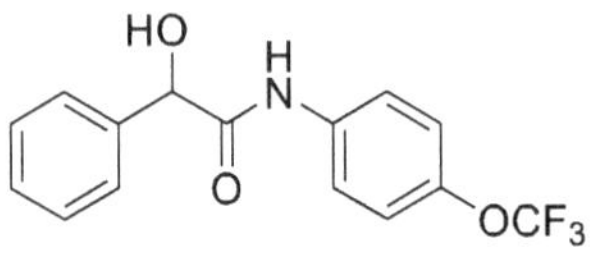

2-Hydroxy-2-phenyl-*N*-(4-(trifluoromethoxy)phenyl)acetamide (13): The representative procedure was followed, using substrate **13a** (0.062 g, 0.201 mmol), 1.0 mL of MeOH:CH_2Cl_2 (4:1) and the reaction mixture was stirred at r.t. for 16 h. Purification by column chromatography on silica gel (petroleum ether/EtOAc: 1/1) yielded **13** (0.040 g, 64%) as a white solid. ^{1}H-NMR (500 MHz, $CDCl_3$): δ = 7.59-7.55 (m, 2H, Ar–H), 7.45 (d, J = 8.4 Hz, 2H, Ar–H), 7.35-7.26 (m, 3H, Ar–H), 7.12 (d, J = 8.5 Hz, 2H, Ar–H), 5.12 (s, 1H, CH). ^{13}C{^{1}H}-NMR (125 MHz, $CDCl_3$) δ = 171.2 (CO), 145.5 (C_q), 139.4 (C_q), 136.1 (C_q), 128.8 (2C, CH), 128.6 (CH), 126.7 (2C, CH), 123.0 (q, $^1J_{C-F}$ = 270.1 Hz, C_q), 121.8 (2C, CH), 121.0 (2C, CH), 74.4 (CH).

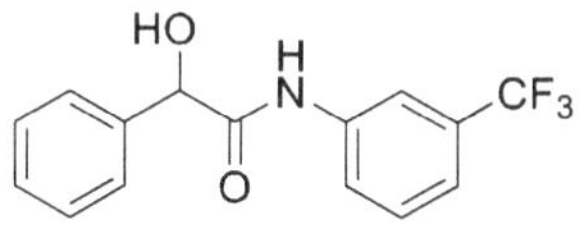

2-Hydroxy-2-phenyl-*N*-(3-(trifluoromethyl)phenyl)acetamide (14): The representative procedure was followed, using substrate **14a** (0.059 g, 0.201 mmol), 1.0 mL of MeOH:CH_2Cl_2 (4:1) and the reaction mixture was stirred at r.t. for 16 h. Purification by column chromatography on silica gel (petroleum ether/EtOAc: 5/1) yielded **14** (0.052 g, 88%) as a white solid. ^{1}H-NMR (400 MHz, $CDCl_3$): δ = 6.22 (s, 1H, Ar–H), 6.06 (d, J = 7.6 Hz, 1H, Ar–H), 5.84-5.87 (m, 7H, Ar–H), 5.75 (s, 1H, CH). ^{13}C{^{1}H}-NMR (100 MHz, $CDCl_3$) δ = 171.8 (CO), 139.4 (C_q), 138.0 (C_q), 131.3 (d, $^2J_{C-F}$ = 32.8 Hz C_q), 129.5 (CH), 128.6 (2C, CH), 128.4 (CH), 126.6 (2C, CH), 123.9 (q, $^1J_{C-F}$ = 272.4 Hz, CF_3), 122.9 (CH), 121.0 (d, $^3J_{C-F}$ = 3.8 Hz, CH), 116.6 (q, $^3J_{C-F}$ = 3.8 Hz, CH), 74.3 (CH). HRMS (ESI): m/z Calcd for $C_{15}H_{12}F_3NO_2 + H^+$ $[M + H]^+$ 296.0893; Found 296.0886.

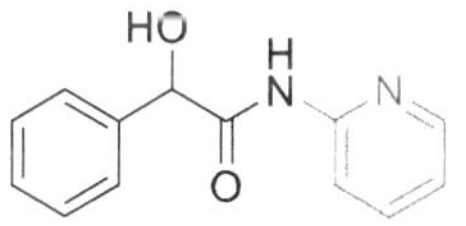

2-Hydroxy-2-phenyl-*N*-(pyridin-2-yl)acetamide (15): The representative procedure was followed, using substrate **15a** (0.046 g, 0.203 mmol), 1.0 mL of MeOH:CH_2Cl_2 (4:1) and the reaction mixture was stirred at r.t. for 16 h. Purification by column chromatography on silica gel (CH_2Cl_2/MeOH: 50/1) yielded **15** (0.044 g, 93%) as a white solid. ^{1}H-NMR (400 MHz, CD_3OD): δ = 8.32-8.31 (m, 1H, Ar–H), 8.14 (d, J = 8.3 Hz, 1H, Ar–H), 7.80 (t, J = 7.8 Hz, 1H, Ar–H), 7.55 (d, J = 7.8 Hz, 2H, Ar–H), 7.41-7.32 (m, 3H, Ar–H), 7.17-7.14 (m, 1H, Ar–H), 5.22 (s, 1H, CH). ^{13}C{^{1}H}-NMR (100 MHz, CD_3OD) δ = 174.6 (CO), 153.0 (C_q), 149.9 (CH), 142.1 (C_q), 140.8 (CH), 130.4 (2C, CH), 130.2 (CH), 128.7 (2C, CH), 122.3 (CH), 116.1 (CH), 76.4 (CH).

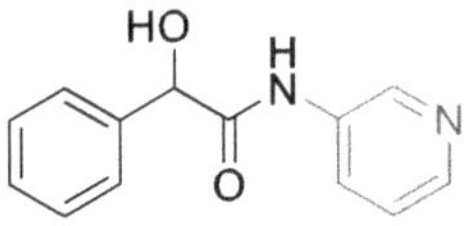

2-Hydroxy-2-phenyl-*N*-(pyridin-3-yl)acetamide (16): The representative procedure was followed, using substrate **16a** (0.046 g, 0.203 mmol), 1.0 mL of MeOH:CH_2Cl_2 (4:1) and the reaction mixture was stirred at r.t. for 16 h. Purification by column chromatography on silica gel (CH_2Cl_2/MeOH: 100/1) yielded **16** (0.046 g, 99%) as a white solid. ^{1}H-NMR (500 MHz, DMSO-d_6): δ = 8.82 (m, 1H, Ar–H), 8.28 (m, 1H, Ar–H), 8.17-8.14 (m, 1H, Ar–H), 7.58-7.56 (m, 2H, Ar–H), 7.41-7.37 (m, 3H, Ar–H), 7.35-7.30 (m, 1H, Ar–H), 5.23 (s, 1H, CH). ^{13}C{^{1}H}-NMR (125 MHz, DMSO-d$_6$) δ = 175.2 (CO), 146.4 (CH), 143.3 (CH), 142.1 (C$_q$), 137.5 (C$_q$), 130.5 (CH), 130.4 (2C, CH), 130.1 (CH), 128.7 (2C, CH), 126.1 (CH), 76.6 (CH). HRMS (ESI): *m/z* Calcd for $C_{13}H_{12}O_2N_2$ + H^+ $[M + H]^+$ 229.0972; Found 229.0969.

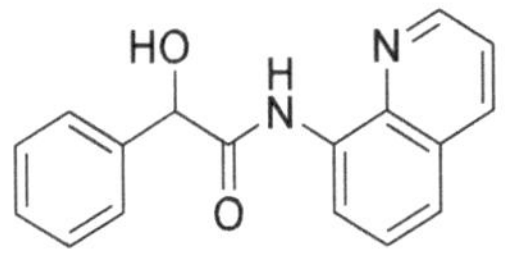

2-Hydroxy-2-phenyl-*N*-(quinolin-8-yl)acetamide (17): The representative procedure was followed, using substrate **17a** (0.055 g, 0.203 mmol), 1.0 mL of MeOH:CH_2Cl_2 (4:1) and the reaction mixture was stirred at r.t. for 16 h. Purification by column chromatography on silica gel (petroleum ether/EtOAc: 2/1) yielded **17** (0.049 g, 87%) as a white solid. ^{1}H-NMR (500 MHz, $CDCl_3$): δ = 10.41 (br s, 1H, *N*H), 8.54-8.45 (m, 2H, Ar–H), 7.89-7.87 (m, 1H, Ar–H), 7.35 (d, J = 7.4 Hz, 2H, Ar–H), 7.28-7.23 (m, 2H, Ar–H), 7.19-7.13 (m, 3H, Ar–H), 7.11-7.02 (m, 1H, Ar–H), 5.11 (s, 1H, CH), 3.89 (br s, 1H, OH). ^{13}C{^{1}H}-NMR (125 MHz, $CDCl_3$) δ = 170.8 (CO), 148.7 (CH), 139.5 (C$_q$), 138.8 (C$_q$), 136.4 (CH), 133.9 (C$_q$), 129.0 (2C, CH), 128.8 (CH), 128.1 (C$_q$), 127.4 (CH), 127.0 (2C, CH), 122.3 (CH), 121.8 (CH), 116.9 (CH), 75.2 (CH). HRMS (ESI): *m/z* Calcd for $C_{17}H_{14}N_2O_2$ + H^+ $[M + H]^+$ 279.1128; Found 279.1125.

4.4.2 ^{1}H and $^{13}C\{^{1}H\}$ NMR Spectra of Selected Hydrogenated Compounds

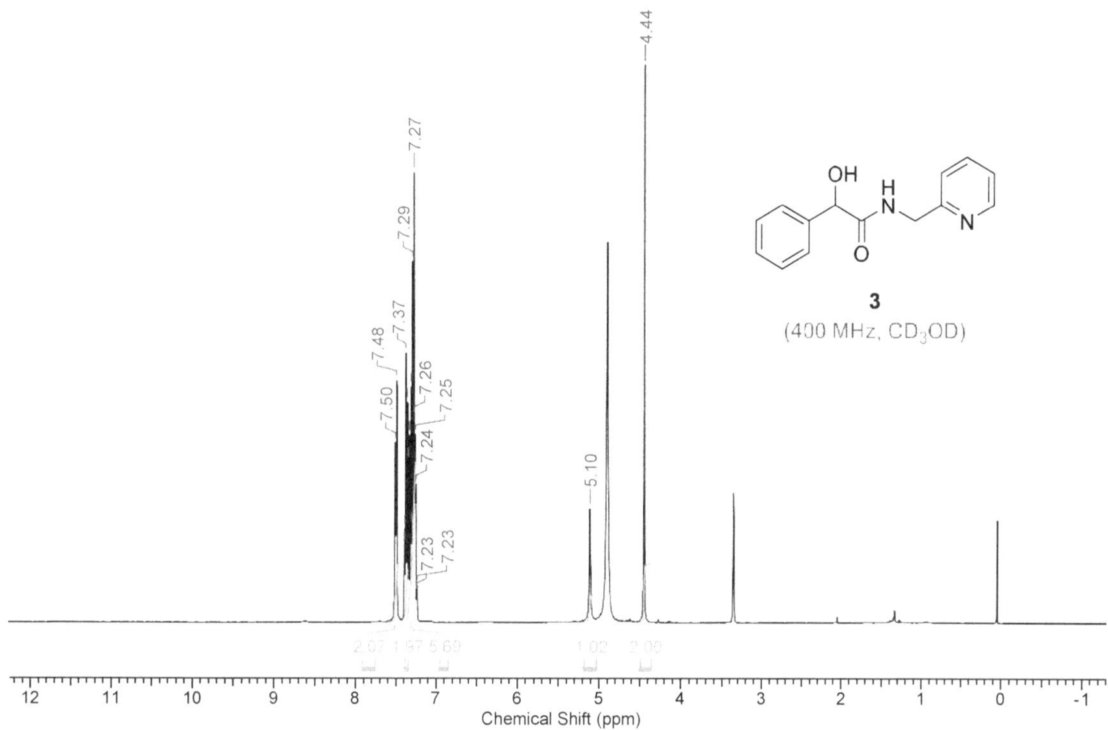

^{1}H-NMR spectrum of **3**

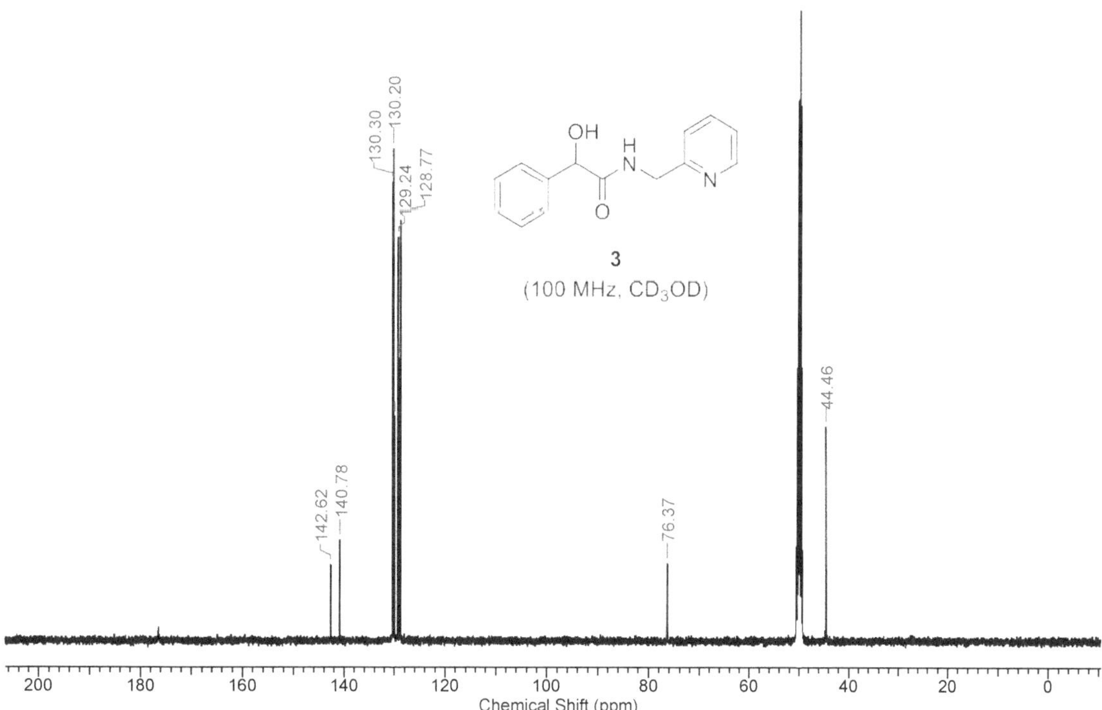

$^{13}C\{^{1}H\}$-NMR spectrum of **3**

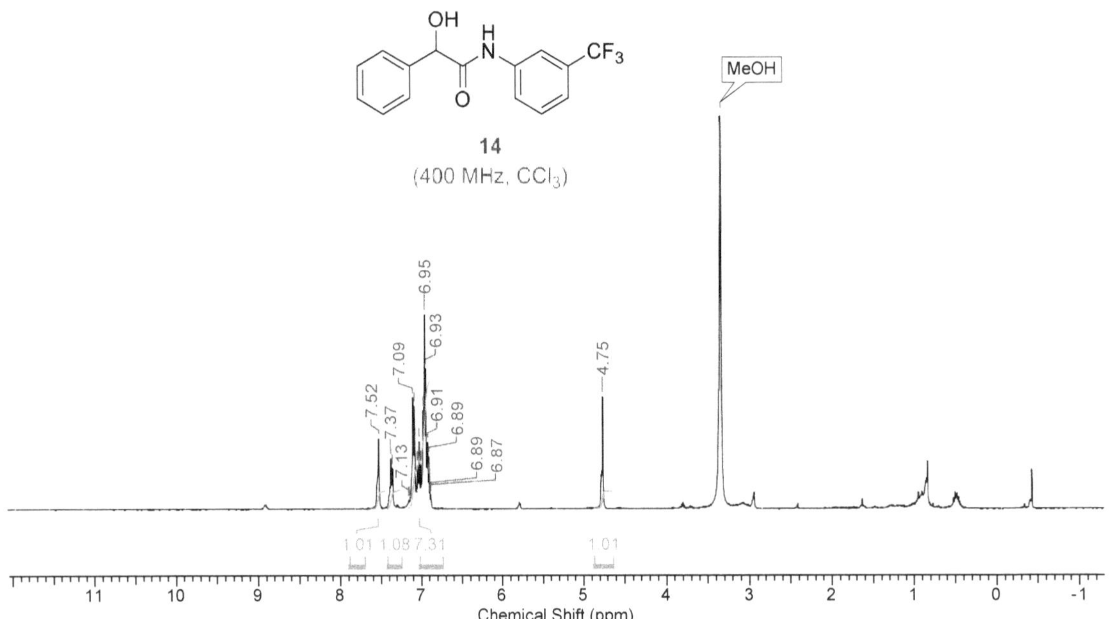

^{1}H-NMR spectrum of **14**

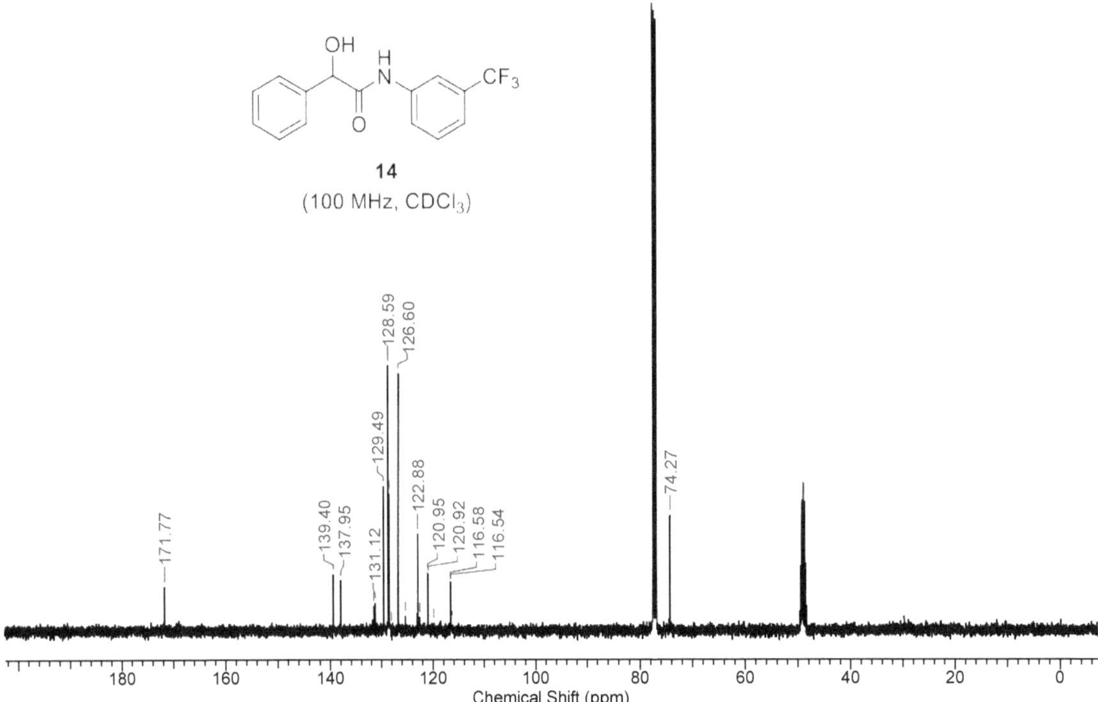

$^{13}C\{^1H\}$-NMR spectrum of **14**

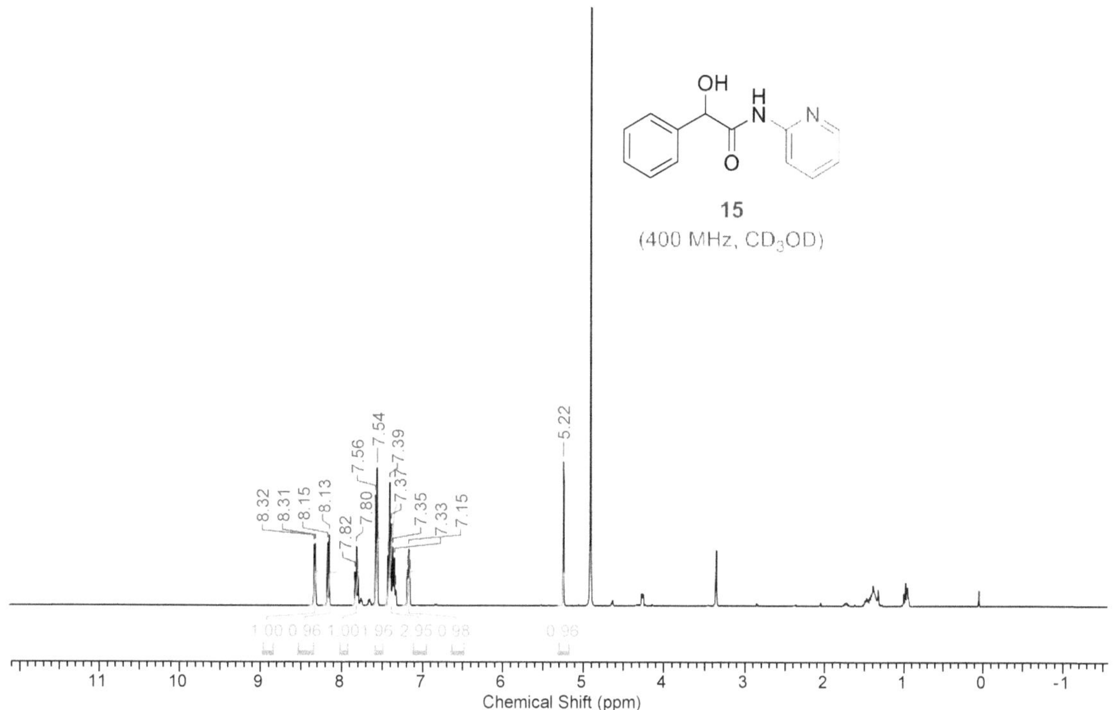

^{1}H-NMR spectrum of **15**

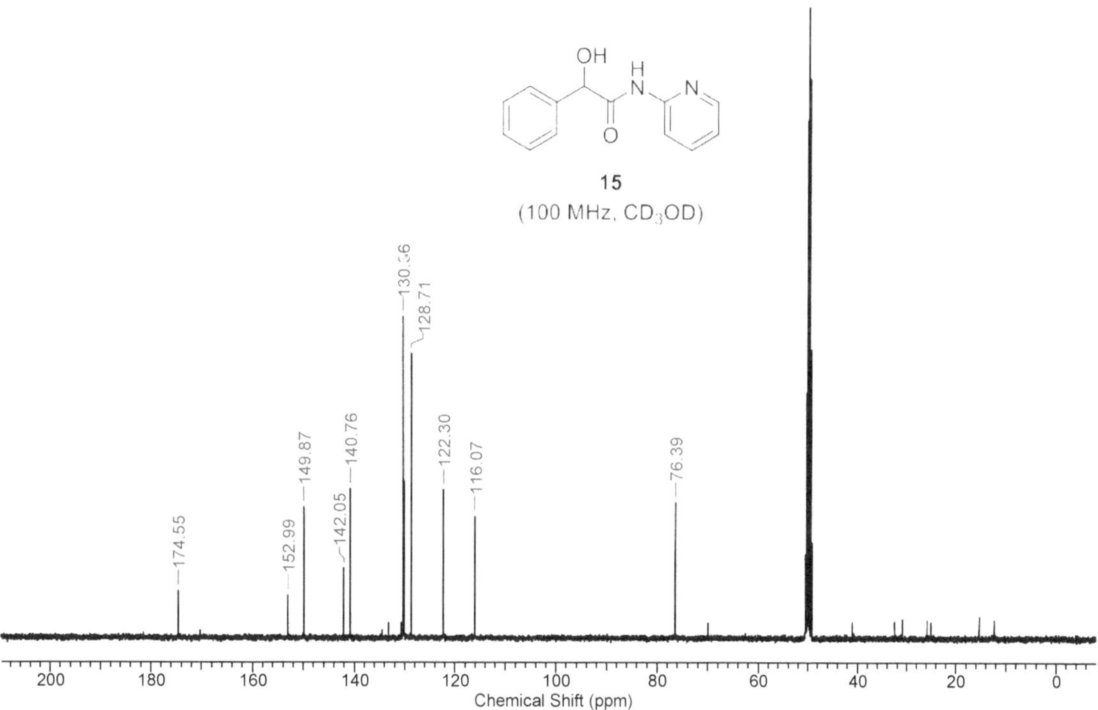

^{13}C{^{1}H}-NMR spectrum of **15**

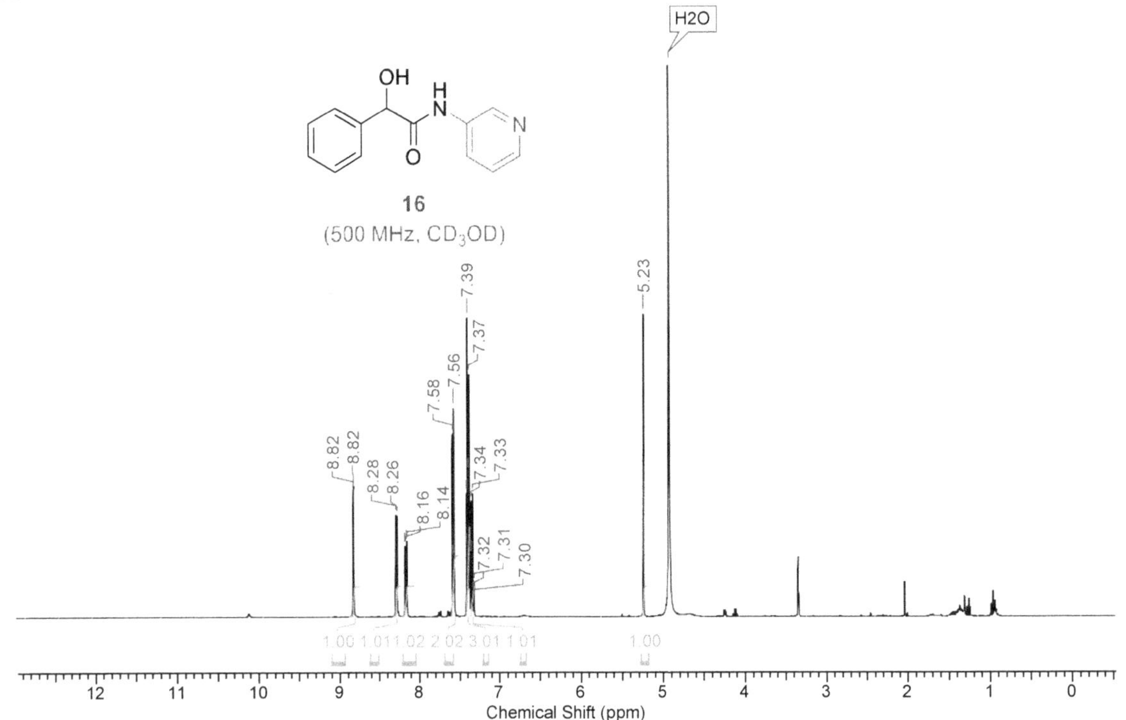

^{1}H-NMR spectrum of **16**

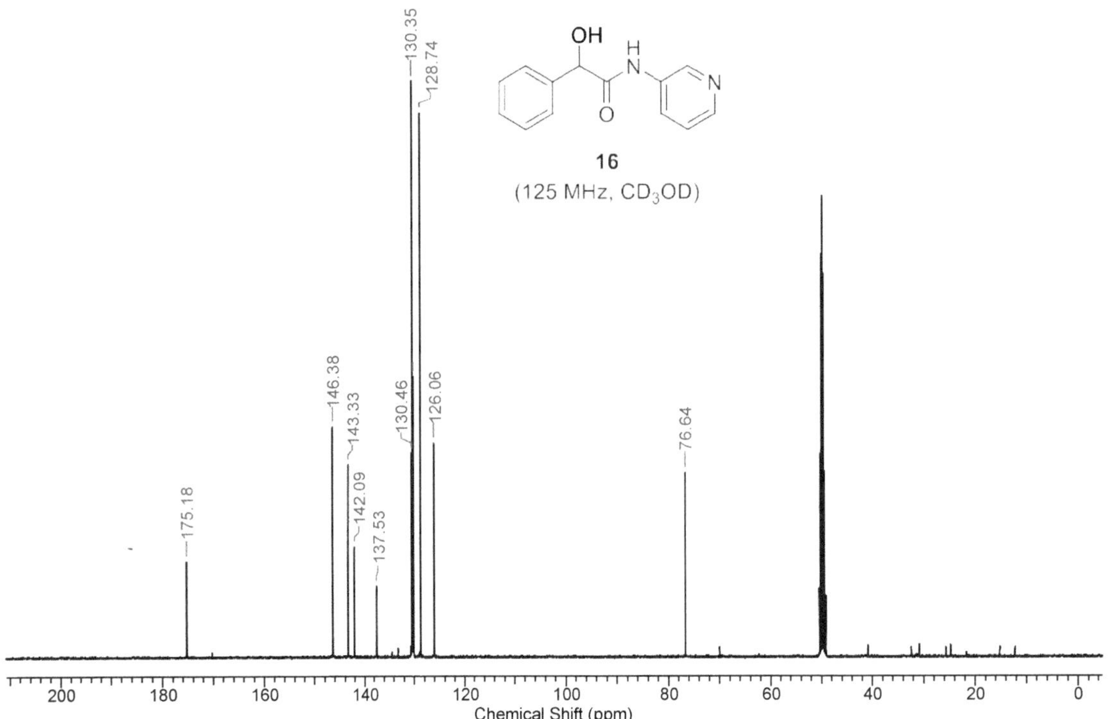

^{13}C{^{1}H}-NMR spectrum of **16**

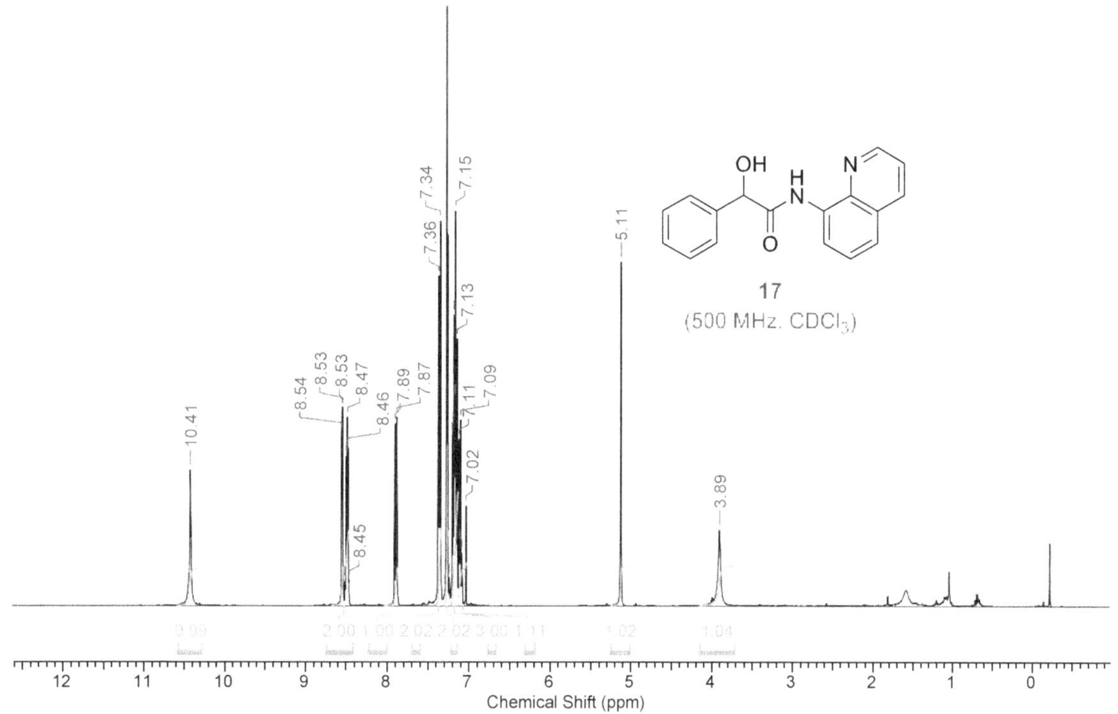

^{1}H-NMR spectrum of **17**

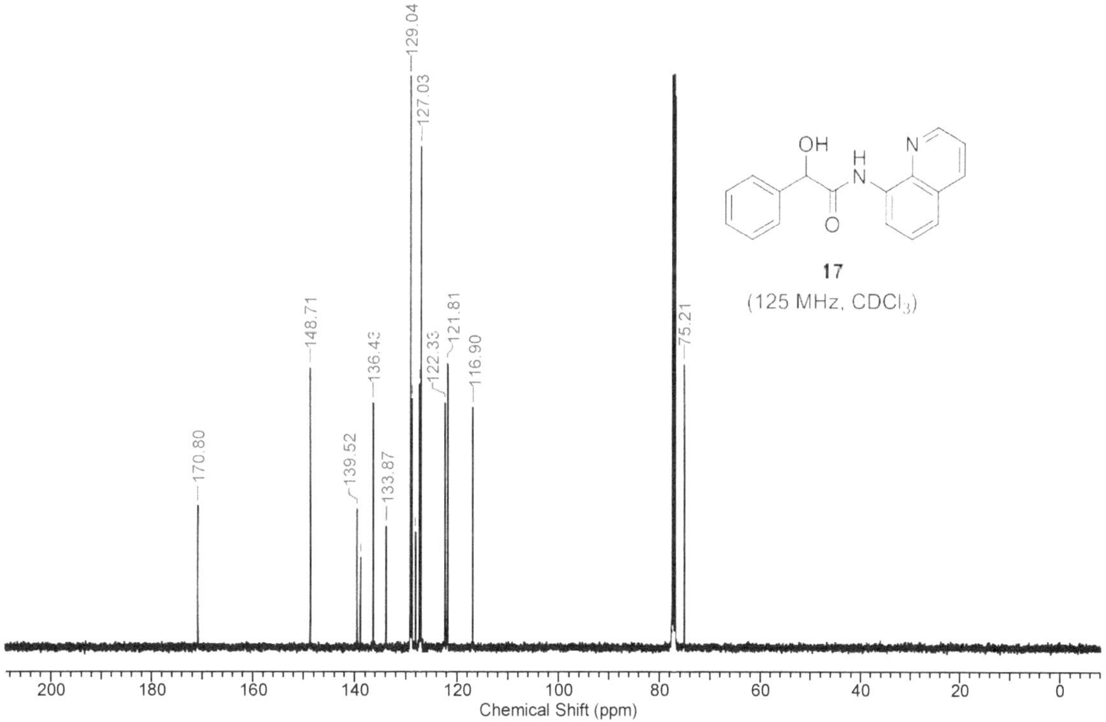

^{13}C{^{1}H}-NMR spectrum of **17**

4.5 REFERENCES

(1) J. G. de Vries; C. J. Elsevier.; de Vries, J. G., Elsevier, C. J., Eds.; Wiley, 2006.

(2) Alig, L.; Fritz, M.; Schneider, S., *Chem. Rev.* **2019**, *119*, 2681–2751.

(3) Wang, Y.; Wang, M.; Li, Y.; Liu, Q., *Chem* **2021**, *7*, 1180–1223.

(4) Tamura, M.; Nakagawa, Y.; Tomishige, K., *J. JPN Petrol. Inst.* **2019**, *62*, 106–119.

(5) Pritchard, J.; Filonenko, G. A.; van Putten, R.; Hensen, E. J. M.; Pidko, E. A., *Chem. Soc. Rev.* **2015**, *44*, 3808–3833.

(6) Das, K.; Waiba, S.; Jana, A.; Maji, B., *Chem. Soc. Rev.* **2022**, *51*, 4386–4464.

(7) Anferov, S. W.; Filatov, A. S.; Anderson, J. S., *ACS Catal.* **2022**, *12*, 9933–9943.

(8) Gholap, S. S.; Dakhil, A. Al; Chakraborty, P.; Li, H.; Dutta, I.; Das, P. K.; Huang, K. W., *Chem. Commun.* **2021**, *57*, 11815–11818.

(9) Vielhaber, T.; Topf, C., *Appl. Catal. A Gen.* **2021**, *623,* 118280.

(10) Zhang, L.; Wang, Z.; Han, Z.; Ding, K., *Angew. Chem. Int. Ed.* **2020**, *59*, 15565–15569.

(11) Glatz, M.; Stöger, B.; Himmelbauer, D.; Veiros, L. F.; Kirchner, K., *ACS Catal.* **2018**, *8*, 4009–4016.

(12) Langer, R.; Leitus, G.; Ben-David, Y.; Milstein, D., *Angew. Chem. Int. Ed.* **2011**, *50*, 2120–2124.

(13) Shapiro, S. L.; Rose, I. M.; Freedman, L., *J. Am. Chem. Soc.* **1959**, *81*, 6322–6329.

(14) Clifton, J. E.; Wicks, P. D., Lunts; L. H. C.; Hartley, D.; Hallett, P.; Collins, I., *J. Med. Chem.* **1982**, *25*, 670–679.

(15) Ablondi, F. B.; Hagen, J. J.; Clarke, R. E., *Proc. Soc. Exp. Biol. Med.* **1957**, *95*, 195–200.

(16) Ye, R.; Hao, F.; Liu, G.; Zuo, Q.; Deng, L.; Jin, Z.; Wu, J., *Org. Chem. Front.* **2019**, *6*, 3562–3565.

(17) Mamillapalli, N. C.; Sekar, G. *RSC Adv.* **2014**, *4*, 61077–61085.

(18) Muthukumar, A.; Mamillapalli, N. C.; Sekar, G., *Adv. Synth. Catal.* **2016**, *358*, 643–652.

(19) Kumar, G.; Muthukumar, A.; Sekar, G., *Eur. J. Org. Chem.* **2017**, *2017*, 4883–4890.

(20) Fang, Z. B.; Yu, R. R.; Hao, F. Y.; Jin, Z. N.; Liu, G. Y.; Dai, G. L.; Yao, W. B.; Wu, J. S., *Tetrahedron Lett.* **2021**, *86,* 153524.

(21) Hao, F.; Gu, Z.; Liu, G.; Yao, W.; Jiang, H.; Wu, J., *Eur. J. Org. Chem.* **2019**, *2019*, 5985–5991.

(22) Haynes, W. M.; Lide, D. R.; Bruno, T. J.; Eds., CRC Press, 2016.

(23) Wieghardt, K., *Angew. Chem. Int. Ed. 1***989**, *28*, 1153–1172.

(24) Christianson, D. W. *Prog. Biophys. Mol. Biol.* **1997**, *67*, 217–252.

Chapter-5

Copper-Catalyzed Regioselective C(2)–H Bond Arylation of Indoles and Related (Hetero)arenes

5.1 INTRODUCTION

Regioselective direct C–H arylation of heteroarenes, particularly indoles and other alkaloids, stands as significant considering the ubiquitous nature of arylated-heteroarenes in pharmaceuticals and biologically relevant compounds.[1,2] Amongst the diverse regio-specific C–H functionalizations of indoles,[3-5] the selective C(2)–H arylation is predominantly explored because of the unique importance of C-2 arylated indole and derivatives.[6-8] Notably, this selective transformation has been studied well with precious transition metal-based catalysts like ruthenium,[9-12] rhodium,[13-15] and palladium (Scheme 5.1a).[16-26] However, excessive dependency on these metals for various coupling reactions, and low availability ofthem in the earth-crust, forces researchers to look for alternate transition-metal catalysts.[27-29] In that direction, Song[30] and Ackermann[31-32] independently manifested the earth-abundant cobalt for the arylation of indoles employing diverse aryl coupling partners (Scheme 5.1b). Recently, indole's C2 arylation via the assistance of a monodentate directing group has been discovered under nickel catalysis (Scheme 5.1c).[33,34] Although these approaches exhibited the capability of 3d metals in the challenging C–H arylation, there is a tremendous scope for the development of such methodology using less-toxicand low-cost metal catalysts under simplified conditions; avoiding the use of expensive ligands and organic solvents.[35-37] This might allow the methodologies to be implemented in large-scale industrial synthesis of probable biologically active intermediates. Furthermore, solvent-free and ligand-free approaches would reduce the generation of hazardous waste that arises from the usage of solvents, and thus, would be economically beneficial.[38]

a) using 4d metal catalyst:

N(DG)–H + X–Ar → [Ru], [Rh], [Pd] / many reports → N(DG)–Ar

DG = directing group
ArX = (Psuedo)halide or organometallic coupling partner

b) using nickel catalyst:

N(DG)–H + $(OH)_2B$–Ar or X–Ar → [Co] / oxidant or base → N(DG)–Ar

c) using nickel catalyst:

N(2-py)–H + Cl/Br/I–Ar → [Ni] / base → N(2-py)–Ar

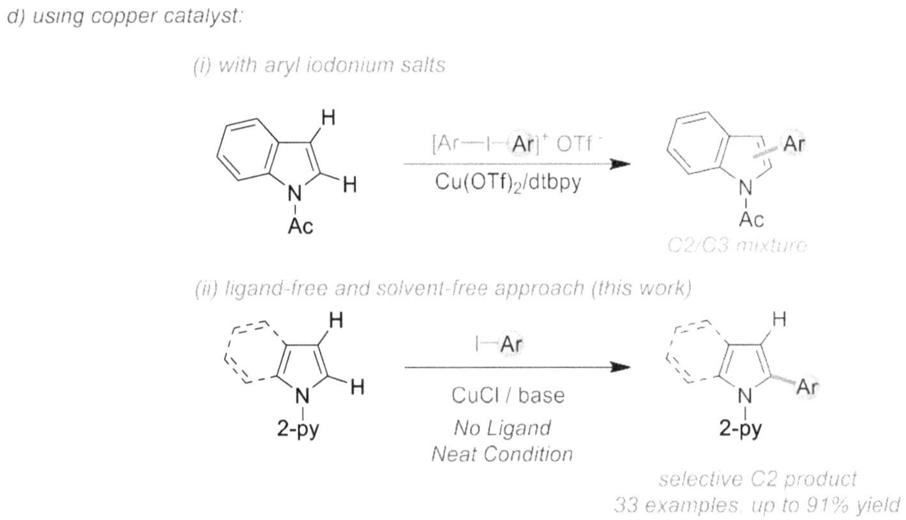

d) using copper catalyst:

(i) with aryl iodonium salts

[Ar—I—Ar]$^+$ OTf$^-$

Cu(OTf)$_2$/dtbpy

H H N Ac → Ar N Ac

C2/C3 mixture

(ii) ligand-free and solvent-free approach (this work)

I—Ar

CuCl / base

No Ligand

Neat Condition

H H N 2-py → H Ar N 2-py

selective C2 product

33 examples, up to 91% yield

Scheme 5.1 Approaches on C(2)–H Arylation of Indoles.

Copper stands as one of the most plentiful metals with low cost with great potential as a catalyst for the coupling reaction. In the last two decades, theeconomically viable copper metal led to significant progress in the development of efficient methods for carbon-carbon, carbon-nitrogen, and carbon-oxygen bond-forming reactions.[39-44] However, this earth-abundant metal seems to be underutilized for efficient C–H bond functionalizations. Recent reports on Cu-catalyzed C–H arylations demonstrate the ability of copper in such challenging transformations.[45-52] Though there has been some progress in the copper-catalyzed C-3 as well as other arylation of indoles,[53-55] a single report demonstrated by Gaunt for the C-2 arylation using activated diaryliodonium salt, [TRIP-I-Ar]OTf, however, with poor regioselectivity (Scheme 5.1d (i); TRIP = 2,4,6-tri-*iso*propylphenyl).[56] Moreover, the activated diaryliodonium salts employed for the coupling are generally prepared from corresponding aryl iodides, thus, enforcing an additional step for the desired arylation. We assumed that the use of aryl iodides for the arylation of indole heteroarenes employing a low-cost and abundant copper catalyst would be highly beneficial for modern chemical synthesis (Scheme 5.1d (ii)). With this hypothesis, herein, we optimized and developed a copper-catalyzed method for indole's C-2 arylation under ligand-free and solvent-free conditions. Additionally, preliminary mechanistic investigation, including external additive experiments, deuterium labeling, and electronic effect studies, has been performed that outlined a single-electron transfer path for the transformation.

5.2 RESULTS AND DISCUSSION

5.2.1 Optimization of Reaction Conditions

We have started the process of screening reaction parameters using the CuCl/dppf catalyst combination for the arylation of **1a** with **2a** as a model reactant (Table 5.1). The employment of relatively mild inorganic bases was unsuccessful in providing the arylated product, though the existence of LiO^{t}Bu and NaO^{t}Bu resulted in the formation of coupled indole (**3aa**) in 32% and 13%, respectively, in 1,4-dioxane at 120 °C (entries 1 and 2). Notably, the arylation led to quantitative conversion (98%) in the presence of $LiN(SiMe_3)_2$ [LiHMDS], which could be partly due to the high solubility of LiHMDS in 1,4-dioxane (entry 4). Surprisingly, the reaction also proceeded in quantitative conversion and afforded 91% of **3aa** by employing only CuCl metal precursor (99.99% pure) in the absence of dppf ligand (entry 5). The use of other copper(I) precursors, such as CuBr, CuI and Cu(OAc) as well as Cu(II) salts like $CuCl_2$, $CuBr_2$, $Cu(OAc)_2$ as catalysts produced **3aa** in 75-96% GC yields (entries 6-11). Even without a solvent, the arylation proceeded without difficulty, yielding arylation product **3aa** in a 91% isolated yield (entry12). Even with 1.3 equiv of LiHMDS, the reaction produced a comparable product yield with neat reaction parameters (entry 14). Furthermore, the 2-pyridinyl group on *N*-atom of indole is necessary, as it coordinates (directly) to copper, and brings the indole's C(2)−H bond in the proximity of copper for selective activation. The use of 4-bromotoluene as electrophile led to trace formation of **3aa**, whereas 4- chlorotoluene could not couple with indole **1a**. The reaction failed to proceed in the absence of CuCl, indicating the role of copper as catalyst (entry 22). The arylation reaction at 100 °C or lower temperature showed incomplete conversion, thus, we have performed further arylations employing CuCl catalyst and aryl iodide electrophiles at 120 °C under neat conditions.

Table 5.1 Optimization of Reaction Parameters.[a]

(1a) + (2a) → [Cu] (5.0 mol %), ligand (5.0 mol %), base, solvent, 120 °C, 16 h → (3aa)

Entry	[Cu]	Ligand	Base	Solvent	Yield of 3aa (%)[b]
1	CuCl	dppf	LiO^{t}Bu	1,4-dioxane	32
2	CuCl	dppf	NaO^{t}Bu	1,4-dioxane	13
3	CuCl	dppf	Na_2CO_3	1,4-dioxane	--
4	CuCl	dppf	LiHMDS	1,4-dioxane	98 (90)
5	**CuCl**	**--**	**LiHMDS**	**1,4-dioxane**	**99 (91)**
6	CuBr	--	LiHMDS	1,4-dioxane	75
7	CuI	--	LiHMDS	1,4-dioxane	87
8	Cu(OAc)	--	LiHMDS	1,4-dioxane	94
9	$CuCl_2$	--	LiHMDS	1,4-dioxane	96
10	$CuBr_2$	--	LiHMDS	1,4-dioxane	93
11	$Cu(OAc)_2$	--	LiHMDS	1,4-dioxane	91
12	CuCl	--	LiHMDS	toluene	26
13	CuCl	--	LiHMDS	--	99 (91)
14[c]	**CuCl**	**--**	**LiHMDS**	**--**	**99 (91)**
15[d]	CuCl	--	LiHMDS	1,4-dioxane	trace
16[d]	CuCl	--	LiHMDS	--	trace
17[e]	CuCl	--	LiHMDS	1,4-dioxane	trace
18[e]	CuCl	bpy	LiHMDS	1,4-dioxane	trace
19[e]	CuCl	dppf	LiHMDS	1,4-dioxane	trace
20[e]	CuCl	--	LiHMDS	--	trace
21[f]	CuCl	--	LiHMDS	--	72
22	--	--	LiHMDS	--	--

[a] Reaction conditions: **1a** (0.097 g, 0.50 mmol), **2a** (0.218 g, 1.0 mmol), base (1.0 mmol), [Cu] precursor (0.025 mmol, 5 mol %), solvent (1.5 mL). [b] GC yield using *n*-hexadecane as internal standard, isolated yield is given in parenthesis. [c] 1.3 equiv of LiHMDS used. [d] 4-Bromotoluene as electrophile. [e] 4-Chlorotoluene as electrophile. [f] 1.2 equiv of LiHMDS used.

5.2.2 Substrate Scope of Arylation of Indoles and Related Heteroarenes

The scope and limitations of arylation with various aryl iodides have been investigated following the successful improvement of reaction parameters for the CuCl-catalyzed arylation of indole with 4-iodotoluene. The developed methodology was applicable to the coupling of 2-pyridinyl indole with a various substituted iodoarenes (Scheme 5.2). Thus, the *para*-alkyl and alkoxy-substituted aryl iodides were reacted efficiently under neat reaction conditions to deliver expected coupled products **3aa-3ah** exclusively. Remarkably, the precedented copper-catalyzed arylation employing iodonium salts was reported with moderate yields for aryl bearing alkyl or alkoxy groups.[56] The arylation proceeded smoothly with iodides bearing bisphenol and thymol (natural monoterpenoid phenol) containing groups (**3ai** and **3aj**). The halide functionalities $^{-}$F, $^{-}$Cl and $^{-}CF_3$ were well tolerated at the *para* position of arene and afforded the arylated products in low to moderate yields (**3ak-3am**). Notably, the $^{-}CF_3$ containing aryl electrophile was unreactive under the nickel-catalysis that was demonstrated earlier.[34] The 1,3-dioxolanederivative, **2n** could be employed in the arylation reaction providing 84% of the coupled product **3an**. This protecting group can be easily removed to generate synthetically important acetyl-functionality. The sterically demanding electrophiles 1-iodonaphthalene and 2-methyl-1-iodobenzene were conveniently reacted with indole **1a** to produce **3ao** and **3ap** in 76% and 91% yields, respectively. A more challenging and sterically bulky electrophile, 2-iodo- 1,3-dimethylbenzene (**2q**) was employed under theneat condition to result **3aq** in 91% yield. Similar sterically demanding electrophiles were not shown in the previous Cu-catalyzed protocol.[56] In addition to the efficient coupling of *ortho*- and *para*- and *meta*-substituted aryl iodides, **2r-2t** reacted to produce desired products **3ar-3at** with moderate to good yields. Interestingly, the aryl iodides containing substituted amine and *N*- heterocycles (*N*-pyrrolyl, indolyl, carbazolyl groups) resulted in the formation of desired arylated products in excellent yields. Such functionalities containing aryls have shown moderate reactivity under nickel catalysis.[34] Unfortunately, the current method failed to toleratethe functional groups, like acetyl, ester, amides and nitro groups. Notably, the employment of *n*-octyl iodide as a coupling partner with indole **1a**, with the standard conditions, showed C-2 alkylated indole, in 40% yield,[57-58] and a substantial side-reaction of *n*- octyl iodide with LiHMDS was observed. Though a copper-catalyzed arylation of indoles is being reported previously,[56] the protocol employed highly reactive aryliodonium(III) electrophiles and a mixture of C2/C3 regio-isomer was observed. Nevertheless, using aryl iodides, we have demonstrated here the ligand-free, solvent-free indole's C-2 arylation.

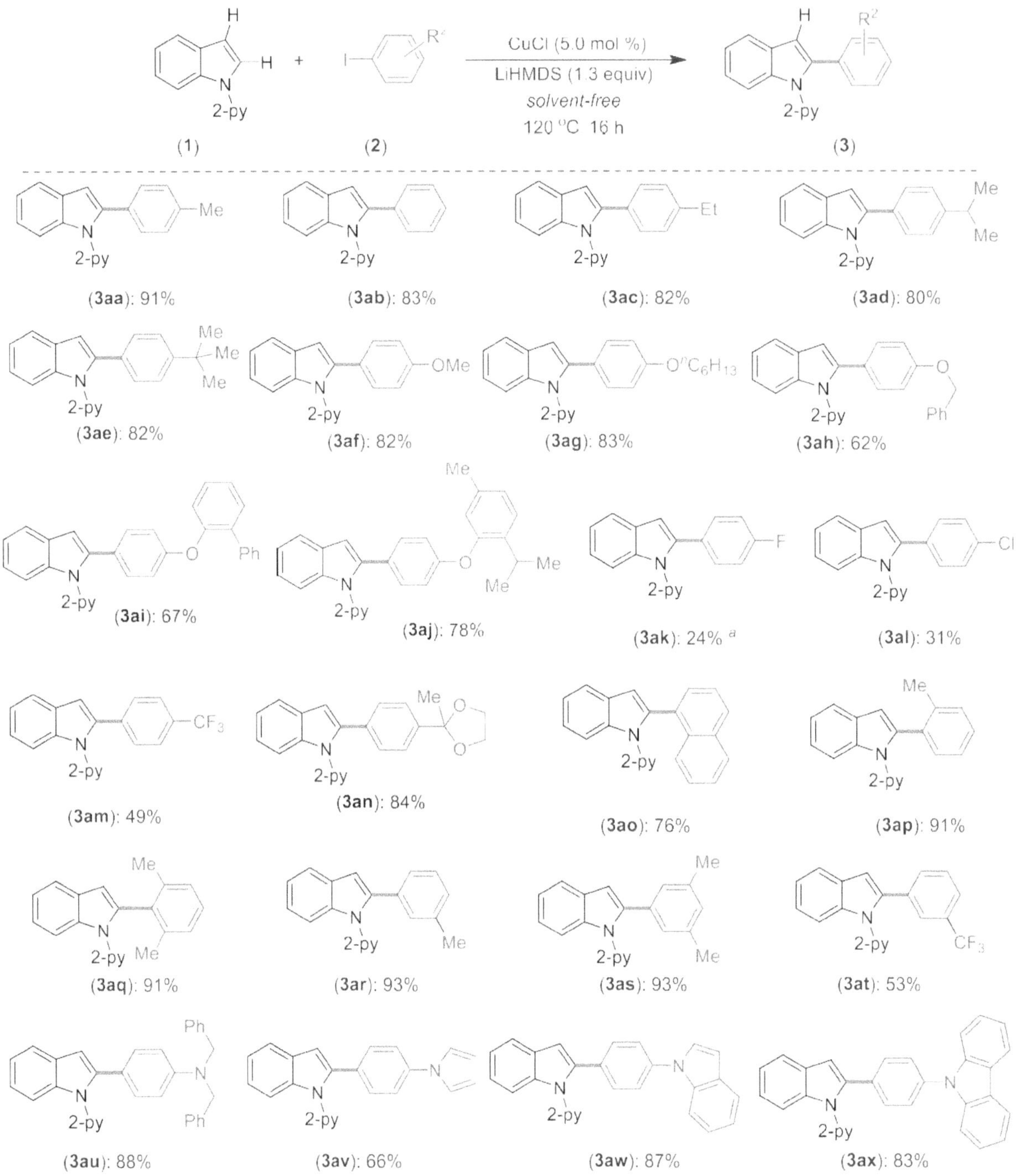

H
H + I—R²
CuCl (5.0 mol %)
LiHMDS (1.3 equiv)
solvent-free
120 °C 16 h
(1) (2) (3)

(3aa): 91% (3ab): 83% (3ac): 82% (3ad): 80%
(3ae): 82% (3af): 82% (3ag): 83% (3ah): 62%
(3ai): 67% (3aj): 78% (3ak): 24% [a] (3al): 31%
(3am): 49% (3an): 84% (3ao): 76% (3ap): 91%
(3aq): 91% (3ar): 93% (3as): 93% (3at): 53%
(3au): 88% (3av): 66% (3aw): 87% (3ax): 83%

Scheme 5.2 Scope for C-2 Arylation of Indoles with Aryl Iodides. Conditions: indole **1a** (0.097 g, 0.50 mmol), aryl iodide **2** (1.0 mmol), LiHMDS (0.109 g, 0.65 mmol), CuCl (0.0025 g, 0.025 mmol, 5.0 mol %). Yield of isolated compound. [a] 10 mol % CuCl, aryl iodide (4.0 equiv) and 1,4-dioxane (1.5 mL) were employed.

Furthermore, the reactivity of substituted indole derivatives was explored with 4-iodotoluene (Scheme 5.3). With electron-donating substituents at the C-5 position, the indole therefore reacted easily to form **3ba** and **3ca** in 77% and 87% yields, respectively. Nevertheless, the halo-substituted indoles reacted moderately with *p*-iodotoluene resulting in the formation of **3da** and **3ea** in 62% and 51% yields, respectively. The sterically bulky indole **1f** upon reaction with 4-iodotoluene afforded 80% of the desired C-2 arylated product **3fa**. This substrate had shown low reactivity for the arylation reaction using a nickel catalyst.[34] The 7-azaindolepyridine could be coupled with 4-iodotoluene in moderate activity affording 32% of **3ga**. Notably, the thiophenyl derivative **1h** reacted efficiently under the copper catalysis and delivered **3ha** in 85% yield. The arylation of such substrate has not been precedented with nickel or copper catalysis, and this example demonstrates the broadness of the presented protocol. Notably, the indole derivatives containing 2-pyrazinyl or 2-pyrimidinyl as the *N*-substituents decomposed under the standard reaction conditions. Nonetheless, the pyridinyl-carbazole, benzo[*h*]quinoline and 2-pyridinyl-arene substrates were unreactive under the optimized reaction conditions (Scheme 5.3).

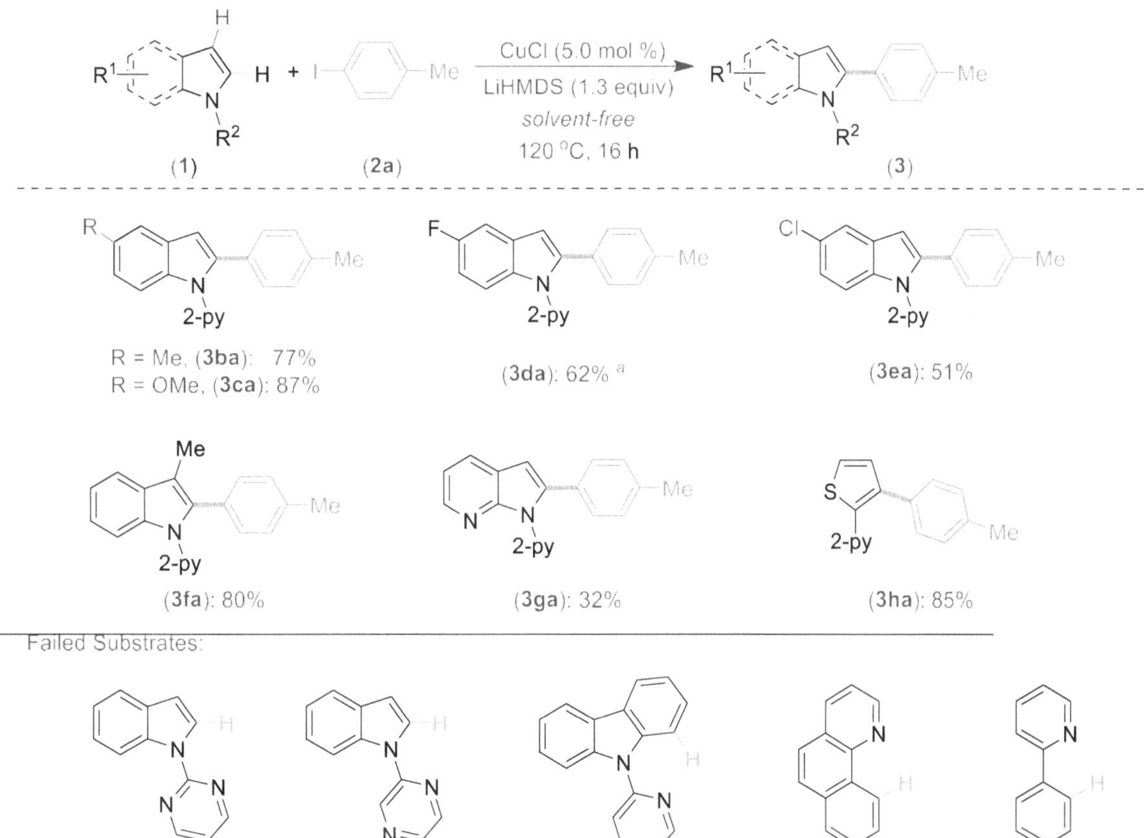

Scheme 5.3 Scope of Substituted-Indoles and Other Heteroarenes. Reaction conditions: Heteroarenes **1** (0.50 mmol), 4-iodotoluene (**2a**; 1.0 mmol), LiHMDS (0.109 g, 0.65 mmol), CuCl (0.0025 g, 0.025 mmol, 5.0 mol

%). Yield of isolated compound. [a] 1,4-dioxane was used as solvent.

The 2-pyrrolpyridine substrate afforded the mixture of mono-arylated (**3ia**) as well as biarylated (**3i'a**) products in 43% and 10% yields, respectively, when treated with 2.0 equiv of 4-iodotoluene (Scheme 5.4). The selective synthesis of mono-arylated product **3ia** was unsuccessful even under various stoichiometries of **1i** and **2a**. However, when **1i** was treated 4.0 equiv of **2a**, the reaction resulted with 10% of **3ia** and 60% of **3i'a**. Although there exist protocols for the regioselective arylation of indoles using various coupling partners, in particular, we have demonstrated a low-cost, solvent-free, ligand-free copper-catalyzed approach, which broadens the scope and utility of this methodology.

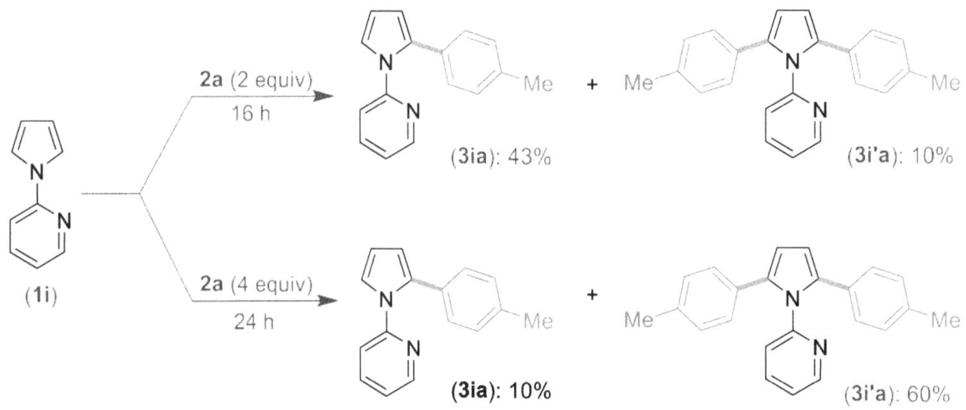

2a (2 equiv)
16 h
(3ia): 43%
(3i'a): 10%
(1i)
2a (4 equiv)
24 h
(3ia): 10%
(3i'a): 60%

Scheme 5.4 Arylation of Pyrrole. Reaction conditions: **1i** (0.30 mmol), LiHMDS (1.3 or 2.6 equiv), CuCl (0.0015 g, 0.015 mmol, 5.0 mol %).

The practical application of present arylation under copper catalysis was showed by performing a gram-scale arylation reaction. Accordingly, the treatment of indole **1a** (1.0 g) with 4-iodotoluene (2.24 g) under the standard reaction conditions, using 5 mol % CuCl and 1,4-dioxane (7.0 mL), gave 1.2 g (82%) of **3aa**. The usefulness of our developed protocol was showcased by the smooth elimination of the directing group to obtain synthetically advantageous C-2 arylated free *NH* indoles (Scheme 5.5).[10] Therefore, MeOTf was applied to the C-2 arylated indoles **3aa**, **3ba**, and **3ca**. This was followed by a reaction in NaOH (2.0 M), which produced the free *NH* indoles **4aa**, **4ba**, and **4ca**, respectively, in excellent yields. The resulting C-2 arylated indoles **4aa** and **4ba** were further functionalized at the C-3 position to synthesize biologically relevant Tryptamine derivatives (Scheme 5.6).[59,60]

5.2.3 Synthetic Utility for C2-Arylation of Indoles

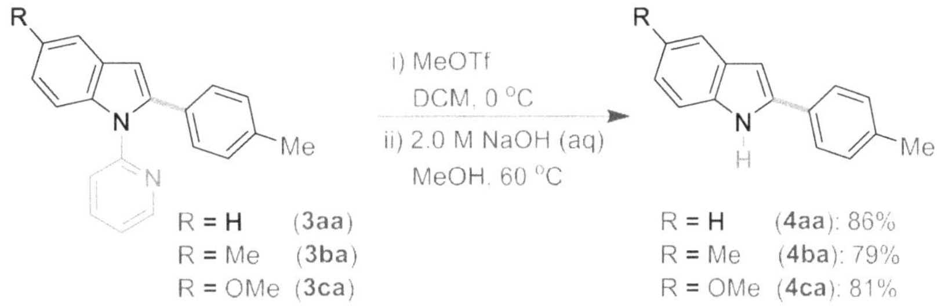

Scheme 5.5 Removal of Directing Group.

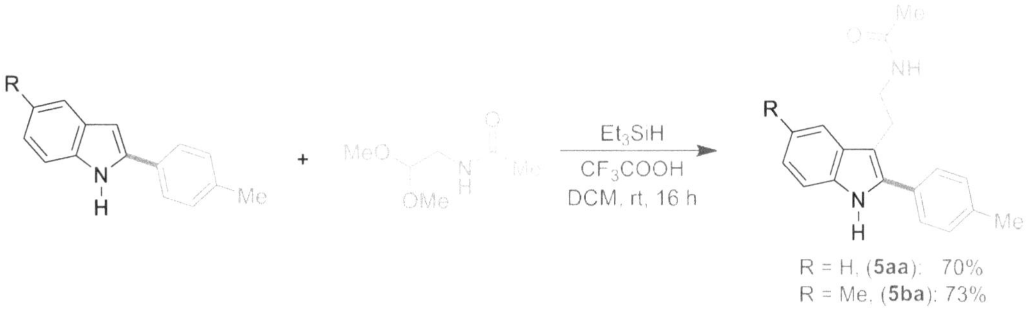

Scheme 5.6 Synthesis of Tryptamine Derivatives.

5.2.4 Mechanistic Aspects

The mechanism of action of the reported Cu-catalyzed arylation is still poorly understood. A small amount of product is formed by the arylation reaction when radical inhibitor TEMPO is present, whereas galvinoxyl or BHT totally suppresses the reaction (Scheme 5.7a). These observations tentatively suggested a radical pathway for the arylation. A competition experiment between electronically distinct substrates **1c** and **1d** showed similar reactivity(Scheme 5.7b), thus, ruling out an electrophilic-type C–H activation.[61] However, the electron-deficient aryl iodide **2m** reacted almost two-fold faster in preference to electron-rich electrophile **2f** (Scheme 5.7c), this suggests that aryl iodide's oxidative addition is likely to have a rate-influencing effect. The independent rate measurements for the arylation of indole **1a** and [2-D]-**1a** did not show an isotope effect (Figure 5.3). Additionally, a noticeable H/D was scrambling between substrates [2-D]-**1a** and **1c** was observed (Scheme 5.8). These findings in addition to the K_H/K_D value (0.66), suggest a reversible C–H metalation process and tentatively ruled out the rate-limiting–H activation (Scheme 5.9).[62]

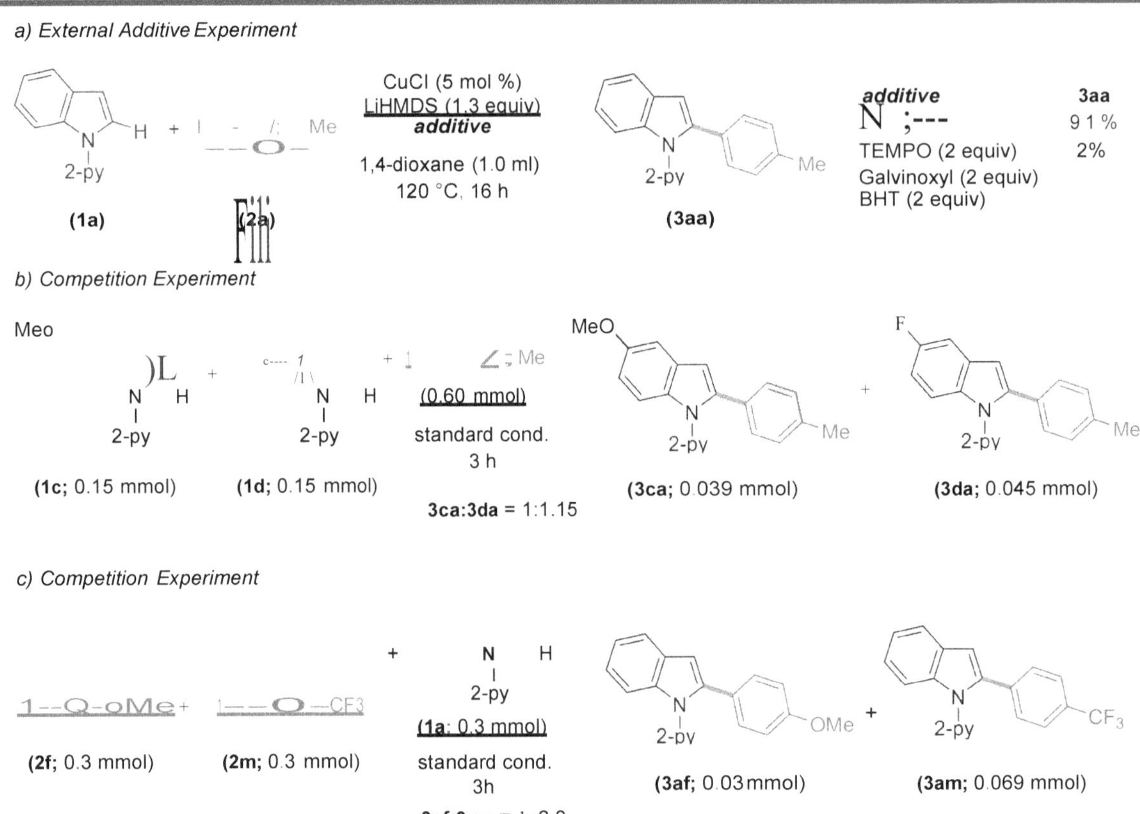

Scheme 5.7 External Additive and Competition Experiments.

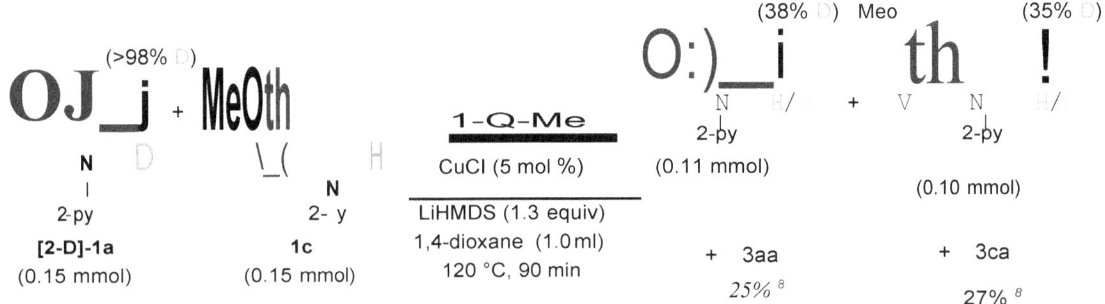

Scheme 5.8 H/D Scrambling Experiment. [a] GC conversion.

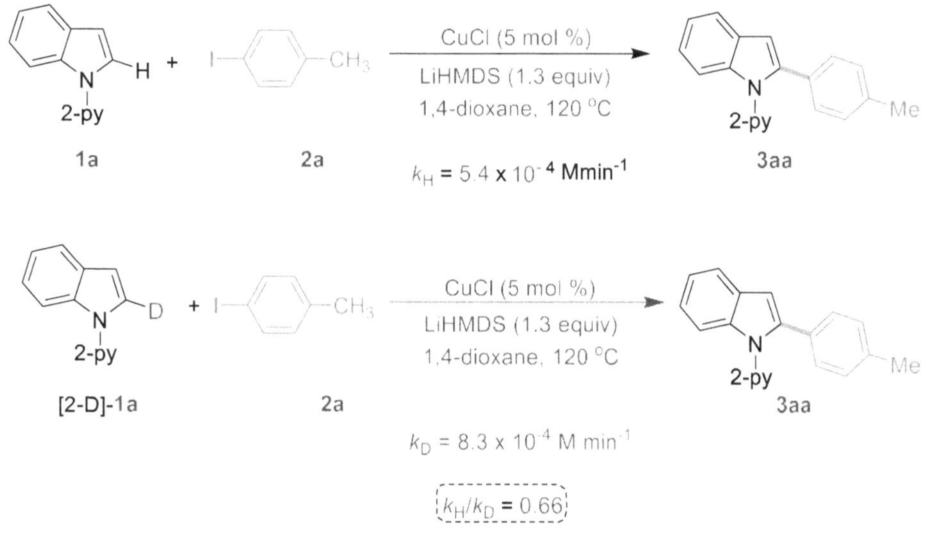

1a + 2a → 3aa
CuCl (5 mol %)
LiHMDS (1.3 equiv)
1,4-dioxane, 120 °C
$k_H = 5.4 \times 10^{-4}$ Mmin^{-1}

[2-D]-1a + 2a → 3aa
CuCl (5 mol %)
LiHMDS (1.3 equiv)
1,4-dioxane, 120 °C
$k_D = 8.3 \times 10^{-4}$ M min^{-1}

$k_H/k_D = 0.66$

Scheme 5.9 Study of the Kinetic Isotopic Effect on the Reaction.

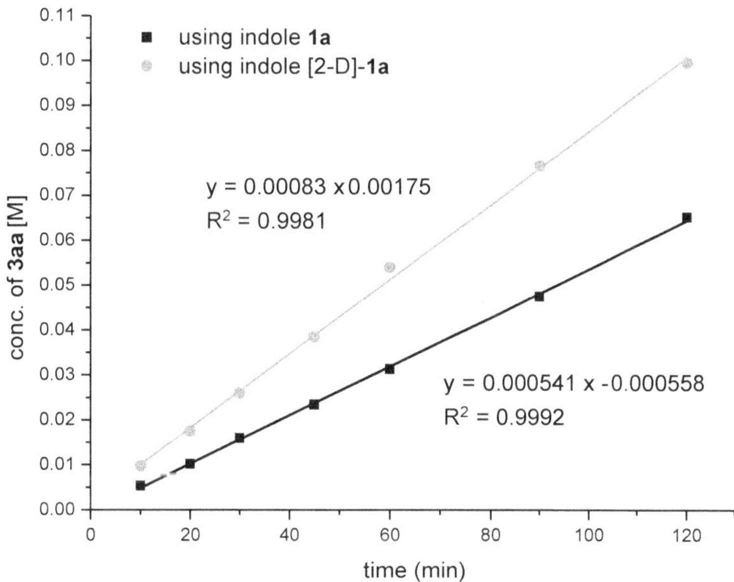

Figure 5.1 Time-Dependent Formation of Product **3aa** Using Indoles **1a** and [2-D]-**1a**.

5.2.5 Plausible Catalytic Cycle

Based on appropriate research precedents and initial mechanistic insights,[33,34,57,58] the Cu(I) complex is thought to function as an active catalyst, wherein indole 1a would coordinate to Cu followed by reversible LiHMDS-mediated C–H activation resulting in the species **A** (Figure 5.2). This species would cause iodoarene to undergo a two-step, one-electron oxidative addition, yielding **B**. Cu(I) species would be regenerated to complete the catalytic cycle upon the reductive elimination of product 3. Further studies are needed to proposea more precise mechanism for the reaction.

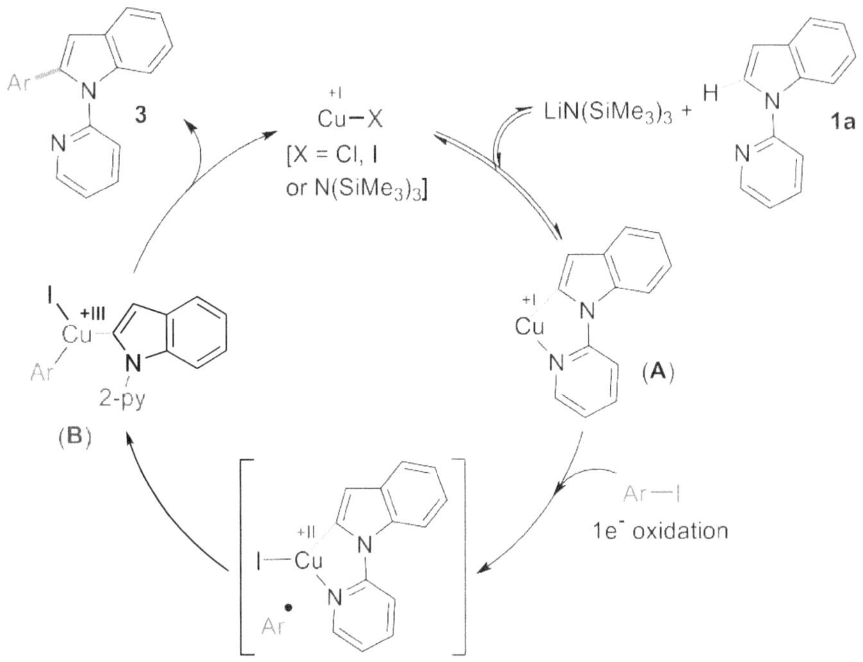

Figure 5.2 Plausible Mechanistic Pathway.

5.3 CONCLUSIONS

In this chapter, the process for regioselective arylation of C-H bonds facilitated by Cu, for the indoles and related heteroarenes with a ligand-free and solvent-free approach was disclosed. A wide range of functionalities, such as halides, ethers, substituted amines, pyrrolyl, indolyl, and carbazolyl groups, are tolerated by the reaction throughout the broad arylation process. Particularly, *ortho*-substituted aryl iodides have a major steric challenge for coupling, and is typically hard. The developed user-friendly protocol is expected to be beneficial for practical use as it employs a simple Cu-salt and avoids the usage of solvent. This approach showed that the 2-pyridinyl directing group could be removed with ease and that tryptamine derivatives could be synthesized. For the arylation, a single-electron transfer (SET) pathway has been postulated based on the initial mechanistic findings.

5.4 EXPERIMENTAL SECTION

All manipulations were conducted under an argon atmosphere either in a glove box or using standard Schlenk techniques in pre-dried glassware. The catalytic reactions were performed in flame-dried reaction vessels with Teflon screw cap. Solvents were dried over Na/benzophenone or CaH_2 and distilled prior to use. Liquid reagents were flushed with argon prior to use. The aryl iodides **2i** and **2j** were synthesized following the procedure similar to the synthesis of other aryl iodides.[63] The

$CuCl$ used was 99.99% pure, and $CuCl_2$ was 99.999% pure obtained from Sigma-Aldrich chemical company. All other chemicals were obtained from commercial sources and were used without further purification. High resolution mass spectrometry (HRMS) mass spectra were recorded on a Thermo Scientific Q-Exactive, Accela 1250 pump. NMR: (^1H and ^{13}C) spectra were recorded at 400 or 500 MHz (^1H), 100 or 125 MHz {^{13}C, DEPT (distortionless enhancement by polarization transfer)}, 377 MHz (^{19}F), respectively in $CDCl_3$ solutions, if not otherwise specified; chemical shifts (δ) are given in ppm. The ^1H and ^{13}C NMR spectra are referenced to residual solvent signals ($CDCl_3$: δ H = 7.26 ppm, δ C = 77.2 ppm).

GC Method. Gas Chromatography analyses were performed using a Shimadzu GC-2010 gas chromatograph equipped with a Shimadzu AOC-20s auto sampler and a Restek RTX-5 capillary column (30 m x 0.25 mm x 0.25 μm). The instrument was set to an injection volume of 1 μL, an inlet split ratio of 10:1, and inlet and detector temperatures of 250 and 320 °C, respectively. UHP-grade argon was used as carrier gas with a flow rate of 30 mL/min. The temperature program used for all the analyses is as follows: 80 °C, 1 min; 30 °C/min to 200 °C, 2 min; 30 °C/min to 260 °C, 3 min; 30 °C/min to 300 °C, 3 min. Response factors for desired compounds were calculated w.r.t. standard *n*-hexadecane from the average of three independent GC runs.

5.4.1 Representative Procedure for Arylation and Characterization Data

Synthesis of 1-(pyridin-2-yl)-2-(*p*-tolyl)-1*H*-indole (3aa): To a flame-dried screw-cap tube equipped with magnetic stir bar were introduced 1-(pyridin-2-yl)-1*H*-indole (**1a**; 0.097 g, 0.50 mmol), 1-iodo-4-methylbenzene (**2a**; 0.218 g, 1.0 mmol), CuCl (0.0025 g, 0.025 mmol, 5 mol %) and LiHMDS (0.109 g, 0.65 mmol, 1.3 equiv) inside the glove box. The resultant reaction mixture in the tube was immersed in preheated oil bath at 120 °C and stirred for 16 h. At ambient temperature the reaction mixture was quenched with distilled H_2O (5 mL) and 1N HCl (2.0 mL), and the crude product was extracted with EtOAc (15 mL x 3). The combined organic layer was dried over anhydrous Na_2SO_4 and volatiles were evaporated in vacuo. The remaining residue was purified by column chromatography on silica gel (petroleum ether/EtOAc: 50/1) to yield **3aa** (0.129 g, 91%) as light yellow solid.

^1H-NMR (500 MHz, $CDCl_3$): δ = 8.67 (d, J = 4.6 Hz, 1H, Ar–H), 7.72-7.68 (m, 2H, Ar–H), 7.62 (td, J = 7.6, 1.9 Hz, 1H, Ar–H), 7.26-7.20 (m, 5H, Ar–H), 7.11 (d, J = 8.0 Hz, 2H, Ar–H), 6.92 (d, J = 8.0 Hz, 1H, Ar–H), 6.81 (s, 1H, Ar–H), 2.36 (s, 3H, CH_3). $^{13}C\{^1H\}$-NMR (125 MHz, $CDCl_3$): δ = 152.3 (C_q), 149.3 (CH), 140.3 (C_q), 138.6 (C_q), 137.9 (CH), 137.4 (C_q), 129.9 (C_q), 129.2 (2C, CH), 128.9 (C_q), 128.8 (2C, CH), 123.0 (CH), 122.2 (CH), 121.7 (CH), 121.4 (CH), 120.6 (CH), 111.6 (CH), 105.3 (CH), 21.4 (CH_3). HRMS (ESI): *m/z* Calcd for $C_{20}H_{16}N_2 + H^+$ $[M + H]^+$ 285.1386;

Found 285.1391. The ^{1}H and ^{13}C{^{1}H} spectra are consistent with those reported in the literature.[34]

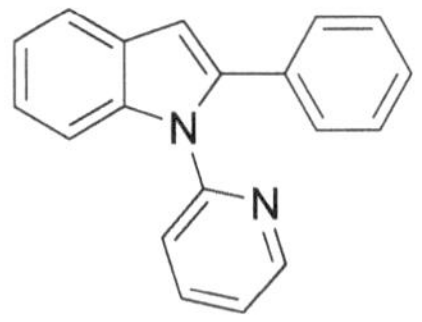

2-Phenyl-1-(pyridin-2-yl)-1*H*-indole (3ab): The representative procedure was followed, using substrate **1a** (0.097 g, 0.50 mmol) and iodobenzene (**2b**; 0.204 g, 1.0 mmol), and the reaction mixture was stirred at 120 °C for 16 h. Purification by column chromatography on silica gel (petroleum ether/EtOAc: 50/1) yielded **3ab** (0.112 g, 83%) as colorless solid. ^{1}H-NMR (400 MHz, $CDCl_3$): δ = 8.65 (d, J = 4.3 Hz, 1H, Ar–H), 7.71-7.67 (m, 2H, Ar–H), 7.61 (vt, J = 7.3 Hz, 1H, Ar–H), 7.28-7.20 (m, 8H, Ar–H), 6.89 (d, J = 8.6 Hz, 1H, Ar–H), 6.82 (s, 1H, Ar–H). ^{13}C{^{1}H}-NMR (100 MHz, $CDCl_3$): δ = 152.3 (C_q), 149.4 (CH), 140.2 (C_q), 138.7 (C_q), 138.0 (CH), 132.9 (C_q), 128.9 (2C, CH), 128.8 (C_q), 128.5 (2C, CH), 127.6 (CH), 123.2 (CH), 122.2 (CH), 121.8 (CH), 121.5 (CH), 120.7 (CH), 111.7 (CH), 105.8 (CH). HRMS (ESI): m/z Calcd for $C_{19}H_{14}N_2 + H^+$ $[M + H]^+$ 271.1230; Found 271.1234. The ^{1}H and ^{13}C{^{1}H} spectra are consistent with those reported in the literature.[34]

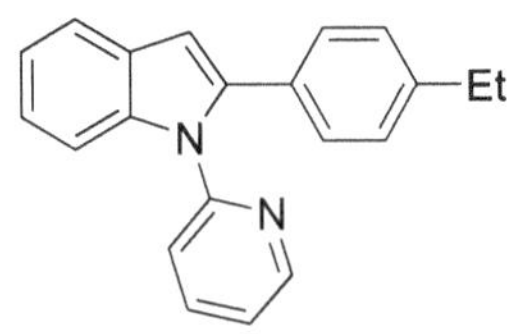

2-(4-Ethylphenyl)-1-(pyridin-2-yl)-1*H*-indole (3ac): The representative procedure was followed, using substrate **1a** (0.097 g, 0.50 mmol) and 1-ethyl-4-iodobenzene (**2c**; 0.232 g, 1.0 mmol), and the reaction mixture was stirred at 120 °C for 16 h. Purification by column chromatography on silica gel (petroleum ether/EtOAc: 50/1) yielded **3ac** (0.122 g, 82%) as brown solid. ^{1}H-NMR (500 MHz, $CDCl_3$): δ = 8.71-8.69 (m, 1H, Ar–H), 7.75-7.71 (m, 2H, Ar–H), 7.65 (t, J = 7.6 Hz, 1H, Ar–H), 7.30-7.24 (m, 5H, Ar–H), 7.17 (d, J = 7.6 Hz, 2H, Ar–H), 6.95 (d, J = 7.6 Hz, 1H, Ar–H), 6.84 (s, 1H, Ar–H), 2.68 (q, J = 7.6 Hz, 2H, Ar–H), 1.28 (t, J = 7.6 Hz, 3H, Ar–H). ^{13}C{^{1}H}-NMR (125 MHz, $CDCl_3$): δ = 152.3 (C_q), 149.3 (CH), 143.7 (C_q), 140.3 (C_q), 138.6 (C_q), 137.9 (CH), 130.1 (C_q), 128.9 (C_q), 128.8 (2C, CH), 128.0 (2C, CH), 123.0 (CH), 122.2 (CH), 121.7 (CH), 121.4 (CH), 120.6 (CH), 111.6 (CH), 105.3 (CH), 28.7 (CH_2), 15.5 (CH_3). HRMS (ESI): m/z Calcd for $C_{21}H_{18}N_2$ + H^+ $[M + H]^+$ 299.1543; Found 299.1548. The ^{1}H and ^{13}C{^{1}H} spectra are consistent with those reported in the literature.[34]

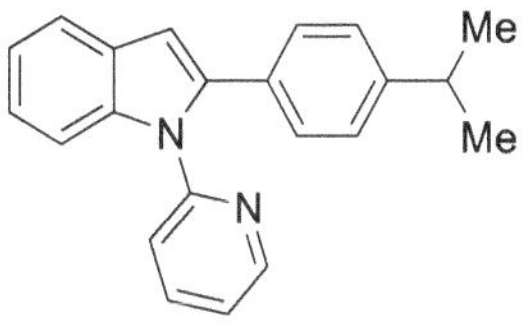

2-(4-Isopropylphenyl)-1-(pyridin-2-yl)-1*H*-indole (3ad): The representative procedure was followed, using substrate **1a** (0.097 g, 0.50 mmol) and 1-iodo-4-isopropylbenzene (**2d**; 0.246 g, 1.00 mmol), and the reaction mixture was stirred at 120 °C for 16 h. Purification by column chromatography on silica gel (petroleum ether/EtOAc: 30/1) yielded **3ad** (0.125 g, 80%) as colorless oil. ^{1}H-NMR (400 MHz, $CDCl_3$): δ = 8.66 (d, J = 3.1 Hz, 1H, Ar–H), 7.69-7.60 (m, 3H, Ar–H), 7.25-7.18 (m, 5H, Ar–H), 7.14 (d, J = 8.4 Hz, 2H, Ar–H), 7.90 (d, J = 7.6 Hz, 1H, Ar–H), 6.79 (s, 1H, Ar–H), 2.89 (sept, J = 6.9 Hz, 1H, CH), 1.24 (d, J = 6.9 Hz, 6H, CH_3). $^{13}C\{^1H\}$-NMR (100 MHz, $CDCl_3$): δ = 152.3 (C_q), 149.3 (CH), 148.4 (C_q), 140.3 (C_q), 138.6 (C_q), 137.9 (CH), 130.2 (C_q), 128.9 (C_q), 128.8 (2C, CH), 126.6 (2C, CH), 123.0 (CH), 122.3 (CH), 121.8 (CH), 121.4 (CH), 120.6 (CH), 111.6 (CH), 105.4 (CH), 33.9 (CH), 24.0 (2C, CH_3). HRMS (ESI): m/z Calcd for $C_{22}H_{20}N_2$ + H^+ $[M + H]^+$ 313.1699; Found 313.1707. The ^{1}H and $^{13}C\{^1H\}$ spectra are consistent with those reported in the literature.[33]

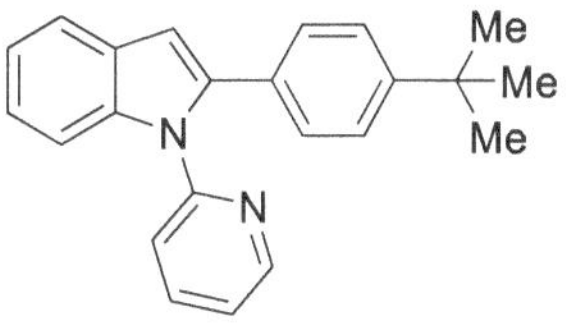

2-(4-(*tert*-Butyl)phenyl)-1-(pyridin-2-yl)-1*H*-indole (3ae): The representative procedure was followed, using substrate **1a** (0.058 g, 0.30 mmol) and 1-(*tert*-butyl)-4-iodobenzene (**2e**; 0.156 g, 0.60 mmol), and the reaction mixture was stirred at 120 °C for 16 h. Purification by column chromatography on silica gel (petroleum ether/EtOAc: 50/1) yielded **3ae** (0.080 g, 82%) as colorless oil. ^{1}H-NMR (500 MHz, $CDCl_3$): δ = 8.62 (d, J = 5.0, 1.1 Hz, 1H, Ar–H), 7.64 (t, J = 8.3 Hz, 2H, Ar–H), 7.57 (td, J = 7.6, 1.5 Hz, 1H, Ar–H), 7.26 (d, J = 8.0 Hz, 2H, Ar–H), 7.20-7.14 (m, 5H, Ar–H), 6.87 (d, J = 8.0 Hz, 1H, Ar–H), 6.75 (s, 1H, Ar–H), 1.27 (s, 9H, CH_3). $^{13}C\{^1H\}$-NMR (125 MHz, $CDCl_3$): δ = 152.3 (C_q), 150.6 (C_q), 149.3 (CH), 140.2 (C_q), 138.6 (C_q), 137.9 (CH), 129.8 (C_q), 128.9 (C_q), 128.5 (2C, CH), 125.4 (2C, CH), 123.0 (CH), 122.3 (CH), 121.8 (CH), 121.4 (CH), 120.6 (CH), 111.6 (CH), 105.4 (CH), 34.7 (C_q), 31.4 (3C, CH_3). HRMS (ESI): m/z Calcd for $C_{23}H_{22}N_2$ + H^+ $[M + H]^+$ 327.1856; Found 327.1866. The ^{1}H and $^{13}C\{^1H\}$ spectra are consistent with those reported in the literature.[64]

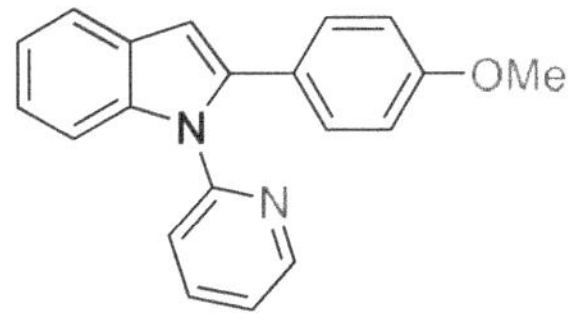

2-(4-Methoxyphenyl)-1-(pyridin-2-yl)-1*H*-indole (3af): The representative procedure was followed, using substrate **1a** (0.097 g, 0.50 mmol) and 4-iodoanisole (**2f**; 0.234 g, 1.00 mmol), and the reaction mixture was stirred at 120 °C for 16 h. Purification by column chromatography on silica gel (petroleum ether/EtOAc: 20/1) yielded **3af** (0.127 g, 85%) as light yellow liquid. ^{1}H-NMR (400 MHz, $CDCl_3$): δ = 8.67 (d, J = 4.6 Hz, 1H, Ar–H), 7.70-7.61 (m, 3H, Ar–H), 7.24-7.21 (m, 5H, Ar–H), 6.90 (d, J = 7.6 Hz, 1H, Ar–H), 6.83 (d, J = 8.4 Hz, 2H, Ar–H), 6.76 (s, 1H, Ar–H), 3.80 (s, 3H, CH_3). $^{13}C\{^{1}H\}$-NMR (100 MHz, $CDCl_3$): δ = 159.2 (C_q), 152.3 (C_q), 149.3 (CH), 140 (C_q), 138.5 (C_q), 137.9 (CH), 130.1 (2C, CH), 128.9 (C_q), 125.3 (C_q), 122.8 (CH), 122.2 (CH), 121.7 (CH), 121.4 (CH), 120.5 (CH), 113.9 (2C, CH), 111.6 (CH), 104.8 (CH), 55.4 (CH_3). HRMS (ESI): *m/z* Calcd for $C_{20}H_{16}N_2O + H^+$ $[M + H]^+$ 301.1335; Found 301.1341. The ^{1}H and $^{13}C\{^{1}H\}$ spectra are consistent with those reported in the literature.[34]

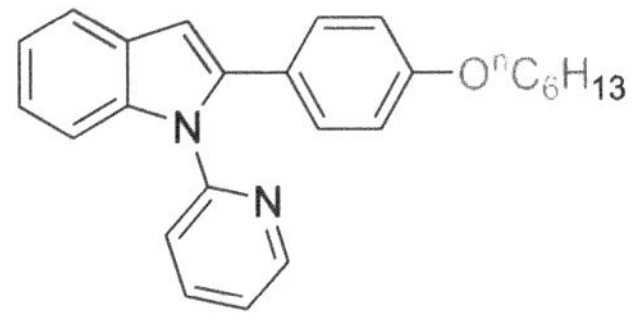

2-(4-(Hexyloxy)phenyl)-1-(pyridin-2-yl)-1*H*-indole (3ag): The representative procedure was followed, using substrate **1a** (0.058 g, 0.30 mmol) and 1-(hexyloxy)-4-iodobenzene (**2g**; 0.183 g, 0.60 mmol), and the reaction mixture was stirred at 120 °C for 16 h. Purification by column chromatography on silica gel (petroleum ether/EtOAc: 50/1) yielded **3ag** (0.092 g, 83%) as yellow liquid. ^{1}H-NMR (400 MHz, $CDCl_3$): δ = 8.62 (d, J = 4.9 Hz, 1H, Ar–H), 7.68-7.62 (m, 2H, Ar–H), 7.56 (vt, J = 7.7, 1.7 Hz, 1H, Ar–H), 7.21-7.15 (m, 5H, Ar–H), 6.85 (d, J = 8.0 Hz, 1H, Ar–H), 6.79 (d, J = 8.8 Hz, 2H, Ar–H), 6.72 (s, 1H, Ar–H), 3.90 (t, J = 6.5 Hz, 2H, CH_2), 1.75 (quint, J = 7.2 Hz, 2H, CH_2), 1.45-1.26 (m, 6H, CH_2), 0.90 (t, J = 6.5 Hz, 3H, CH_3). $^{13}C\{^{1}H\}$-NMR (100 MHz, $CDCl_3$): δ = 158.8 (C_q), 152.3 (C_q), 149.2 (CH), 140.1 (C_q), 138.4 (C_q), 137.8 (CH), 130.1 (2C, CH), 128.9 (C_q), 125.0 (C_q),122.8 (CH), 122.2 (CH), 121.6 (CH), 121.4 (CH), 120.4 (CH), 114.5 (2C, CH), 111.6 (CH), 104.7 (CH), 68.1 (CH_2), 31.7 (CH_2), 29.3 (CH_2), 25.8 (CH_2), 22.7 (CH_2), 14.2 (CH_3). HRMS (ESI): *m/z* Calcd for $C_{25}H_{26}N_2O + H^+$ $[M + H]^+$ 371.2118; Found 371.2123.

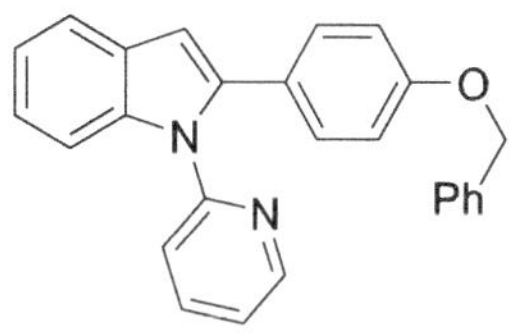

2-(4-(Benzyloxy)phenyl)-1-(pyridin-2-yl)-1*H*-indole (3ah): The representative procedure was followed, using substrate **1a** (0.058 g, 0.30 mmol) and 1-(benzyloxy)-4-iodobenzene (**2h**; 0.186 g, 0.60 mmol), and the reaction mixture was stirred at 120 °C for 16 h. Purification by column chromatography on silica gel (petroleum ether/EtOAc: 20/1) yielded **3ah** (0.070 g, 62%) as yellow solid. ^{1}H-NMR (400 MHz, $CDCl_3$): δ = 8.64 (d, J = 3.8 Hz, 1H, Ar–H), 7.66-7.59 (m, 3H, Ar–H), 7.42-7.30 (m, 5H, Ar–H), 7.23-7.18 (m, 5H, Ar–H), 6.90-6.87 (m, 3H, Ar–H), 6.73 (s, 1H, Ar–H), 5.02 (s, 2H, CH_2). $^{13}C\{^1H\}$-NMR (100 MHz, $CDCl_3$): δ = 158.5 (C_q), 152.3 (C_q), 149.3 (CH), 140.0 (C_q), 138.5 (C_q), 138.0 (CH), 136.9 (C_q), 130.2 (2C, CH), 129.0 (C_q), 128.8 (2 CH), 128.2 (CH), 127.7 (2C, CH), 125.6 (C_q), 122.9 (CH), 122.3 (CH), 121.8 (CH), 121.5 (CH), 120.5 (CH), 114.9 (2C, CH), 111.6 (CH), 104.9 (CH), 70.2 (CH_2). HRMS (ESI): m/z Calcd for $C_{26}H_{20}N_2O + H^+$ $[M + H]^+$ 377.1648; Found 377.1659. The ^{1}H and $^{13}C\{^1H\}$ spectra are consistent with those reported in the literature.[65]

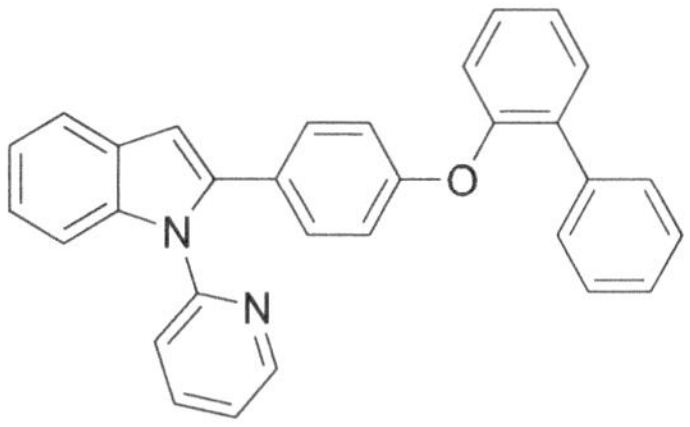

2-(4-([1,1'-Biphenyl]-2-yloxy)phenyl)-1-(pyridin-2-yl)-1*H*-indole (3ai): The representative procedure was followed, using substrate **1a** (0.058 g, 0.30 mmol) and 2-(4-iodophenoxy)-1,1'-biphenyl (**2i**; 0.223 g, 0.60 mmol), and the reaction mixture was stirred at 120 °C for 16 h. Purification by column chromatography on silica gel (petroleum ether/EtOAc: 20/1) yielded **3ai** (0.088 g, 67%) as yellow liquid. ^{1}H-NMR (500 MHz, $CDCl_3$): δ = 8.61 (d, J = 3.1 Hz, 1H, Ar–H), 7.66-7.62 (m, 2H, Ar–H), 7.57 (vt, J = 7.7, 1.8 Hz, 1H, Ar–H), 7.50 (d, J = 7.6 Hz, 2H, Ar–H), 7.44 (dd, J = 1.5, 7.6 Hz, 1H, Ar–H), 7.36-7.28 (m, 4H, Ar–H), 7.23-7.12 (m, 6H, Ar–H), 7.03 (d, J = 8.4 Hz, 1H, Ar–H), 6.83 (d, J= 8.4 Hz, 1H, Ar–H), 6.80 (d, J = 8.4 Hz, 2H, Ar–H), 6.72 (s, 1H, Ar–H). $^{13}C\{^1H\}$-NMR (125 MHz, $CDCl_3$): δ = 157.5 (C_q), 153.3 (C_q), 152.2 (C_q), 149.4 (CH), 139.7 (C_q), 138.6 (C_q), 137.9 (CH), 137.8 (C_q), 134.0 (C_q), 131.6 (CH), 130.2 (2 CH), 129.4 (2C, CH), 128.9 (CH), 128.9 (C_q), 128.3 (2C, CH), 127.4 (CH), 127.2 (C_q), 124.6 (CH), 123.0 (CH), 122.2 (CH), 121.8 (CH), 121.5 (CH), 120.6 (CH), 120.5 (CH), 117.9 (2C, CH), 111.6 (CH), 105.2 (CH).

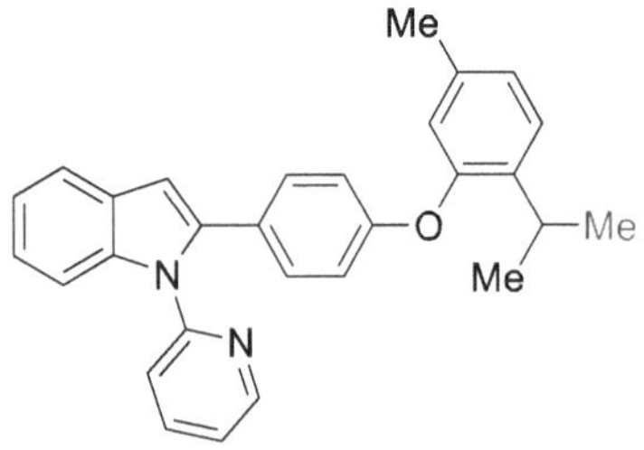

2-(4-(2-Isopropyl-5-methylphenoxy)phenyl)-1-(pyridin-2-yl)-1*H*-indole (3aj): The representative procedure was followed, using substrate **1a** (0.039 g, 0.20 mmol) and 2-(4-iodophenoxy)-1-isopropyl-4-methylbenzene (**2j**; 0.141 g, 0.40 mmol), and the reaction mixture was stirred at 120 °C for 16 h. Purification by column chromatography on silica gel (petroleum ether/EtOAc: 50/1) yielded **3aj** (0.065 g, 78%) as colorless liquid. ^{1}H-NMR (400 MHz, $CDCl_3$): δ = 8.70 (d, J = 3.0 Hz, 1H, Ar–H), 7.73-7.68 (m, 3H, Ar–H), 7.29-7.25 (m, 6H, Ar–H), 7.03-6.98 (m, 2H, Ar–H), 6.88 (d, J = 8.4 Hz, 2H, Ar–H), 6.83 (s, 1H, Ar–H), 6.80 (s, 1H, Ar–H), 3.25 (sept, J = 6.9 Hz, 1H, CH), 2.34 (s, 3H, CH_3), 1.25 (d, J = 6.9 Hz, 6H, CH_3). $^{13}C\{^{1}H\}$-NMR (100 MHz, $CDCl_3$): δ = 158.3 (C_q), 153.0 (C_q), 152.3 (CH), 149.3 (C_q), 139.8 (C_q), 138.5 (C_q), 137.9 (CH), 137.4 (C_q), 137.0 (C_q), 130.2 (2C, CH), 128.9 (C_q), 127.0 (CH), 126.8 (C_q), 125.6 (CH), 123.0 (CH), 122.3 (CH), 121.8 (CH), 121.5 (CH), 121.0 (CH),120.6 (CH), 117.2 (2C, CH), 111.6 (CH), 105.1 (CH), 26.9 (CH), 23.3 (2C, CH_3), 21.1 (CH_3). HRMS (ESI): *m/z* Calcd for $C_{29}H_{26}N_2O + H^+$ $[M + H]^+$ 419.2118; Found 419.2128.

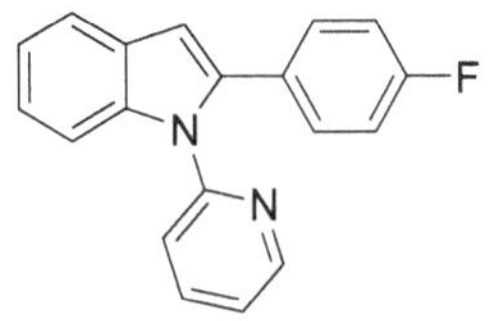

2-(4-Fluorophenyl)-1-(pyridin-2-yl)-1*H*-indole (3ak): The representative procedure was followed, using CuCl (0.005 g, 0.05 mmol, 10 mol %), **1a** (0.097 g, 0.50 mmol), 1-fluoro-4-iodobenzene (**2k**; 0.444 g, 2.0 mmol), LiHMDS (0.22 g, 1.3 mmol) and 1,4-dioxane (1.5 mL). The reaction mixture was stirred at 120 °C for 24 h. Purification by column chromatography on silica gel (petroleum ether/EtOAc: 100/1) yielded **3ak** (0.035 g, 24%) as yellow solid. ^{1}H-NMR (400 MHz, $CDCl_3$): δ = 8.62 (d, J = 3.8 Hz, 1H, Ar–H), 7.67-7.62 (m, 3H, Ar–H), 7.25-7.19 (m, 5H, Ar–H), 6.99-6.94 (m, 2H, Ar–H), 6.91 (d, J = 8.4 Hz, 1H, Ar–H), 6.77 (s, 1H, Ar–H). $^{13}C\{^{1}H\}$-NMR (100 MHz, $CDCl_3$): δ = 162.4 (d, J_{C-F} = 247.3 Hz, C_q), 152.1 (C_q), 149.5 (CH), 139.1 (C_q), 138.6 (C_q), 138.1 (CH), 130.6 (d, J_{C-F} = 7.7 Hz, 2C, CH), 129.0 (d, J_{C-F} = 3.8 Hz, C_q), 128.8 (C_q), 123.3 (CH), 122.1 (CH), 122.0 (CH), 121.6 (CH), 120.8 (CH), 115.6 (d, J_{C-F} = 21.1 Hz, 2C, CH), 111.6 (CH), 105.7 (CH). ^{19}F-NMR (377 MHz, $CDCl_3$): δ = −114.2 (s). HRMS (ESI): *m/z* Calcd for $C_{19}H_{13}FN_2 + H^+$ $[M + H]^+$ 289.1136; Found 289.1135. The ^{1}H and $^{13}C\{^{1}H\}$ spectra are consistent with those reported in the literature.[66]

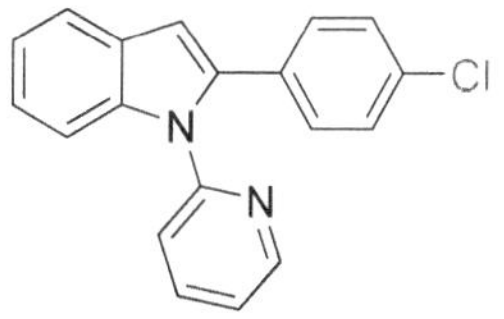

2-(4-Chlorophenyl)-1-(pyridin-2-yl)-1H-indole (3al): The representative procedure was followed, using substrate **1a** (0.097 g, 0.50 mmol) and 1-chloro-4-iodobenzene (**2l**; 0.238 g, 1.0 mmol), and the reaction mixture was stirred at 120 °C for 16 h. Purification by column chromatography on silica gel (petroleum ether/EtOAc: 100/1) yielded **3al** (0.047 g, 31%) as light yellow liquid. ^{1}H-NMR (500 MHz, $CDCl_3$): δ = 8.64 (d, J = 3.8 Hz, 1H, Ar–H), 7.71-7.62 (m, 3H, Ar–H), 7.77-7.17 (m, 7H, Ar–H), 6.93 (d, J = 7.6 Hz, 1H, Ar–H), 6.80 (s, 1H, Ar–H). $^{13}C\{^1H\}$-NMR (125 MHz, $CDCl_3$): δ = 152.0 (C_q), 149.5 (CH), 138.9 (C_q), 138.8 (C_q), 138.2 (CH), 133.6 (C_q), 131.4 (C_q), 130.0 (2C, CH), 128.8 (2C, CH), 123.5 (CH), 122.1 (CH), 122.0 (CH), 121.7 (CH), 120.9 (CH), 111.6 (CH), 106.1 (CH). HRMS (ESI): m/z Calcd for $C_{19}H_{13}ClN_2 + H^+$ $[M + H]^+$ 305.0840; Found 305.0848. The ^{1}H and $^{13}C\{^1H\}$ spectra are consistent with those reported in the literature.[34]

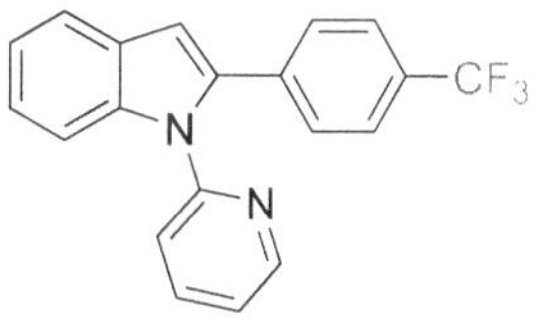

1-(Pyridin-2-yl)-2-(4-(trifluoromethyl)phenyl)-1*H*-indole (3am): The representative procedure was followed, using substrate **1a** (0.097 g, 0.50 mmol) and 1-iodo-4-(trifluoromethyl)benzene (**2m**; 0.272 g, 1.0 mmol), and the reaction mixture was stirred at 120 °C for 16 h. Purification by column chromatography on silica gel (petroleum ether/EtOAc: 50/1) yielded **3am** (0.083 g, 49%) as white solid. ^{1}H-NMR (500 MHz, $CDCl_3$): δ = 8.62 (d, J = 4.8 Hz, 1H, Ar–H), 7.71-7.63 (m, 3H, Ar–H), 7.52 (d, J = 8.4 Hz, 2H, Ar–H), 7.36 (d, J = 8.0 Hz, 2H, Ar–H), 7.26-7.20 (m, 3H, Ar–H), 6.97 (d, J = 8.0 Hz, 1H, Ar–H), 6.88 (s, 1H, Ar–H). $^{13}C\{^1H\}$-NMR (125 MHz, $CDCl_3$): δ = 151.9 (C_q), 149.7 (CH), 139.1 (C_q), 138.6 (C_q), 138.3 (CH), 136.5 (C_q), 129.3 (q, J_{C-F} = 32.4 Hz, C_q), 128.9 (2C, CH), 128.7 (C_q), 125.5 (q, J_{C-F} = 3.8 Hz, 2C, CH), 124.3 (q, J_{C-F} = 271.8 Hz, CF_3), 123.9 (CH), 122.2 (CH), 122.1 (CH), 121.8 (CH), 121.1 (CH), 111.6 (CH), 107.1 (CH). ^{19}F-NMR (377 MHz, $CDCl_3$): δ = −62.5 (s). HRMS (ESI): m/z Calcd for $C_{20}H_{13}F_3N_2 + H^+$ $[M + H]^+$ 339.1104; Found 339.1111. The ^{1}H and $^{13}C\{^1H\}$ spectra are consistent with those reported in the literature.[67]

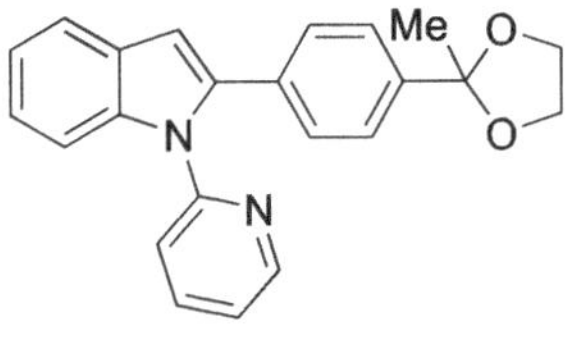

2-(4-(2-Methyl-1,3-dioxolan-2-yl)phenyl)-1-(pyridin-2-yl)-1*H*-indole (3an): The representative procedure was followed, using substrate **1a** (0.58 g, 0.30 mmol) and 2-(4-iodophenyl)-2-methyl-1,3-dioxolane (**2n**; 0.174 g, 0.60 mmol), and the reaction mixture was stirred at 120 °C for 16 h. Purification by column chromatography on silica gel (petroleum ether/EtOAc: 10/1) yielded **3an** (0.090 g, 84%) as yellow liquid. ^{1}H-NMR (400 MHz, $CDCl_3$): δ = 8.64-8.62 (m, 1H, Ar–H), 7.67-7.60 (m, 3H, Ar–H), 7.38 (d, J = 8.4 Hz, 2H, Ar–H), 7.25-7.17 (m, 5H, Ar–H), 6.92 (d, J = 7.6 Hz, 1H, Ar–H), 6.80 (s, 1H, Ar–H), 4.03-4.00 (m, 2H, CH_2), 3.79-3.75 (m, 2H, CH_2), 1.64 (s, 3H, CH_3). $^{13}C\{^1H\}$-NMR (100 MHz, $CDCl_3$): δ = 152.2 (C_q), 149.4 (CH), 142.6 (C_q), 139.8 (C_q), 138.8 (C_q), 138.0 (CH), 132.3 (C_q), 128.8 (C_q), 128.6 (2C, CH), 125.4 (2C, CH), 123.2 (CH), 122.2 (CH), 121.9 (CH), 121.5 (CH), 120.7 (CH), 111.6 (CH), 108.8 (C_q), 105.9 (CH), 64.7 (2C, CH_2), 27.6 (CH_3). HRMS (ESI): *m/z* Calcd for $C_{23}H_{20}N_2O_2 + H^+$ $[M + H]^+$ 357.1598; Found 357.1609.

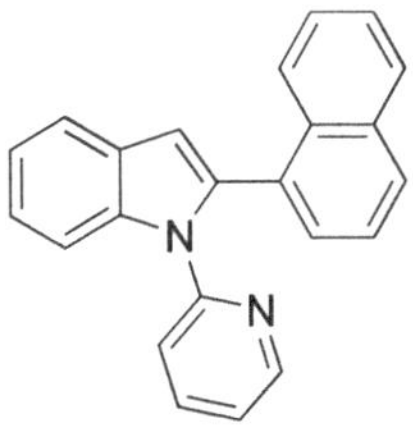

2-(Naphthalen-1-yl)-1-(pyridin-2-yl)-1*H*-indole (3ao): The representative procedure was followed, using substrate **1a** (0.097 g, 0.50 mmol) and 1-iodonaphthalene (**2o**; 0.254 g, 1.00 mmol), and the reaction mixture was stirred at 120 °C for 16 h. Purification by column chromatography on silica gel (petroleum ether/EtOAc: 50/1) yielded **3ao** (0.121 g, 76%) as yellow solid. ^{1}H-NMR (400 MHz, $CDCl_3$): δ = 8.49 (d, J = 4.3 Hz, 1H, Ar–H), 8.01 (d, J = 7.9 Hz, 1H, Ar–H), 7.88 (d, J = 7.9 Hz, 1H Ar–H), 7.79 (d, J = 7.9 Hz, 2H, Ar–H), 7.72 (d, J = 7.3 Hz, 1H, Ar–H), 7.44-7.29 (m, 5H, Ar–H), 7.25-7.20 (m, 2H, Ar–H), 6.94 (t, J = 6.2 Hz, 1H, Ar–H), 6.85 (s, 1H, Ar–H), 6.63 (d, J = 7.9 Hz, 1H, Ar–H). $^{13}C\{^1H\}$-NMR (100 MHz, $CDCl_3$): δ= 152.0 (C_q), 148.9 (CH), 137.8 (C_q), 137.7 (C_q), 137.6 (CH), 133.6 (C_q), 132.3 (C_q), 130.8 (C_q), 129.1 (CH), 128.8 (C_q), 128.7 (CH), 128.2 (CH), 126.6 (CH), 126.1 (CH), 126.1 (CH), 125.2 (CH), 123.2 (CH), 121.5 (CH), 121.2 (CH), 120.7 (2C, CH), 112.2 (CH), 107.7 (CH). HRMS (ESI): *m/z* Calcd for $C_{23}H_{16}N_2 + H^+$ $[M + H]^+$ 321.1386; Found 321.1395. The ^{1}H and $^{13}C\{^1H\}$ spectra are consistent with those reported in the literature.[34]

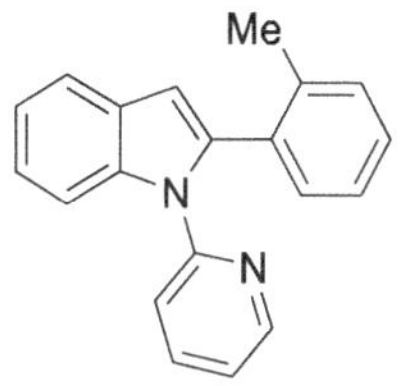

1-(Pyridin-2-yl)-2-(o-tolyl)-1*H*-indole (3ap): The representative procedure was followed, using substrate **1a** (0.097 g, 0.50 mmol) and 1-iodo-2-methylbenzene (**2p**; 0.220 g, 1.0 mmol), and the reaction mixture was stirred at 120 °C for 16 h. Purification by column chromatography on silica gel (petroleum ether/EtOAc: 100/1) yielded **3ap** (0.129 g, 91%) as light yellow solid. ^{1}H-NMR (500 MHz, $CDCl_3$): δ = 8.61-8.60 (m, 1H, Ar–H), 7.89 (d, J = 8.4 Hz, 1H, Ar–H), 7.71 (d, J = 6.9 Hz, 1H, Ar–H), 7.51 (td, J = 1.5 Hz, 7.6 Hz, 1H, Ar–H), 7.34 (d, J = 8.4 Hz, 1H, Ar–H), 7.30-7.24 (m, 3H, Ar–H), 7.23-7.13 (m, 3H, Ar–H), 6.73 (d, J = 8.4 Hz, 1H, Ar–H), 6.69 (s, 1H, Ar–H), 2.09 (s, 3H, CH_3). $^{13}C\{^1H\}$-NMR (125 MHz, $CDCl_3$): δ = 152.1 (C_q), 149.0 (CH), 139.2 (C_q), 137.7 (CH), 137.6 (C_q), 137.3 (C_q), 132.9 (C_q), 131.2 (CH), 130.4 (CH), 128.8 (C_q), 128.5 (CH), 125.8 (CH), 123.0 (CH), 121.4 (CH), 121.2 (CH), 120.7 (CH), 120.6 (CH), 112.1 (CH), 106.4 (CH), 20.3 (CH_3). HRMS (ESI): *m/z* Calcd for $C_{20}H_{16}N_2$ + H^+ $[M + H]^+$ 285.1386; Found 285.1392. The ^{1}H and $^{13}C\{^1H\}$ spectra are consistent with those reported in the literature.[34]

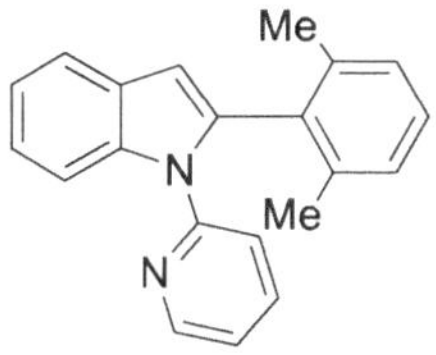

2-(2,6-Dimethylphenyl)-1-(pyridin-2-yl)-1*H*-indole (3aq): The representative procedure was followed, using substrate **1a** (0.097 g, 0.50 mmol) and 2-iodo-1,3-dimethylbenzene (**2q**; 0.232 g, 1.00 mmol), and the reaction mixture was stirred for 16 h at 120 °C. Purification by column chromatography on silica gel (petroleum ether/EtOAc: 50/1) yielded **3aq** (0.136 g, 91%) as light yellow liquid. ^{1}H-NMR (400 MHz, $CDCl_3$): δ = 8.54 (d, J = 4.6 Hz, 1H, Ar–H), 7.88 (d, J = 7.6 Hz, 1H, Ar–H), 7.69 (d, J = 7.6 Hz, 1H, Ar–H), 7.49 (td, J = 1.5 Hz, 1H, Ar–H), 7.29-7.21 (m, 2H, Ar–H), 7.17 (t, J = 7.6 Hz, 1H, Ar–H), 7.12-7.09 (m, 1H, Ar–H), 7.03 (d, J = 7.6 Hz, 2H, Ar–H), 6.71 (d, J = 8.4 Hz, 1H, Ar–H), 6.57 (s, 1H, Ar–H), 2.09 (s, 6H, CH_3). $^{13}C\{^1H\}$-NMR (100 MHz, $CDCl_3$): δ = 151.8 (C_q), 149.0 (CH), 138.6 (2C, C_q), 137.7 (CH), 137.5 (C_q), 136.8 (C_q), 132.7 (C_q), 129.0 (C_q), 128.6 (CH), 127.4 (2C, CH), 122.7 (CH), 121.3 (CH), 121.2 (CH), 120.5 (CH), 119.5 (CH), 112.4 (CH), 105.6 (CH), 20.8 (2C, CH_3). HRMS (ESI): *m/z* Calcd for $C_{21}H_{18}N_2$ + H^+ $[M + H]^+$ 299.1543; Found 299.1551. The ^{1}H and $^{13}C\{^1H\}$ spectra are consistent with those reported in the literature.[34]

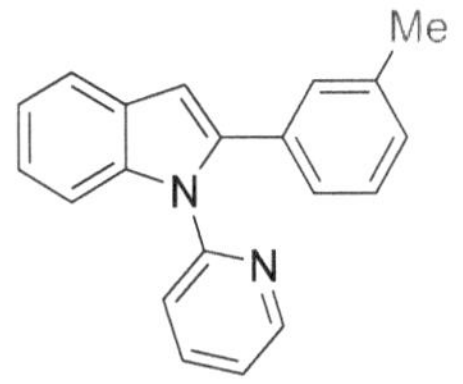

1-(Pyridin-2-yl)-2-(*m*-tolyl)-1*H*-indole (3ar): The representative procedure was followed, using substrate **1a** (0.097 g, 0.50 mmol) and 1-iodo-3-methylbenzene (**2r**; 0.220 g, 1.0 mmol), and the reaction mixture was stirred at 120 °C for 16 h. Purification by column chromatography on silica gel (petroleum ether/EtOAc: 100/1) yielded **3ar** (0.132 g, 93%) as light yellow solid. ^{1}H-NMR (500 MHz, $CDCl_3$): δ = 8.65 (dd, J = 4.6, 1.5 Hz, 1H, Ar–H), 7.71-7.67 (m, 2H, Ar–H), 7.62 (td, J = 7.6, 2.3 Hz, 1H, Ar–H), 7.26-7.53 (m, 5H, Ar–H), 7.08 (d, J = 6.9 Hz, 1H, Ar–H), 7.01 (d, J = 7.6 Hz, 1H, Ar–H), 6.91 (d, J = 7.6 Hz, 1H, Ar–H), 6.81 (s, 1H, Ar–H), 2.30 (s, 3H, CH_3). $^{13}C\{^1H\}$-NMR (125 MHz, $CDCl_3$): δ = 152.3 (C_q), 149.3 (CH), 140.3 (C_q), 138.6 (C_q), 138.2 (C_q), 137.9 (CH), 132.7 (C_q), 129.6 (CH), 128.9 (C_q), 128.4 (CH), 128.3 (CH), 126.1 (CH), 123.1 (CH), 122.2 (CH), 121.7 (CH), 121.5 (CH), 120.7 (CH), 111.7 (CH), 105.6 (CH), 21.6 (CH_3). HRMS (ESI): *m/z* Calcd for $C_{20}H_{16}N_2 + H^+$ $[M + H]^+$ 285.1386; Found 285.1390. The ^{1}H and $^{13}C\{^1H\}$ spectra are consistent with those reported in the literature.[34]

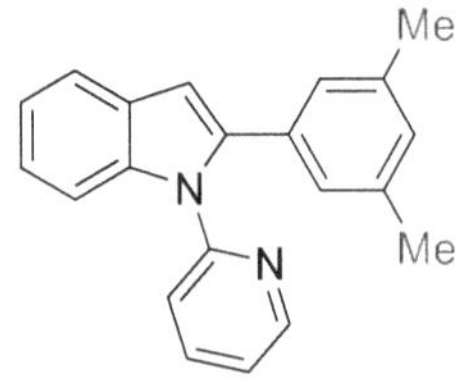

2-(3,5-Dimethylphenyl)-1-(pyridin-2-yl)-1*H*-indole (3as): The representative procedure was followed, using substrate **1a** (0.97 g, 0.50 mmol) and 1-iodo-3,5-dimethylbenzene (**2s**; 0.232 g, 1.0 mmol), and the reaction mixture was stirred at 120 °C for 16 h. Purification by column chromatography on silica gel (petroleum ether/EtOAc: 50/1) yielded **3as** (0.139 g, 93%) as colorless liquid. ^{1}H-NMR (500 MHz, $CDCl_3$): δ = 8.76 (d, J = 3.8 Hz, 1H, Ar–H), 7.83-7.77 (m, 2H, Ar–H), 7.73 (td, J = 7.6, 2.3 Hz, 1H, Ar–H), 7.36-7.30 (m, 3H, Ar–H), 7.03-7.02 (m, 4H, Ar–H), 6.91 (s, 1H, Ar–H), 2.35 (s, 6H, CH_3). $^{13}C\{^1H\}$-NMR (125 MHz, $CDCl_3$): δ = 152.4 (C_q), 149.2 (CH), 140.4 (C_q), 138.6 (C_q), 137.9 (2C, C_q), 137.8 (CH), 132.6 (C_q), 129.3 (CH), 128.9 (C_q), 126.8 (2C, CH), 123.0 (CH), 122.2 (CH), 121.6 (CH), 121.4 (CH), 120.6 (CH), 111.7 (CH), 105.5 (CH), 21.4 (CH_3). HRMS (ESI): *m/z* Calcd for $C_{21}H_{18}N_2 + H^+$ $[M + H]^+$ 299.1543; Found 299.1549. The ^{1}H and $^{13}C\{^1H\}$ spectra are consistent with those reported in the literature.[66]

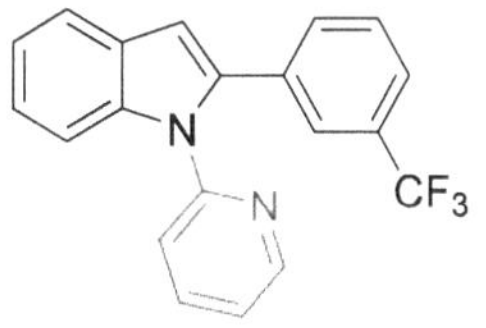

1-(Pyridin-2-yl)-2-(3-(trifluoromethyl)phenyl)-1*H*-indole (3at): The representative procedure was followed, using substrate **1a** (0.097 g, 0.50 mmol) and 1-iodo-3-(trifluoromethyl)benzene (**2t**; 0.272 g, 1.0 mmol), and the reaction mixture was stirred at 120 °C for 16 h. Purification by column chromatography on silica gel (petroleum ether/EtOAc: 50/1) yielded **3at** (0.090 g, 53%) as yellow oil. ^{1}H-NMR (500 MHz, $CDCl_3$): δ = 8.60 (d, J = 3.8 Hz, 1H, Ar–H), 7.69-7.66 (m, 2H, Ar–H), 7.62 (d, J = 8.4 Hz, 1H, Ar–H), 7.52 (s, 1H, Ar–H), 7.48 (d, J = 7.6 Hz, 1H, Ar–H), 7.40-7.33 (m, 2H, Ar–H), 7.25-7.18 (m, 3H, Ar–H), 6.98 (d, J = 8.4 Hz, 1H, Ar–H), 6.88 (s, 1H, Ar–H). ^{13}C{^{1}H}-NMR (125 MHz, $CDCl_3$): δ = 151.7 (C_q), 149.6 (CH), 138.9 (C_q), 138.5 (C_q), 138.3 (CH), 133.6 (C_q), 131.9 (CH), 130.9 (q, J_{C-F} = 32.6 Hz, C_q), 128.9 (CH), 128.6 (C_q), 125.4 (q, J_{C-F} = 3.8 Hz, CH), 124.0 (q, J_{C-F} = 272.2 Hz, CF_3), 124.1 (q, J_{C-F} = 3.8 Hz, CH), 123.7 (CH), 122.2 (CH), 122.1 (CH), 121.8 (CH), 121.1 (CH), 111.5 (CH), 106.6 (CH). ^{19}F-NMR (377 MHz, $CDCl_3$): δ = −62.9 (s). HRMS (ESI): m/z Calcd for $C_{20}H_{13}F_3N_2 + H^+$ $[M + H]^+$ 339.1104; Found 339.1112. The ^{1}H and ^{13}C{^{1}H} spectra are consistent with those reported in the literature.[68]

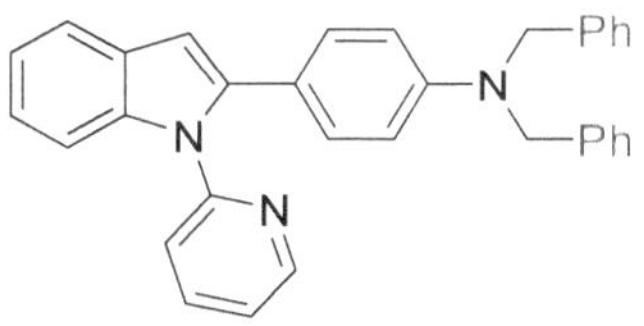

***N,N*-Dibenzyl-4-(1-(pyridin-2-yl)-1*H*-indol-2-yl)aniline (3au):** The representative procedure was followed, using substrate **1a** (0.058 g, 0.30 mmol) and *N,N*-dibenzyl-4-iodoaniline (**2u**; 0.240 g, 0.6 mmol), and the reaction mixture was stirred at 120 °C for 16 h. Purification by column chromatography on silica gel (petroleum ether/EtOAc: 20/1) yielded **3au** (0.123 g, 88%) as colorless solid. ^{1}H-NMR (400 MHz, $CDCl_3$): δ = 8.63 (dd, J = 5.3, 1.5 Hz, 1H, Ar–H), 7.64-7.59 (m, 3H, Ar–H), 7.33-7.30 (m, 4H, Ar–H), 7.25-7.13 (m, 9H, Ar–H), 7.06 (d, J = 8.4 Hz, 2H, Ar–H), 6.92 (d, J = 8.4 Hz, 1H, Ar–H), 6.66 (s, 1H, Ar–H), 6.64-6.60 (m, 2H, Ar–H), 4.62 (s, 4H, CH_2). ^{13}C{^{1}H}-NMR (100 MHz, $CDCl_3$): δ = 152.5 (C_q), 149.2 (CH), 148.7 (C_q), 140.7 (C_q), 138.4 (C_q), 138.3 (2C, C_q), 137.8 (CH), 129.9 (2C, CH), 129.1 (C_q), 128.8 (4C, CH), 127.1 (2C, CH), 126.8 (4C, CH), 122.4 (2C, CH), 121.6 (CH), 121.3 (CH), 120.9 (C_q), 120.2 (CH), 112.3 (2C, CH), 111.4 (CH), 103.9 (CH), 54.2 (2C, CH_2). HRMS (ESI): m/z Calcd for $C_{33}H_{27}N_3 + H^+$ $[M + H]^+$ 466.2278; Found 466.2292.

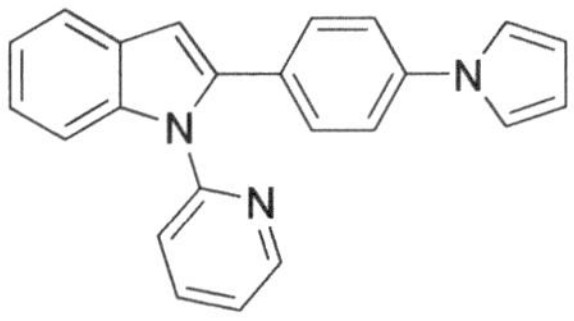

2-(4-(1*H*-Pyrrol-1-yl)phenyl)-1-(pyridin-2-yl)-1*H*-indole (3av): The representative procedure was followed, using substrate **1a** (0.097 g, 0.50 mmol) and 1-(4-iodophenyl)-1*H*-pyrrole (**2v**; 0.269 g, 1.0 mmol), and the reaction mixture was stirred at 120 °C for 16 h. Purification by column chromatography on silica gel (petroleum ether/EtOAc: 20/1) yielded **3av** (0.110 g, 66%) as light yellow liquid. ^{1}H-NMR (500 MHz, $CDCl_3$): δ = 8.64 (d, J = 3.8 Hz, 1H, Ar–H), 7.68-7.62 (m, 3H, Ar–H), 7.31-7.26 (m, 4H, Ar–H), 7.25-7.18 (m, 3H, Ar–H), 7.06 (vt, J = 2.3 Hz, 2H, Ar–H), 6.95 (d, J = 7.63 Hz, 1H, Ar–H), 6.82 (s, 1H, Ar–H), 6.33 (vt, J = 2.3 Hz, 2H, Ar–H). ^{13}C{^{1}H}-NMR (125 MHz, $CDCl_3$): δ = 152.1 (C_q), 149.5 (CH), 139.9 (C_q), 139.2 (C_q), 138.7 (C_q), 138.1 (CH), 130.1 (C_q), 129.9 (2C, CH), 128.8 (C_q), 123.3 (CH), 122.2 (CH), 122.0 (CH), 121.6 (CH), 120.8 (CH), 120.2 (2C, CH), 119.2 (2C, CH), 111.6 (CH), 110.8 (2C, CH), 105.8 (CH). HRMS (ESI): *m/z* Calcd for $C_{23}H_{17}N_3 + H^+$ $[M + H]^+$ 336.1495; Found 336.1504. The ^{1}H and ^{13}C{^{1}H} spectra are consistent with those reported in the literature.[34]

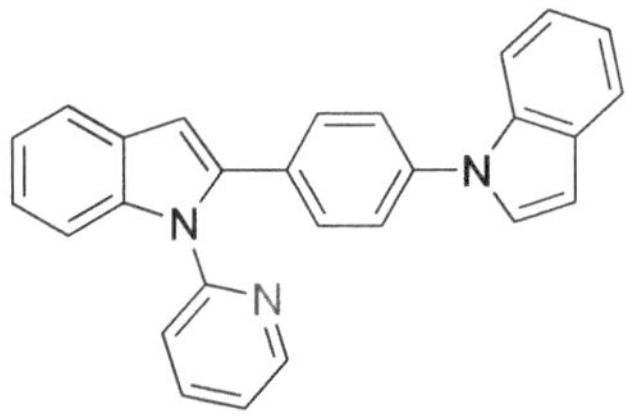

2-(4-(1*H*-Indol-1-yl)phenyl)-1-(pyridin-2-yl)-1*H*-indole (3aw): The representative procedure was followed, using substrate **1a** (0.039 g, 0.20 mmol) and 1-(4-iodophenyl)-1*H*-indole (**2w**; 0.128 g, 0.40 mmol), and the reaction mixture was stirred at 120 °C for 16 h. Purification by column chromatography on silica gel (petroleum ether/EtOAc: 20/1) yielded **3aw** (0.067 g, 87%) as light yellow solid. ^{1}H-NMR (400 MHz, $CDCl_3$): δ = 8.66 (dd, J = 5.0, 1.2 Hz, 1H, Ar–H), 7.72-7.67 (m, 4H, Ar–H), 7.58 (d, J = 8.4 Hz, 1H, Ar–H), 7.41 (br s, 4H, Ar–H), 7.33 (d, J = 3.3 Hz, 1H, Ar–H), 7.28-7.16 (m, 5H, Ar–H), 7.04 (d, J = 8.0 Hz, 1H, Ar–H), 6.89 (s, 1H, Ar–H), 6.69 (d, J = 3.3 Hz, 1H, Ar–H). ^{13}C{^{1}H}-NMR (100 MHz, $CDCl_3$): δ = 152.1 (C_q), 149.6 (CH), 139.2 (C_q), 139.1 (C_q), 138.8 (C_q), 138.2 (CH), 135.8 (C_q), 130.8 (C_q), 129.9 (2C, CH), 129.6 (C_q), 128.8 (C_q), 127.8 (CH), 124.0 (2C, CH), 123.4 (CH), 122.6 (CH), 122.2 (CH), 122.1 (CH), 121.7 (CH), 121.4 (CH), 120.9 (CH), 120.7 (CH), 111.6 (CH), 110.7 (CH), 106.1 (CH), 104.1 (CH). HRMS (ESI): *m/z* Calcd for $C_{27}H_{19}N_3 + H^+$ $[M + H]^+$ 386.1652; Found 386.1661. The ^{1}H and ^{13}C{^{1}H} spectra are consistent with those reported in the literature.[34]

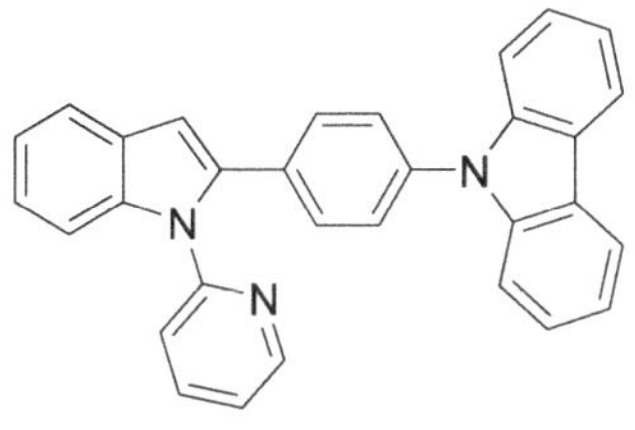

9-(4-(1-(Pyridin-2-yl)-1*H*-indol-2-yl)phenyl)-9*H*-carbazole (3ax): The representative procedure was followed, using substrate **1a** (0.97 g, 0.50 mmol) and 9-(4-iodophenyl)-9*H*-carbazole (**2x**; 0.369 g, 1.0 mmol), and the reaction mixture was stirred at 120 °C for 16 h. Purification by column chromatography on silica gel (petroleum ether/EtOAc: 30/1) yielded **3ax** (0.181 g, 83%) as light yellow solid. ^{1}H-NMR (500 MHz, $CDCl_3$): δ = 8.70 (d, J = 5.3 Hz, 1H, Ar–H), 8.16 (d, J = 7.6 Hz, 2H, Ar–H), 7.77-7.71 (m, 3H, Ar–H), 7.50 (s, 4H, Ar–H), 7.46-7.41 (m, 4H, Ar–H), 7.33-7.24 (m, 5H, Ar–H), 7.10 (d, J = 8.4 Hz, 1H, Ar–H), 6.95 (s, 1H, Ar–H). ^{13}C{^{1}H}-NMR (125 MHz, $CDCl_3$): δ = 152.1 (C_q), 149.6 (CH), 140.7 (2C, C_q), 139.3 (C_q), 138.9 (C_q), 138.2 (CH), 136.9 (C_q), 131.8 (C_q), 130.1 (2C, CH), 128.9 (C_q), 126.9 (2C, CH), 126.1 (2C, CH), 123.6 (2C, C_q), 123.5 (CH), 122.2 (CH), 122.1 (CH), 121.7 (CH), 120.9 (CH), 120.5 (2C, CH), 120.3 (2C, CH), 111.6 (CH), 109.9 (2C, CH), 106.2 (CH). HRMS (ESI): m/z Calcd for $C_{31}H_{21}N_3 + H^+$ $[M + H]^+$ 436.1808; Found 436.1821. The ^{1}H and ^{13}C{^{1}H} spectra are consistent with those reported in the literature.[34]

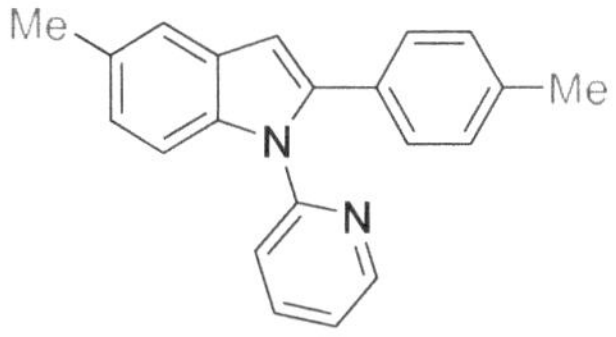

5-Methyl-1-(pyridin-2-yl)-2-(*p*-tolyl)-1*H*-indole (3ba): The representative procedure was followed, using substrate **1b** (0.104 g, 0.50 mmol) and 1-iodo-4-methylbenzene (**2a**; 0.220 g, 1.0 mmol), and the reaction mixture was stirred at 120 °C for 16 h. Purification by column chromatography on silica gel (petroleum ether/EtOAc: 50/1) yielded **3ba** (0.115 g, 77%) as light yellow liquid. ^{1}H-NMR (500 MHz, $CDCl_3$): δ = 8.60 (dd, J = 4.6, 1.5 Hz, 1H, Ar–H), 7.57-7.53 (m, 2H, Ar–H), 7.42 (s, 1H, Ar–H), 7.17-7.13 (m, 3H, Ar–H), 7.06-7.01 (m, 3H, Ar–H), 6.84 (d, J = 8.4 Hz, 1H, Ar–H), 6.68 (s, 1H, Ar–H), 2.44 (s, 3H, CH_3), 2.30 (s, 3H, CH_3). ^{13}C{^{1}H}-NMR (125 MHz, $CDCl_3$): δ = 152.5 (C_q), 149.2 (CH), 140.2 (C_q), 137.8 (CH), 137.3 (C_q), 137.0 (C_q), 130.7 (C_q), 130.1 (C_q), 129.1 (C_q), 129.2 (2C, CH), 128.7 (2C, CH), 124.5 (CH), 122.0 (CH), 121.5 (CH), 120.3 (CH), 111.4 (CH), 105.1 (CH), 21.6 (CH_3), 21.4 (CH_3). HRMS (ESI): m/z Calcd for $C_{21}H_{18}N_2 + H^+$ $[M + H]^+$ 299.1543; Found 299.1548. The ^{1}H and ^{13}C{^{1}H} spectra are consistent with those reported in the literature.[34]

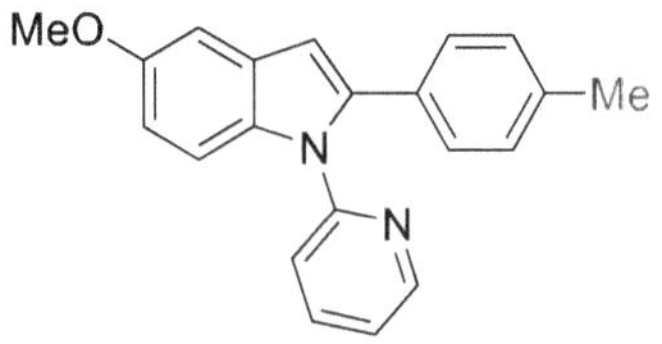

5-Methoxy-1-(pyridin-2-yl)-2-(*p*-tolyl)-1*H*-indole (3ca): The representative procedure was followed, using substrate **1c** (0.112 g, 0.50 mmol) and 1-iodo-4-methylbenzene (**2a**; 0.220 g, 1.0 mmol), and the reaction mixture was stirred at 120 °C for 16 h. Purification by column chromatography on silica gel (petroleum ether/EtOAc: 100/1) yielded **3ca** (0.136 g, 87%) as light yellow liquid. ^{1}H-NMR (400 MHz, $CDCl_3$): δ = 8.64 (dd, J = 4.6, 1.5 Hz, 1H, Ar–H), 7.64-7.57 (m, 2H, Ar–H), 7.21-7.16 (m, 3H, Ar–H), 7.13 (d, J = 2.3 Hz, 1H, Ar–H), 7.09 (d, J = 8.4 Hz, 2H, Ar–H), 6.90-6.84 (m, 2H, Ar–H), 6.71 (s, 1H, Ar–H), 3.88 (s, 3H, CH_3), 2.34 (s, 3H, CH_3). $^{13}C\{^{1}H\}$-NMR (100 MHz, $CDCl_3$): δ = 155.3 (C_q), 152.4 (C_q), 149.2 (CH), 140.7 (C_q), 137.8 (CH), 137.5 (C_q), 133.8 (C_q), 130.0 (C_q), 129.4 (C_q), 129.2 (2C, CH), 128.7 (2C, CH), 122.0 (CH), 121.5 (CH), 112.8 (CH), 112.6 (CH), 105.2 (CH), 102.4 (CH), 56.0 (CH_3), 21.4 (CH_3). HRMS (ESI): *m/z* Calcd for $C_{21}H_{18}N_2O + H^+$ $[M + H]^+$ 315.1492; Found 315.1499. The ^{1}H and $^{13}C\{^{1}H\}$ spectra are consistent with those reported in the literature.[34]

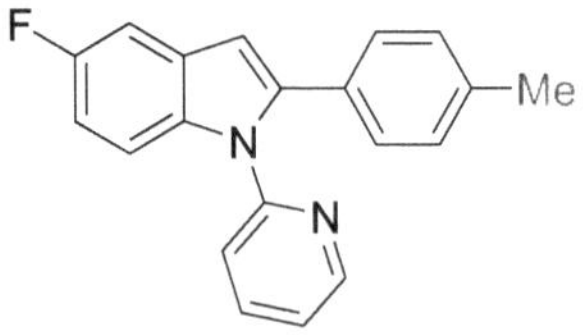

5-Fluoro-1-(pyridin-2-yl)-2-(*p*-tolyl)-1*H*-indole (3da): The representative procedure was followed, using substrate **1d** (0.106 g, 0.50 mmol) and 1-iodo-4-methylbenzene (**2a**; 0.220 g, 1.0 mmol), and the reaction mixture was stirred at 120 °C for 16 h. Purification by column chromatography on silica gel (petroleum ether/EtOAc: 100/1) yielded **3da** (0.094 g, 62%) as light yellow solid. ^{1}H-NMR (400 MHz, $CDCl_3$): δ = 8.65 (dd, J = 4.6, 1.5 Hz, 1H, Ar–H), 7.64-7.60 (m, 2H, Ar–H), 7.28 (dd, J = 9.2, 2.3 Hz, 1H, Ar–H), 7.25-7.22 (m, 1H, Ar–H), 7.15 (d, J = 7.6 Hz, 2H, Ar–H), 7.09 (d, J = 8.4 Hz, 2H, Ar–H), 6.95 (td, J = 9.2, 3.1 Hz, 1H, Ar–H), 6.84 (d, J = 8.4 Hz, 1H, Ar–H), 6.72 (s, 1H, Ar–H), 2.33 (s, 3H, CH_3). $^{13}C\{^{1}H\}$-NMR (100 MHz, $CDCl_3$): δ = 158.9 (d, J_{C-F} = 236 Hz, C_q), 152.0 (C_q), 149.2 (CH), 141.8 (C_q), 138.2 (CH), 137.9 (C_q), 135.1 (C_q), 129.5 (C_q), 129.4 (C_q), 129.3 (2C, CH), 128.8 (2C, CH), 122.2 (CH), 121.9 (CH), 112.6 (d, J_{C-F} = 9.6 Hz, CH), 111.1 (d, J_{C-F} = 25.0 Hz, CH), 105.5 (d, J_{C-F} = 24.0 Hz, CH), 105.2 (d, J_{C-F} = 3.8 Hz, CH), 21.4 (CH_3). ^{19}F-NMR (377 MHz, $CDCl_3$): δ = −123.1 (s). HRMS (ESI): *m/z* Calcd for $C_{20}H_{15}FN_2 + H^+$ $[M + H]^+$ 303.1292; Found 303.1299. The ^{1}H and $^{13}C\{^{1}H\}$ spectra are consistent with those reported in the literature.[34]

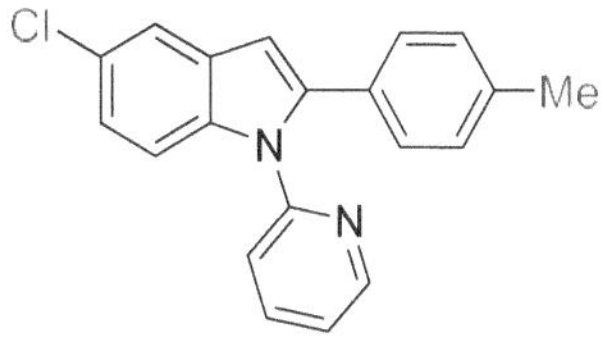

5-Chloro-1-(pyridin-2-yl)-2-(*p*-tolyl)-1*H*-indole (3ea): The representative procedure was followed, using substrate **1e** (0.114 g, 0.50 mmol) and 1-iodo-4-methylbenzene (**2a**; 0.220 g, 1.0 mmol), and the reaction mixture was stirred at 120 °C for 16 h. Purification by column chromatography on silica gel (petroleum ether/EtOAc: 50/1) yielded **3ea** (0.082 g, 51%) as colorless solid. ^{1}H-NMR (400 MHz, $CDCl_3$): δ = 8.63 (dd, J = 4.6, 1.5 Hz, 1H, Ar–H), 7.62-7.57 (m, 3H, Ar–H), 7.22 (dd, J = 7.6, 3.8 Hz, 1H, Ar–H), 7.15-7.12 (m, 3H, Ar–H), 7.07 (d, J = 8.4 Hz, 2H, Ar–H), 6.83 (d, J = 7.6 Hz, 1H, Ar–H), 6.68 (s, 1H, Ar–H), 2.32 (s, 3H, CH_3). $^{13}C\{^1H\}$-NMR (100 MHz, $CDCl_3$): δ = 151.9 (C_q), 149.4 (CH), 141.5 (C_q), 138.1 (CH), 137.9 (C_q), 136.9 (C_q), 129.9 (C_q), 129.4 (C_q), 129.3 (2C, CH), 128.8 (2C, CH), 126.9 (C_q), 123.1 (CH), 122.2 (CH), 122.0 (CH), 119.9 (CH), 112.8(CH), 104.6 (CH), 21.4 (CH_3). HRMS (ESI): *m/z* Calcd for $C_{20}H_{15}ClN_2 + H^+$ $[M + H]^+$ 319.0997; Found 319.1009.

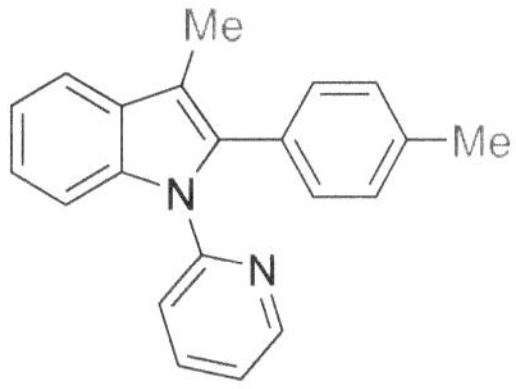

3-Methyl-1-(pyridin-2-yl)-2-(*p*-tolyl)-1*H*-indole (3fa): The representative procedure was followed, using substrate **1f** (0.062 g, 0.30 mmol) and 1-iodo-4-methylbenzene (**2a**; 0.131 g, 0.6 mmol), and the reaction mixture was stirred at 120 °C for 16 h. Purification by column chromatography on silica gel (petroleum ether/EtOAc: 50/1) yielded **3fa** (0.072 g, 80%) as light yellow solid. ^{1}H-NMR (500 MHz, $CDCl_3$): δ = 8.58 (d, J = 5.3 Hz, 1H, Ar–H), 7.75 (dd, J = 6.1, 2.3 Hz, 1H, Ar–H), 7.63-7.61 (m, 1H Ar–H), 7.50 (td, J = 8.4, 2.3 Hz, 1H, Ar–H), 7.25-7.19 (m, 2H, Ar–H), 7.14-7.09 (m, 5H, Ar–H), 7.73 (d, J = 7.6, 1H, Ar–H), 2.39 (s, 3H, CH_3), 2.33 (s, 3H, CH_3). $^{13}C\{^1H\}$-NMR (125 MHz, $CDCl_3$): δ = 152.5 (C_q), 149.0 (CH), 137.6 (CH), 137.3 (C_q), 137.1 (C_q), 135.8 (C_q), 130.3 (2C, CH), 129.9 (C_q), 129.5 (C_q), 129.1 (2C, CH), 132.2 (CH), 121.7 (CH), 121.0 (CH), 120.9 (CH), 118.9 (CH), 112.5 (C_q), 111.6 (CH), 21.4 (CH_3), 9.7 (CH_3). HRMS (ESI): *m/z* Calcd for $C_{21}H_{18}N_2 + H^+$ $[M + H]^+$ 299.1543; Found 299.1550. The ^{1}H and $^{13}C\{^1H\}$ spectra are consistent with those reported in the literature.[34]

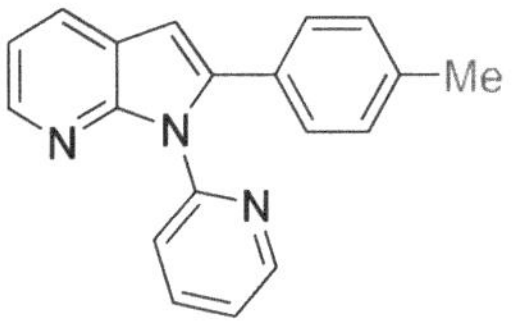

1-(Pyridin-2-yl)-2-(*p*-tolyl)-1*H*-pyrrolo[2,3-b]pyridine (3ga): The representative procedure was followed, using substrate **1g** (0.097 g, 0.50 mmol) and 1-iodo-4-methylbenzene (**2a**; 0.220 g, 1.0 mmol), and the reaction mixture was stirred at 120 °C for 16 h. Purification by column chromatography on silica gel (petroleum ether/EtOAc: 20/1) yielded **3ga** (0.045 g, 32%) as yellow solid. ^{1}H-NMR (500 MHz, $CDCl_3$): δ = 8.55 (d, J = 3.4 Hz, 1H, Ar–H), 8.32 (d, J = 4.2 Hz, 1H, Ar–H), 7.93 (d, J = 7.6 Hz, 1H, Ar–H), 7.81 (td, J = 7.6, 1.9 Hz, 1H, Ar–H), 7.48 (d, J = 7.6 Hz, 1H, Ar–H), 7.27-7.24 (m, 1H, Ar–H), 7.16-7.11 (m, 3H, Ar–H), 7.06 (d, J = 8.0 Hz, 2H, Ar–H), 6.70 (s, 1H, Ar–H), 2.31 (s, 3H, CH_3). ^{13}C{^{1}H}-NMR (125 MHz, $CDCl_3$): δ = 150.8 (C_q), 150.0 (C_q), 149.5 (CH), 143.6 (CH), 141.3 (C_q), 138.1 (CH), 137.8 (C_q), 129.6 (C_q), 129.2 (2C, CH), 128.5 (2C, CH), 128.4 (CH), 123.2 (CH), 122.5 (CH), 121.4 (C_q), 117.5 (CH), 102.1 (CH), 21.4 (CH_3). HRMS (ESI): *m/z* Calcd for $C_{19}H_{15}N_3 + H^+$ $[M + H]^+$ 286.1339; Found 286.1345.

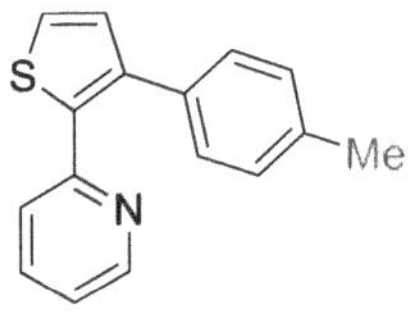

2-(3-(*p*-Tolyl)thiophen-2-yl)pyridine (3ha): The representative procedure was followed, using substrate **1i** (0.048 g, 0.30 mmol) and 1-iodo-4-methylbenzene (**2a**; 0.130 g, 0.48 mmol), and 1,4-dioxane (1.0 mL). The reaction mixture was stirred at 120 °C for 16 h. Purification by column chromatography on silica gel (petroleum ether/EtOAc: 100/1) yielded **3ia** (0.064 g, 85%) as yellow solid. ^{1}H-NMR (400 MHz, $CDCl_3$): δ = 8.58 (d, J = 4.6 Hz, 1H, Ar–H), 7.71-7.65 (m, 2H, Ar–H), 7.57 (d, J = 8.4 Hz, 2H, Ar–H), 7.54 (d, J = 3.8 Hz, 1H, Ar–H), 7.29 (d, J = 3.8 Hz, 1H, Ar–H), 7.21 (d, J = 7.6 Hz, 2H, Ar–H), 7.16-7.14 (m, 1H, Ar–H), 2.38 (s, 3H, CH_3). ^{13}C{^{1}H}-NMR (125 MHz, $CDCl_3$): δ = 152.8 (C_q), 149.8 (CH), 146.6 (C_q), 143.4 (C_q), 138.0 (C_q), 136.8 (CH), 131.7 (C_q), 129.8 (2C, CH), 125.8 (2C, CH), 125.6 (CH), 123.7 (CH), 121.9 (CH), 118.6 (CH), 21.4 (CH_3). HRMS (ESI): *m/z* Calcd for $C_{16}H_{13}NS + H^+$ $[M + H]^+$ 252.0841; Found 252.0841.

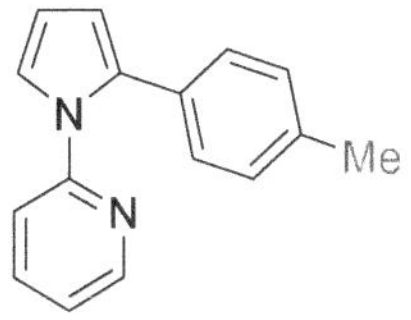

2-(2-(*p*-Tolyl)-1*H*-pyrrol-1-yl)pyridine (3ia): The representative procedure was followed, using substrate **1i** (0.043 g, 0.30 mmol) and 1-iodo-4-methylbenzene (**2a**; 0.131 g, 0.60 mmol), and the reaction mixture was stirred at 120 °C for 16 h. Purification by column chromatography on silica gel (petroleum ether/EtOAc: 100/1) yielded **3ia** (0.030 g, 43%) and **3i'a** (10%) as colorless oil. For **3ai**: ^{1}H-NMR (400 MHz, $CDCl_3$): δ = 8.51 (dd, J = 4.6, 1.5 Hz, 1H, Ar–H), 7.51 (td, J = 7.6, 1.5 Hz, 1H, Ar–H), 7.36 (vt, J = 2.3 Hz, 1H, Ar–H), 7.14 (dd, J = 7.6, 3.8 Hz, 1H, Ar–H), 7.08 (s, 4H, Ar–H), 6.79 (d, J = 8.4 Hz, 1H, Ar–H), 6.41-6.37 (m, 2H, Ar–H), 3.16 (s, 3H, CH_3). $^{13}C\{^1H\}$-NMR (100 MHz, $CDCl_3$): δ = 152.4 (C_q), 149.1 (CH), 137.6 (CH), 136.5 (C_q), 133.4 (C_q), 130.5 (C_q), 129.1 (2C, CH), 128.5 (2C, CH), 123.4 (CH), 121.3 (CH), 119.4 (CH), 112.3 (CH), 110.0 (CH), 21.3 (CH_3). HRMS (ESI): *m/z* Calcd for $C_{16}H_{14}N_2 + H^+$ $[M + H]^+$ 235.1235; Found 235.1236. The ^{1}H and $^{13}C\{^1H\}$ spectra are consistent with those reported in the literature.[34]

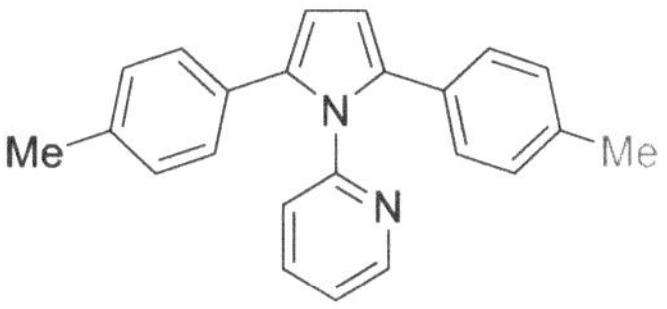

2-(2,5-Di-*p*-tolyl-1*H*-pyrrol-1-yl)pyridine (3i'a): The representative procedure was followed, using substrate **1i** (0.072 g, 0.50 mmol), 1-iodo-4-methylbenzene (**2a**; 0.440 g, 2.00 mmol) and LiHMDS (0.22 g, 1.3 mmol), and the reaction mixture was stirred at 120 °C for 24 h. Purification by column chromatography on silica gel (petroleum ether/EtOAc: 100/1) yielded **3i'a** (0.098 g, 60%) as yellow solid. ^{1}H-NMR (500 MHz, $CDCl_3$): δ = 8.45 (d, J = 3.4 Hz, 1H, Ar–H), 7.55 (td, J = 7.6, 1.5 Hz, 1H, Ar–H), 7.17 (dd, J = 6.9, 1.9 Hz, 1H, Ar–H), 7.98-7.96 (m, 9H, Ar–H), 6.43 (s, 2H, Ar–H), 2.27 (s, 6H, CH_3). $^{13}C\{^1H\}$-NMR (125 MHz, $CDCl_3$): δ = 152.7 (C_q), 149.1 (CH), 137.8 (CH), 136.1 (2C, C_q), 136.0 (2C, C_q), 130.6 (2C, C_q), 128.9 (4C, CH), 128.5 (4C, CH), 124.0 (CH), 122.7 (CH), 110.1 (2C, CH), 21.3 (2C, CH_3). HRMS (ESI): *m/z* Calcd for $C_{23}H_{20}N_2 + H^+$ $[M + H]^+$ 325.1699; Found 325.1701. The ^{1}H and $^{13}C\{^1H\}$ spectra are consistent with those reported in the literature.[34]

5.4.2 Procedure for Removal of Directing Group

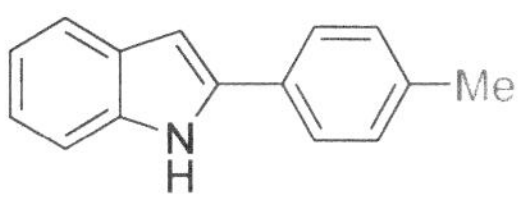

Representative procedure for synthesis of 2-(*p*-Tolyl)-1*H*-indole (4aa): In an oven dried round bottom flask, 1-(pyridin-2-yl)-2-(*p*-tolyl)-1*H*-indole (**3aa**; 0.125 g, 0.44 mmol) was introduced and CH_2Cl_2 (5 mL) was added into it. Methyl trifluoromethanesulfonate (MeOTf; 0.087 g, 0.53 mmol) was added drop wise via a syringe to the reaction mixture at 0 °C and the resultant reaction mixture was stirred at room temperature for 12 h. Then the volatiles were removed under vacuum and the residue was redissolved in MeOH (3 mL). To the resultant mixture, NaOH (2 mL, 2M aqueous) solution was added and the reaction mixture was stirred at 60 °C for 10 h. At ambient temperature, the volatiles were evaporated under reduced pressure, and the resulting residue was extracted with EtOAc (15 mL x 3). The combined organic extract was washed with brine, dried over Na_2SO_4 and the volatiles were evaporated in vacuo. The remaining residue was purified by column chromatography on silica gel (petroleum ether/EtOAc/Et_3N: 50/1/0.5) to yield **4aa** (0.078 g, 86%) as a colorless solid. ^1H NMR (500 MHz, $CDCl_3$): δ= 8.32 (br s, 1H, NH), 7.62 (d, J = 7.6 Hz, 1H, Ar–H), 7.56 (d, J = 7.6 Hz, 2H, Ar–H), 7.39 (d, J = 7.6 Hz, 1H, Ar–H), 7.25 (d, J = 7.6 Hz, 2H, Ar–H), 7.19 (t, J = 7.6 Hz, 1H, Ar–H), 7.12 (t, J = 7.6 Hz, 1H, Ar–H), 6.78 (s, 1H), 2.39 (s, 3H, CH_3). $^{13}C\{^1H\}$-NMR (125 MHz, $CDCl_3$): δ = 138.2 (C_q), 137.8 (C_q), 136.8 (C_q), 129.9 (2C, CH), 129.7 (C_q), 129.5 (C_q), 125.2 (2C, CH), 122.3 (CH), 120.7 (CH), 120.4 (CH), 111.0 (CH), 99.6 (CH), 21.4 (CH_3). HRMS (ESI): *m/z* Calcd for $C_{15}H_{13}N + H^+$ $[M + H]^+$ 208.1126; Found 208.1127. The ^1H and $^{13}C\{^1H\}$ spectra are consistent with those reported in the literature.[69]

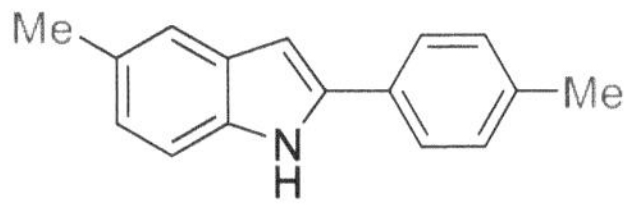

Me, Me, N, H

5-Methyl-2-(*p*-tolyl)-1*H*-indole (4ba): The representative procedure was followed, using substrate **3ba** (0.100 g, 0.33 mmol), and methyl trifluoromethanesulfonate (MeOTf; 0.065 g, 0.40 mmol). Purification by column chromatography on silica gel (petroleum ether/EtOAc/Et_3N): 50/1/0.5) yielded **4ba** (0.058 g, 79%) as colorless solid. ^1H NMR (500 MHz, $CDCl_3$): δ= 8.21 (br s, 1H, NH), 7.54 (d, J = 7.6 Hz, 2H, Ar–H), 7.40 (s, 1H, Ar–H), 7.28-7.23 (m, 3H, Ar–H), 7.01 (d, J = 8.4 Hz, 1H, Ar–H), 6.70 (s, 1H, Ar–H), 2.45 (s, 3 H, CH_3), 2.39 (s, 3H, CH_3). $^{13}C\{^1H\}$-NMR (125 MHz, $CDCl_3$): δ = 138.3 (C_q), 137.7 (C_q), 135.2 (C_q), 129.9 (C_q), 129.8 (2C, CH), 129.7 (C_q), 129.6 (C_q), 125.2 (2C, CH), 123.9 (CH), 120.3 (CH), 110.6 (CH), 99.1 (CH), 21.7 (CH_3), 21.4 (CH_3). HRMS (ESI): *m/z* Calcd for $C_{16}H_{15}N + H^+$ $[M + H]^+$ 222.1277; Found 222.1276. The ^1H and $^{13}C\{^1H\}$ spectra are consistent with those reported in the literature.[70]

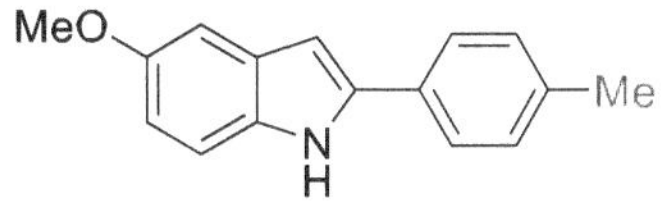

MeO Me N H

5-Methoxy-2-(*p*-tolyl)-1*H*-indole (4ca): The representative procedure was followed, using substrate **4ba** (0.093 g, 0.30 mmol), and methyl trifluoromethanesulfonate (MeOTf; 0.058 g, 0.35 mmol). Purification by column chromatography on silica gel (petroleum ether/EtOAc/Et_3N: 50/1/0.5) yielded **4ba** (0.058 g, 81%) as colorless solid. ^1H NMR (400 MHz, $CDCl_3$): δ= 8.22 (br s, 1H, N–H), 7.51 (d, J = 8.4 Hz, 2H, Ar–H), 7.22 (t, J = 8.4 Hz, 3H, Ar–H), 7.07 (d, J = 2.3 Hz, 1H, Ar–H), 6.83 (dd, J = 2.3, 8.4 Hz, 1H, Ar–H), 6.69 (s, 1H, Ar–H), 3.85 (s, 3 H, CH_3), 2.37 (s, 3H, CH_3). $^{13}C\{^1H\}$-NMR (100 MHz, $CDCl_3$): δ = 154.6 (C_q), 139.0 (C_q), 137.7 (C_q), 132.1 (C_q), 129.9 (C_q), 129.8 (2C, CH), 129.7 (C_q), 125.1 (2C, CH), 112.4 (CH), 111.7 (CH), 102.3 (CH), 99.4 (CH), 56.0 (CH_3), 21.4 (CH_3). HRMS (ESI): m/z Calcd for $C_{16}H_{15}NO + H^+$ $[M + H]^+$ 238.1226; Found 238.1224. The ^1H and $^{13}C\{^1H\}$ spectra are consistent with those reported in the literature.[71]

5.4.3 Synthesis of Tryptamine Derivatives:

Representative Procedure: A solution of **4aa** (0.050 g, 0.24 mmol) and *N*-(2,2-dimethoxyethyl)acetamide (0.047 g, 0.32 mmol) in CH_2Cl_2 (2.0 mL) was added to the mixture of trifluoroacetic acid (0.160 g, 1.43 mmol) and triethylsilane (0.100 g, 0.87 mmol) in CH_2Cl_2 (3 mL), and the resulted reaction mixture was stirred at room temperature for 12 h. The reaction mixture was cooled to 0 °C, neutralized with saturated aqueous solution of $NaHCO_3$ and diluted with CH_2Cl_2. Aqueous layer was extracted with CH_2Cl_2 (20 mL x 3). The combined organic extract was washed with brine, dried over Na_2SO_4 and the volatiles were evaporated in vacuo. The remaining residue was purified by column chromatography on silica gel (petroleum ether/EtOAc/Et_3N: 50/1/0.5) to yield the compound **5aa** (0.049 g, 70%).

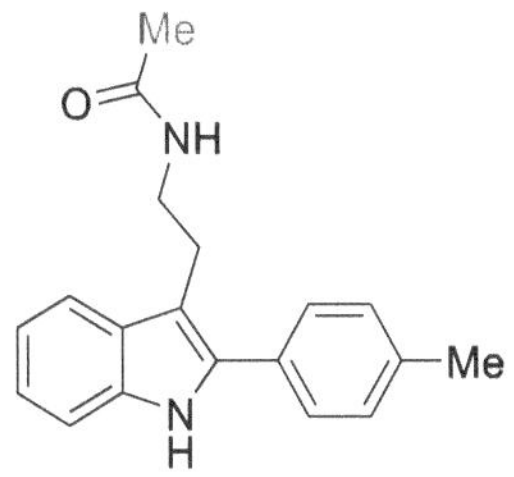

Me O NH Me N H

***N*-(2-(2-(*p*-Tolyl)-1*H*-indol-3-yl)ethyl)acetamide (5aa):** ^1H-NMR (500 MHz, $CDCl_3$): δ = 8.40 (br s, 1H, *NH*), 7.61 (d, J = 7.6 Hz, 1H, Ar–H), 7.44 (d, J = 8.4 Hz, 2H, Ar–H), 7.37 (d, J = 7.6 Hz, 1H, Ar–H), 7.25 (d, J = 8.4 Hz, 2H, Ar–H), 7.20 (t, J = 6.9 Hz, 1H, Ar–H), 7.13 (t, J = 8.4 Hz, 1H, Ar–H), 5.53 (br s, 1H, *NH*), 3.52 (vq, J = 6.9 Hz, 2H, CH_2), 3.09 (t, J = 6.9 Hz, 2H, CH_2), 2.38 (s, 3H, CH_3), 1.74 (s, 3H,CH_3). $^{13}C\{^1H\}$-NMR (125 MHz, $CDCl_3$): δ = 170.3 (CO), 138.0 (C_q), 136.0 (C_q),

135.6 (C_q), 130.2 (C_q), 129.9 (2C, CH), 129.2 (C_q), 128.0 (2C, CH), 122.4 (CH), 119.9 (CH), 118.9 (CH), 110.1 (CH), 109.5 (C_q), 40.3 (CH_2), 24.6 (CH_2), 23.3 (CH_3), 21.4 (CH_3). HRMS (ESI): m/z Calcd for $C_{19}H_{20}N_2O + H^+$ $[M + H]^+$ 293.1648; Found 293.1644. The ^1H and $^{1313}C\{^1H\}$ C spectra are consistent with those reported in the literature.[72]

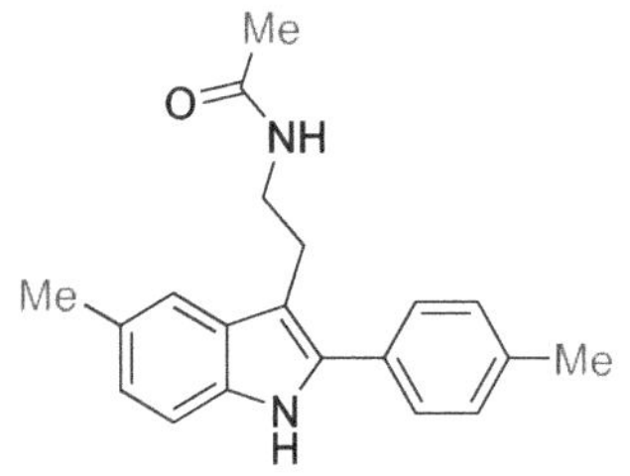

***N*-(2-(5-Methyl-2-(*p*-tolyl)-1*H*-indol-3-yl)ethyl)acetamide (5ba):** The representative procedure was followed, using **4ba** (0.045 g, 0.20 mmol), *N*-(2,2-dimethoxyethyl)acetamide (0.033 g, 0.22 mmol), trifluoroacetic acid (0.113 g, 0.99 mmol) and triethylsilane (0.07 g, 0.60 mmol). Column chromatography on silica gel (petroleum ether/EtOAc/Et_3N: 50/1/0.5) to yielded compound **5ba** (0.045 g, 73%). ^1H-NMR (400 MHz, $CDCl_3$): δ = 8.19 (br s, 1H, N–H), 7.44 (d, J = 8.4 Hz, 2H, Ar–H), 7.39 (s, 1H, Ar–H), 7.26 (d, J = 8.4 Hz, 3H, Ar–H), 7.03 (d, J = 7.6 Hz, 1H, Ar–H), 5.54 (br s, 1H, N–H), 3.53 (q, J = 6.1 Hz, 2H, CH_2), 3.08 (t, J = 6.9 Hz, 2H, CH_2), 2.47 (s, 3H, CH_3), 2.39 (s, 3H, CH_3), 1.76 (s, 3H, CH_3). 13C{^{1}H}-NMR (100 MHz, $CDCl_3$): δ = 170.3 (CO), 137.9 (C_q), 135.8 (C_q), 134.3 (C_q), 130.3 (C_q), 129.9 (2C, CH), 129.5 (C_q), 129.2 (CH), 128.0 (2C, CH), 124.1 (CH), 118.6 (CH), 110.7 (CH), 109.1 (C_q), 40.4 (CH_2), 24.6 (CH_2), 23.3 (CH_3), 21.7 (CH_3), 21.4 (CH_3). HRMS (ESI): m/z Calcd for $C_{20}H_{22}N_2O + H^+$ $[M + H]^+$ 307.1805; Found 307.1803.

5.4.4 Procedure for External Additive Experiments

The representative procedure of the alkylation reaction was followed, using indole (**1a**; 0.097 g, 0.50 mmol), 1-iodo-4-methylbenzene (**2a**; 0.218 g, 1.0 mmol), CuCl (0.0025 g, 0.025 mmol, 5 mol %), LiHMDS (0.109 g, 0.65 mmol), and TEMPO (0.156 g, 1.0 mmol) or galvinoxyl (0.422 g, 1.0 mmol) or BHT (0.220 g, 1.0 mmol) and the reaction mixture was stirred at 120 °C for 16 h. The ^1H NMR analyses of the reaction mixture did not show the formation of product **3aa**.

5.4.5 Deuterium Labeling Experiment

Procedure for Kinetic Isotope Effect (KIE) Study: To a Teflon-screw capped tube equipped with magnetic stir bar were introduced CuCl (0.0015 g, 0.015 mmol, 0.015 M), LiHMDS (0.065 g, 0.39 mmol), indole **1a** (0.058 g, 0.30 mmol, 0.3 M), or [2-D]-**1a** (0.059 g, 0.30 mmol, 0.3

M), 1-iodo-4-methylbenzene (0.133 g, 0.60 mmol, 0.6 M) and *n*-hexadecane (0.025 mL, 0.085 mmol, 0.085 M, internal standard), and 1,4-dioxane (0.9 mL) was added to make the total volume to 1.0 mL. The reaction mixture was then stirred at 120 °C in a pre-heated oil bath. At regular intervals (10, 20, 30, 45, 60, 90, 120) the reaction vessel was cooled to ambient temperature and an aliquot of sample was withdrawn to the GC vial. The sample was diluted with EtOAC and subjected to GC analysis. The concentration of product **3aa** obtained in each sample was determined with respect to the internal standard *n*-hexadecane. The data of the concentration of the product *vs* time (min) plot was drawn (Figure 5.2) with Origin Pro 8, and the rate was determined by initial rate method (up to 120 minutes). The data's were taken from the average of two independent experiments.

Procedure for *H/D* scrambling experiment (intermolecular): To a screw-capped tube equipped with magnetic stir bar were introduced 1-(pyridine-2-yl)-1*H*-indole-2-*d* ([2-D]-**1a**; 0.029 g, 0.15 mmol), 5-methoxy-1-(pyridine-2-yl)-1*H*-indole (**1c**; 0.034 g, 0.15 mmol), 1-iodo-4-methylbenzene (**2a**; 0.131 g, 0.60 mmol), LiHMDS (0.065 g, 0.39 mmol) inside the glove-box. To the above mixture 1,4-dioxane (1.0 mL) was added and the resultant reaction mixture was stirred at 120 °C in a preheated oil bath for 90 min. At ambient temperature, the reaction mixture was quenched with distilled H_2O (10 mL). The crude product was then extracted with EtOAc (15 mL x 3). The combined organic extract was dried over Na_2SO_4 and the volatiles were evaporated in *vacuo*. The remaining residue was subjected to column chromatography on neutral alumina (petroleum ether/EtOAc: 50/1) to recover the starting compounds. The ^{1}H NMR analysis of the recovered compound **1c** shows 35% incorporation of deuterium at the C(2)–H, whereas compound [2-D]-**1a** shows a significant loss of deuterium.

5.4.6 ¹H and ¹³C{¹H} NMR Spectra of Selected Arylated Compounds

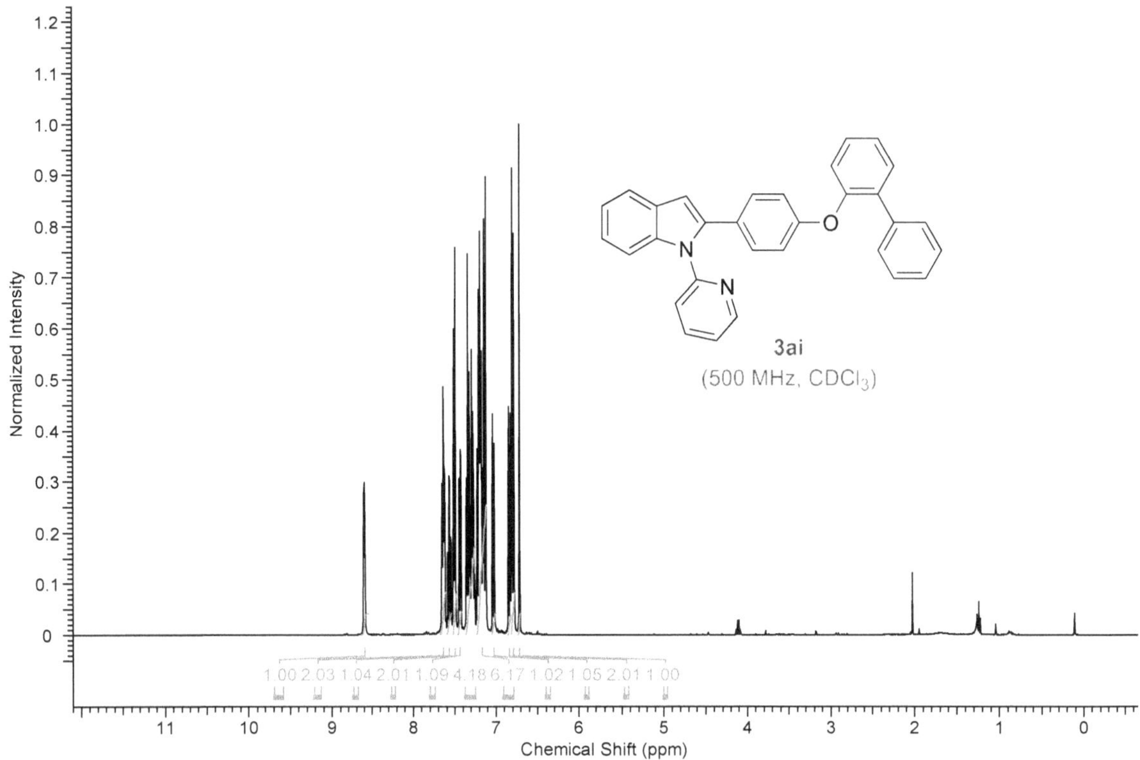

¹H-NMR spectrum of **3ai**

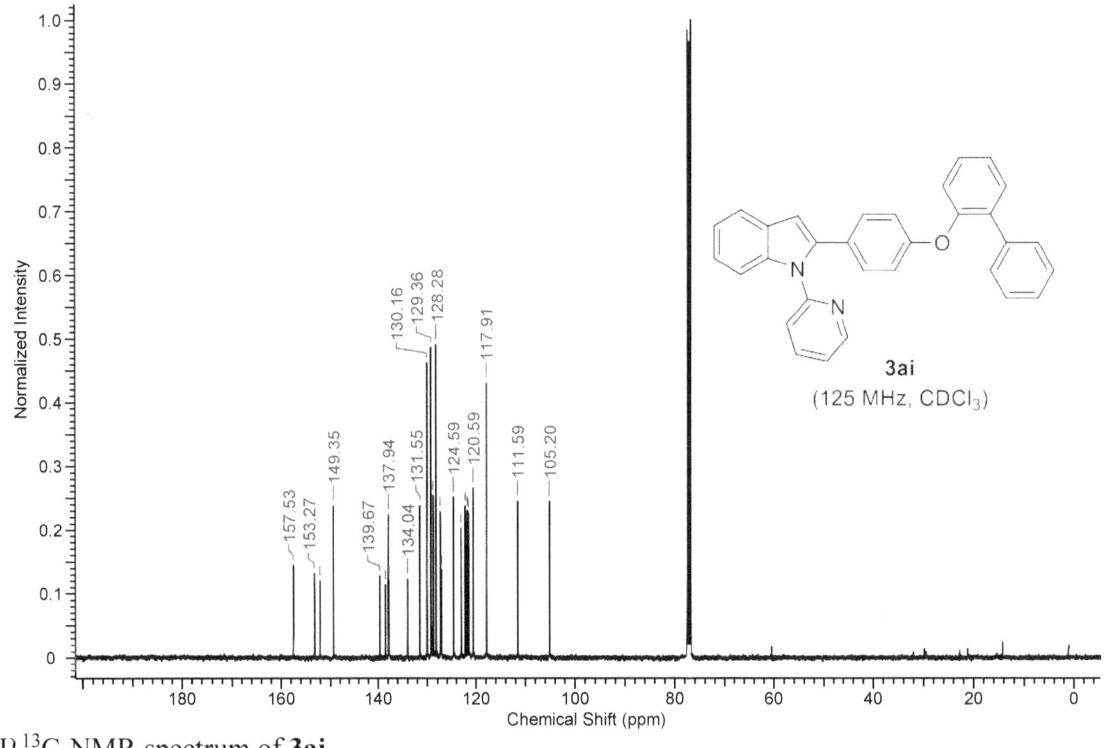

{¹H}¹³C-NMR spectrum of **3ai**.

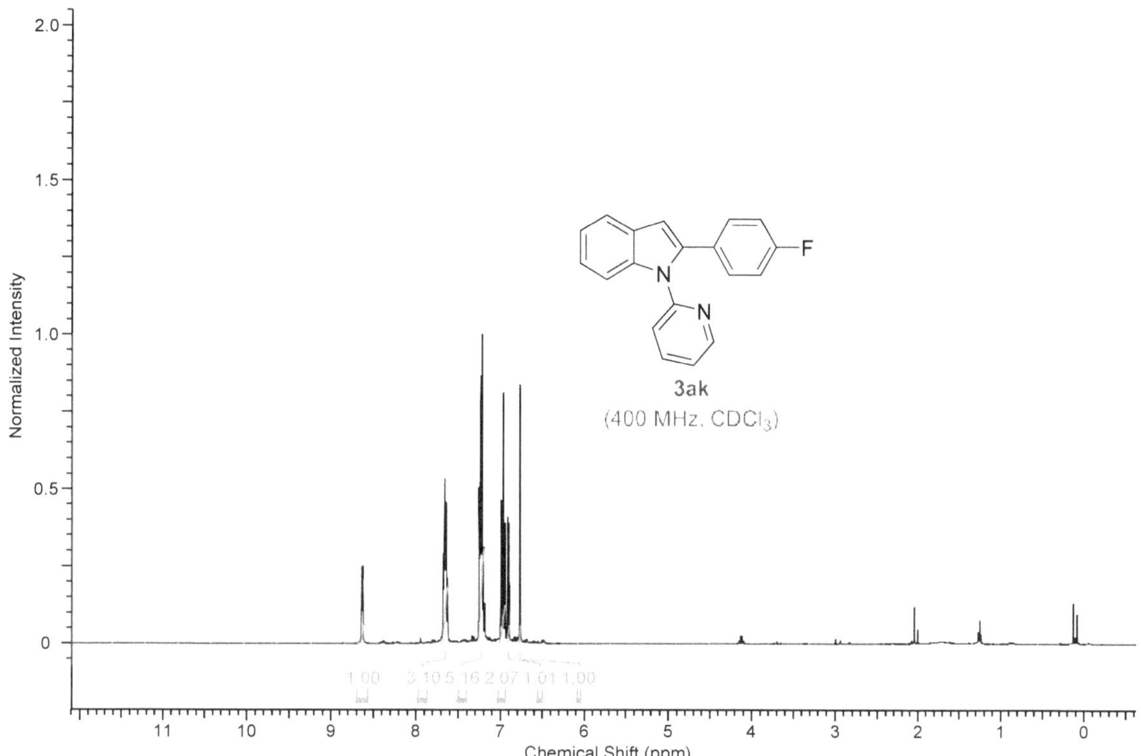

^{1}H-NMR spectrum of **3ak**.

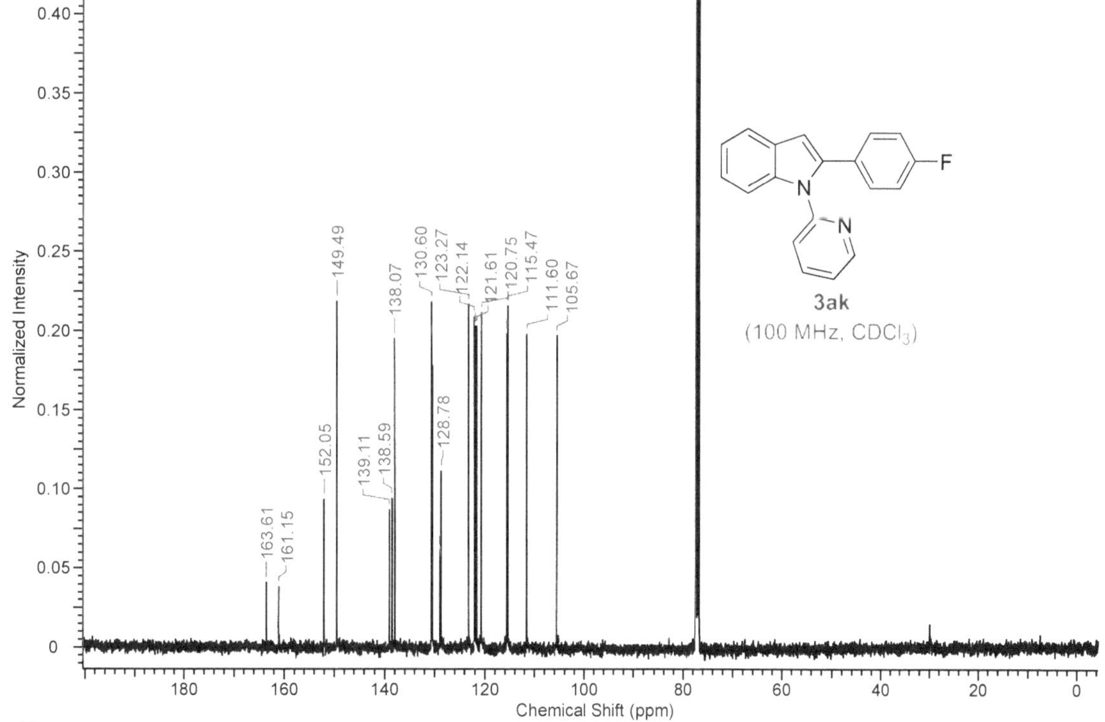

{^{1}H}^{13}C-NMR spectrum of **3ak**.

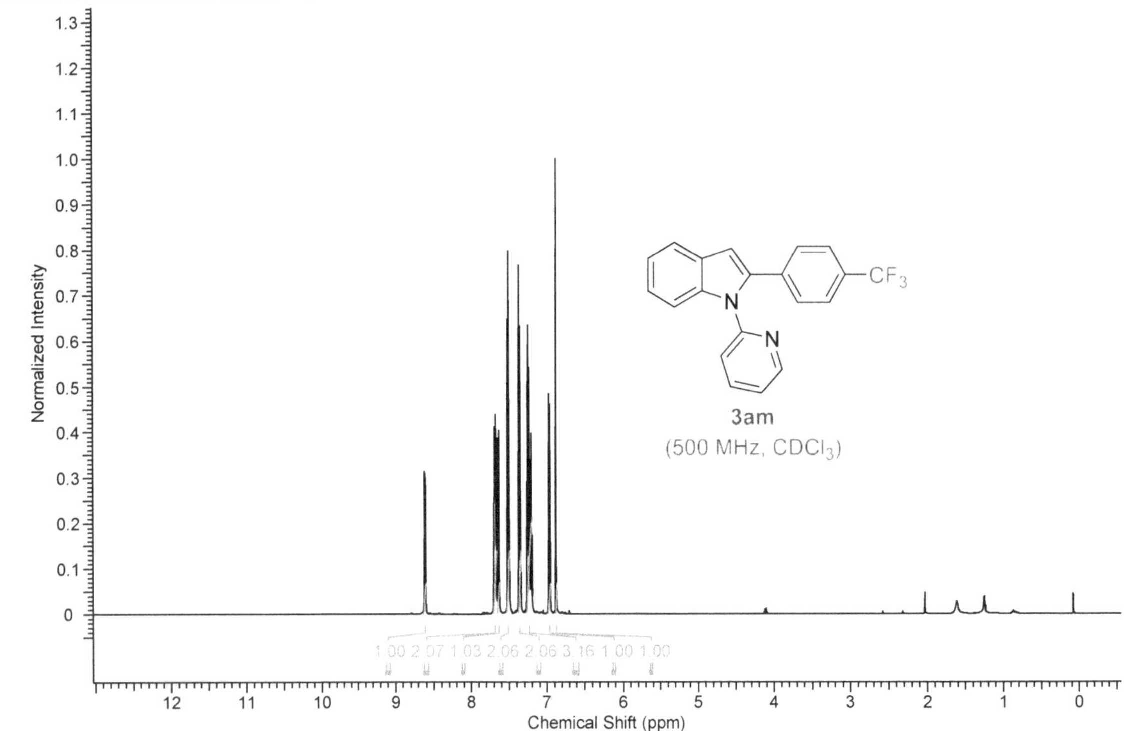

^{1}H-NMR spectrum of **3am**.

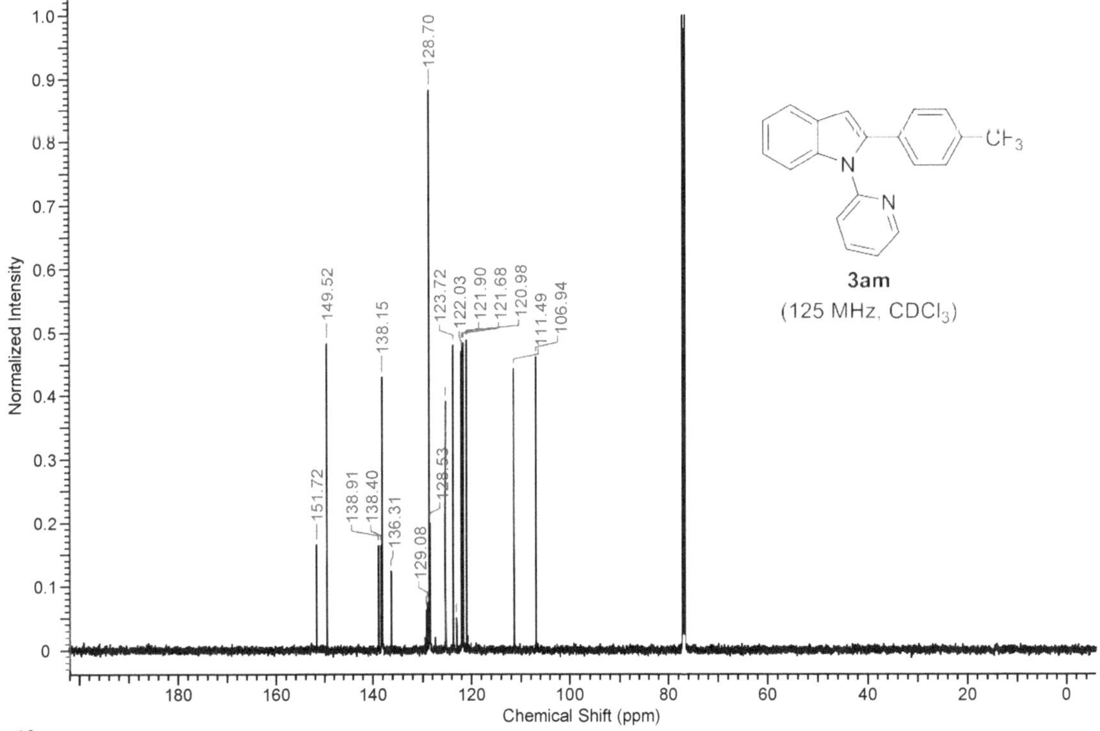

{^{1}H}^{13}C-NMR spectrum of **3am**.

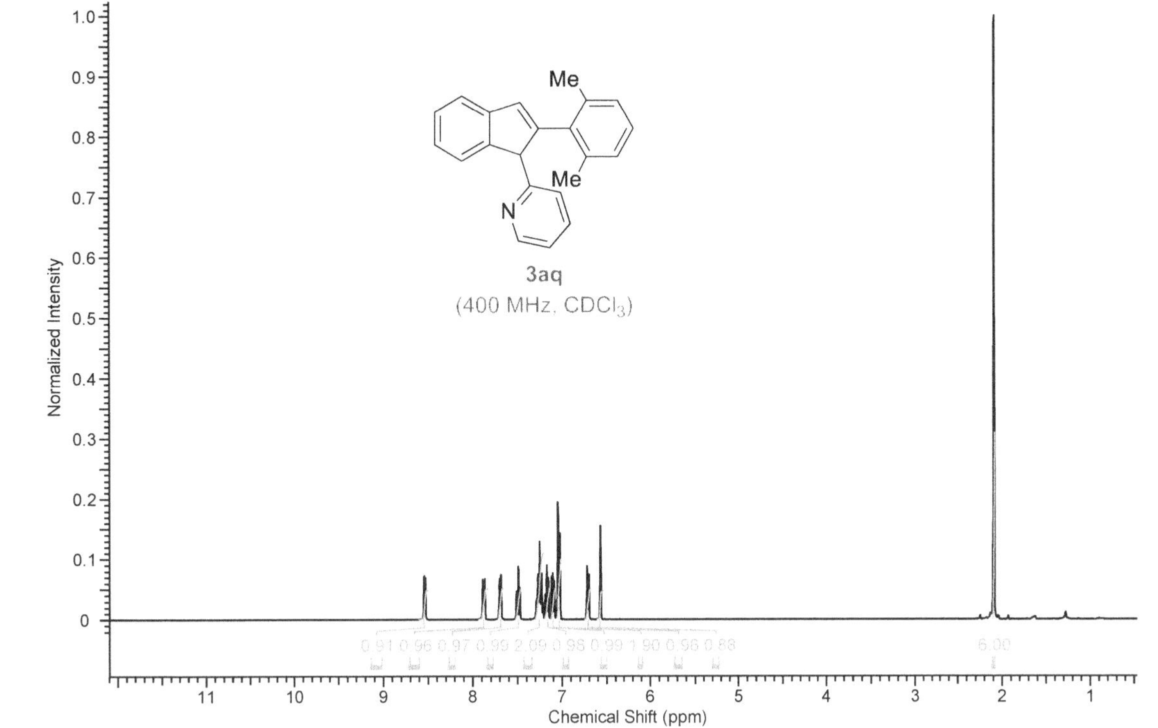

^{1}H-NMR spectrum of **3aq**

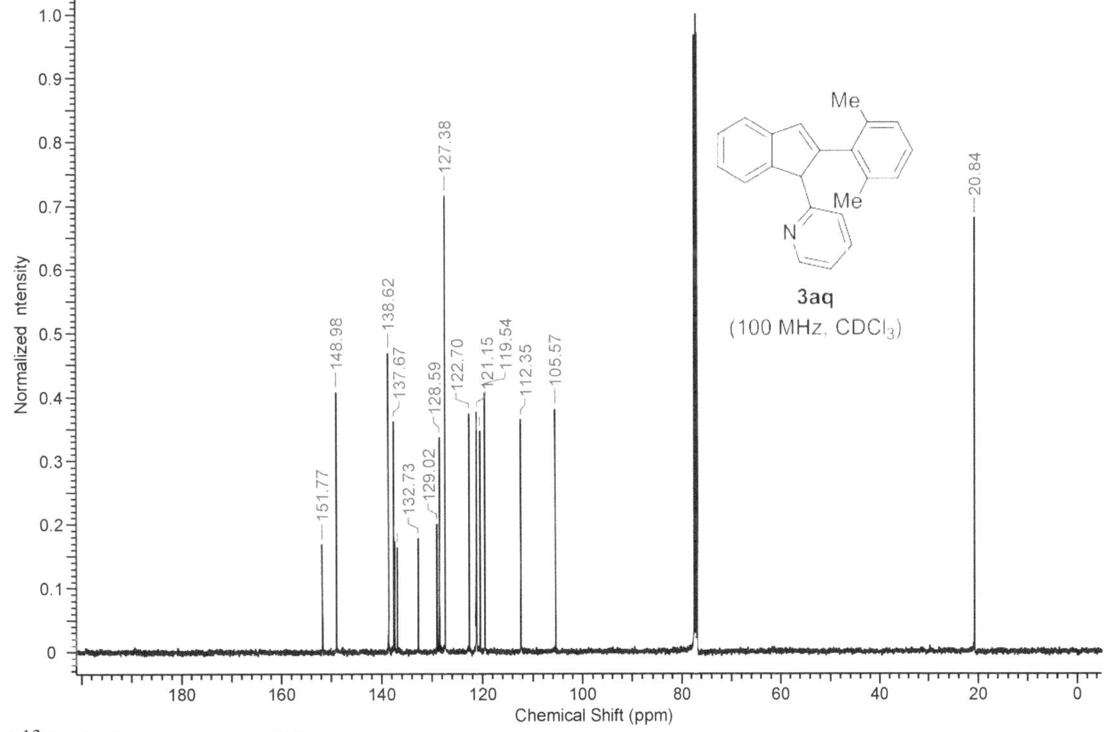

{^{1}H}^{13}C-NMR spectrum of **3aq**.

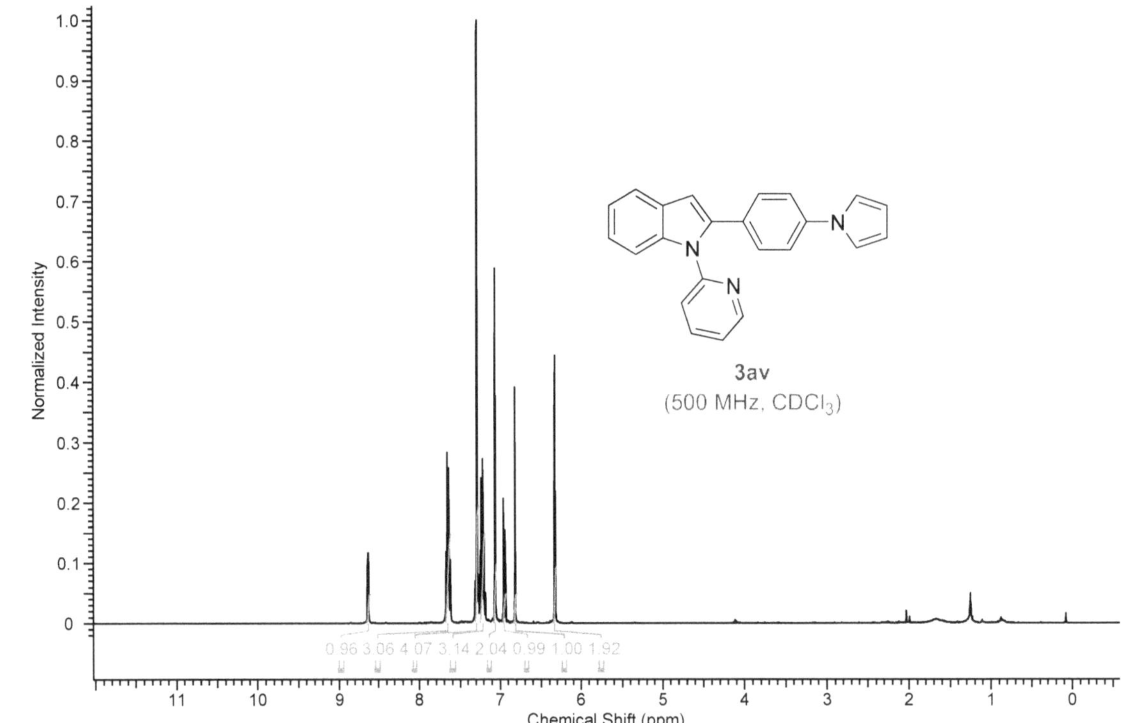

^{1}H-NMR spectrum of **3av**.

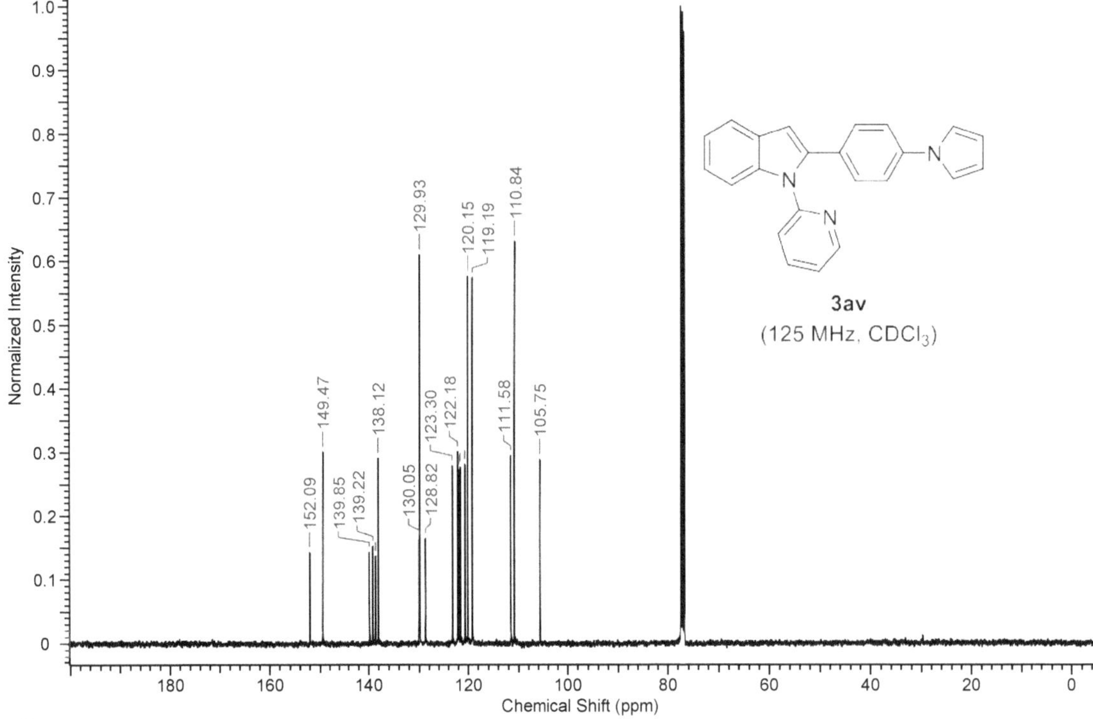

{^{1}H}^{13}C-NMR spectrum of **3av**.

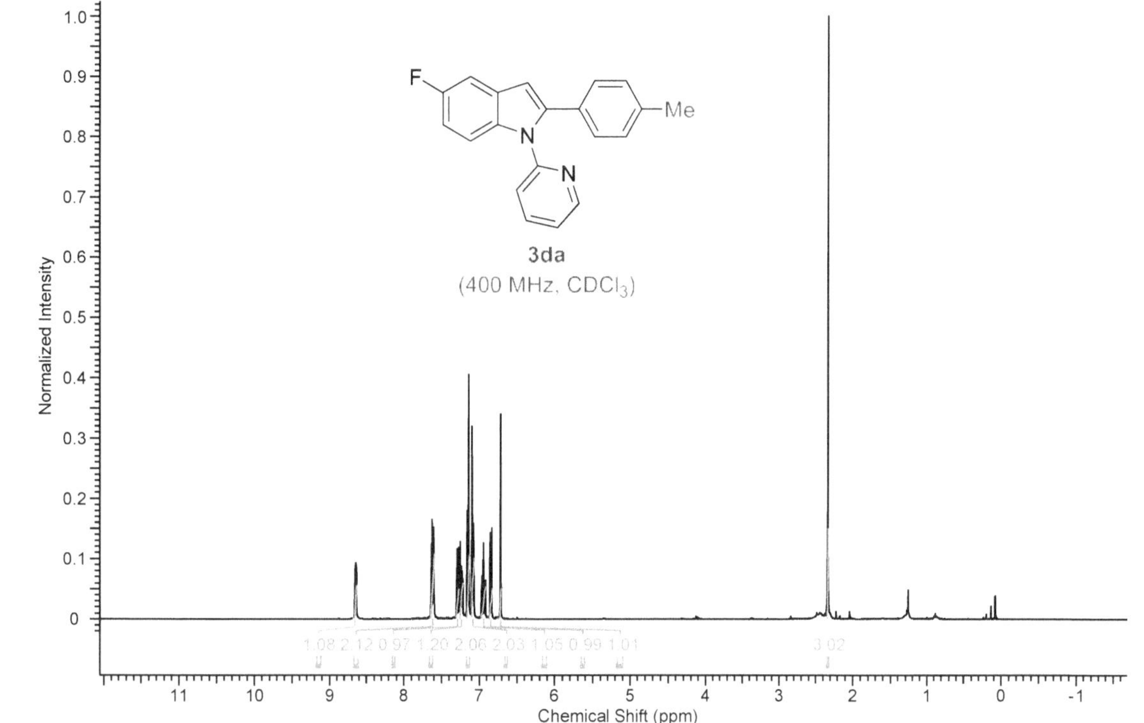

^{1}H-NMR spectrum of **3da**.

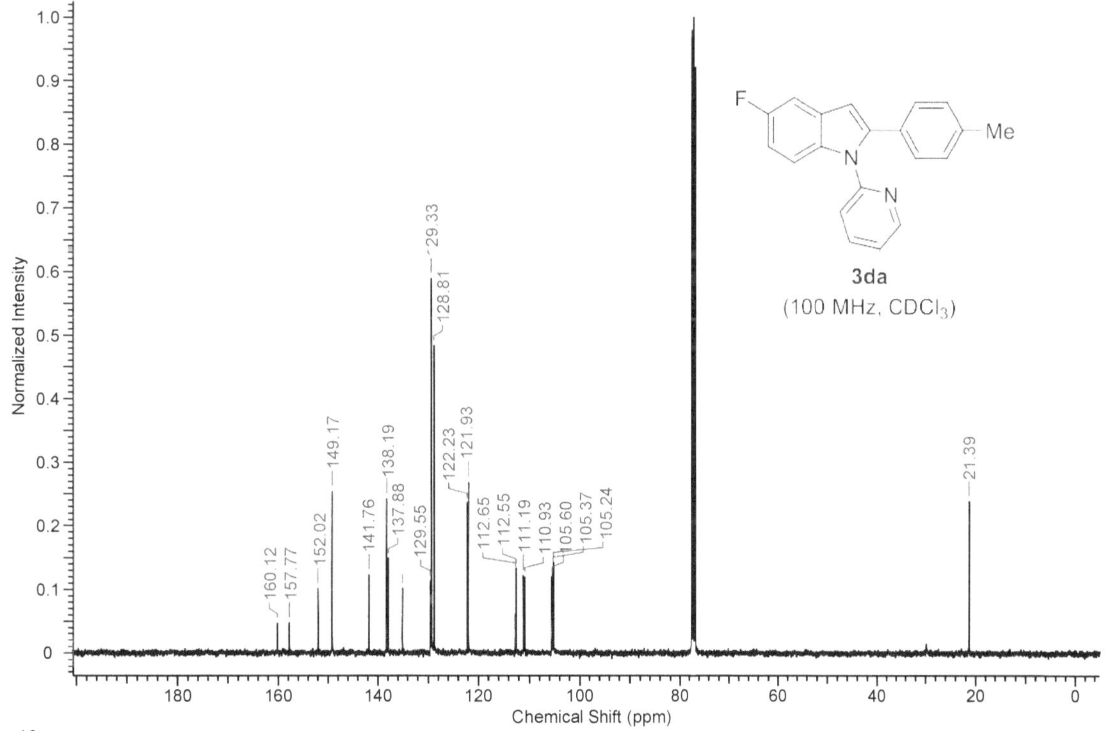

{^{1}H}^{13}C-NMR spectrum of **3da**.

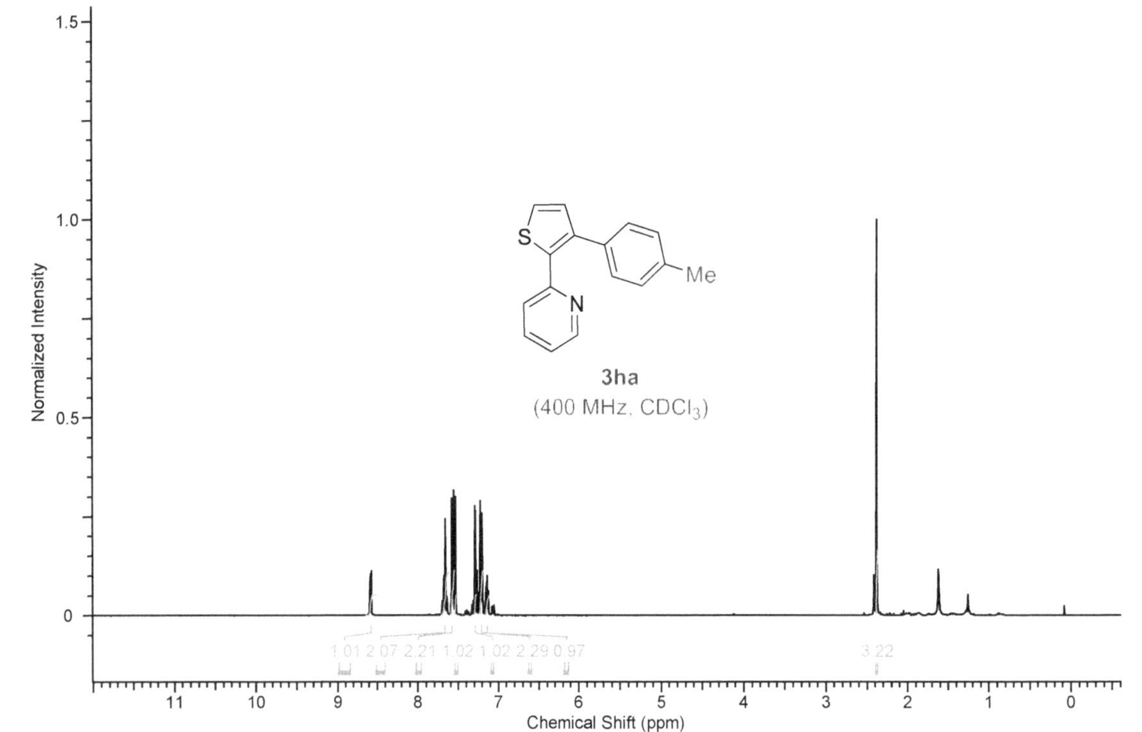

^{1}H-NMR spectrum of **3ha**.

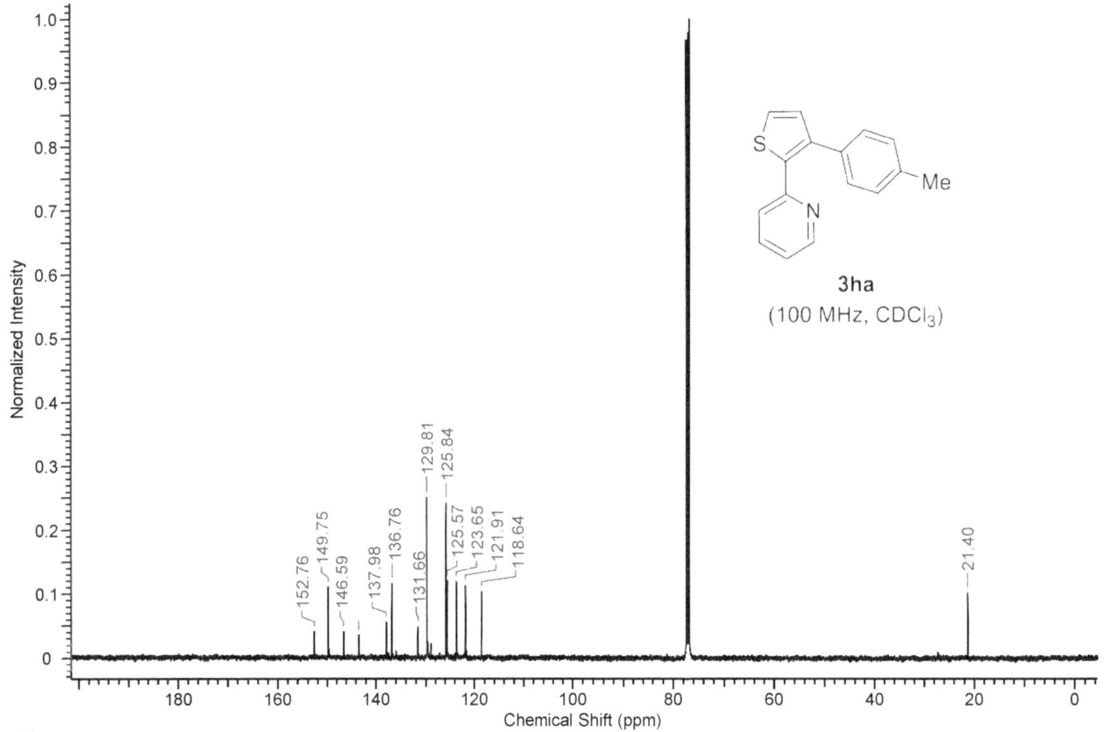

{^{1}H}^{13}C-NMR spectrum of **3ha**.

5.5 REFERENCES

(1) Sharma, V.; Kumar, P.; Pathak, D.; *J. Heterocycl. Chem.* **2010**, *47*, 491–502.

(2) Kaushik, N. K.; Kaushik, N.; Attri, P.; Kumar, N.; Kim, C. H.; Verma, A. K.; Choi, E. H., *Molecules* **2013**, *18*, 6620–6662.

(3) Joucla, L.; Djakovitch, L., *Adv. Synth. Catal.* **2009**, *351*, 673–714.

(4) Leitch, J.; Bhonoah, Y.; Frost, C. G., *ACS Catal.* **2017**, *7*, 5618–5627.

(5) Jagtap, R. A.; Punji, B., *Asian J. Org. Chem.* **2020**, *9*, 326–342; and references cited therein.

(6) Kher, S.; Lake, K.; Sircar, I.; Pannala, M.; Bakir, F.; Zapf, J.; Xu, K.; Zhang, S. H.; Liu, J.; Morera, L.; Sakurai, N.; Jack, R.; Cheng, J. F., *Bioorg. Med. Chem. Lett.* **2007**, *17*, 4442–4446.

(7) Sharma, P.; Kumar, A.; Sahu, V.; Upadhyay, S.; Singh, J., *Med. Chem. Res.* **2009**, *18*, 383–395.

(8) Hadimani, M. B.; MacDonough, M. T.; Ghatak, A.; Strecker, T. E.; Lopez, R.; Sriram, M.; Nguyen, B. L.; Hall, J. J.; Kessler, R. J.; Shirali, A. R.; Liu, L.; C. Garner, M.; Pettit, G. R.; Hamel, E.; Chaplin, D. J.; Mason, R. P.; Trawick, M. L.; Pinney, K. G., *J. Nat. Prod.* **2013**, *76*, 1668–1678; and references cited therein.

(9) Ackermann, L.; Lygin, A. V., *Org. Lett.* **2011**, *13*, 3332–3335.

(10) Tiwari, V. K.; Kamal, N.; Kapur, M., *Org. Lett.* **2015**, *17*, 1766–1769.

(11) Sollert, C.; Devaraj, K.; Orthaber, A.; Gates, P. J.; Pilarski, L. T., *Chem. Eur. J.* **2015**, *21*, 5380–5386.

(12) Nareddy, P.; Jordan, F.; Szostak, M., *Org. Lett.* **2018**, *20*, 341–344.

(13) Wang, X.; Lane, B. S.; Sames, D., *J. Am. Chem. Soc.* **2005**, *127*, 4996–4997.

(14) Wang, L.; Qu, X.; Li, Z.; Peng, W. M., *Tetrahedron Lett.* **2015**, *56*, 3754–3757.

(15) Wu, H.; Liu, T.; Cui, M.; Li, Y.; Jian, J.; Wang, H.; Zeng, Z., *Org. Biomol. Chem.* **2017**, *15*, 536–540.

(16) Deprez, N. R.; Kalyani, D.; Krause, D. A.; Sanford, M. S., *J. Am. Chem. Soc.* **2006**, *128*, 4972–4973.

(17) Wang, X.; Gribkov, D. V.; Sames, D., *J. Org. Chem.* **2007**, *72*, 1476–1479.

(18) Lebrasseur, N.; Larrosa, I., *J. Am. Chem. Soc.* **2008**, *130*, 2926–2927.

(19) Liang, Z.; Yao, B.; Zhang, Y., *Org. Lett.* **2010**, *12*, 3185–3187.

(20) Potavathri, S.; Pereira, K. C.; Gorelsky, S. I.; Pike, A.; LeBris, A. P.; DeBoef, B., *J. Am. Chem. Soc.* **2010**, *132*, 14676–14681.

(21) Joucla, L.; Batail, N.; Djakovitch, L., *Adv. Synth. Catal.* **2010**, *352*, 2929–2936.

(22) Zhou, J.; Hu, P.; Zhang, M.; Huang, S.; Wang, M.; Su, W., *Chem. Eur. J.* **2010**, *16*, 5876–5881.

(23) Nadres, E. T.; Lazareva, A.; Daugulis, O., *J. Org. Chem.* **2011**, *76*, 471–483.

(24) Wang, L.; Yi, W. B.; Cai, C., *Chem. Commun.* **2011**, *47*, 806–808.

(25) Islam, S.; Larrosa, I., *Chem. Eur. J.* **2013**, *19*, 15093–15096.

(26) Ghobrial, M.; Mihovilovic, M. D.; Schnürch, M., *Beilstein J. Org. Chem.* **2014**, *10*, 2186–2199.

(27) Kulkarni, A. A.; Daugulis, O., *Synthesis* **2009**, 4087–4109.

(28) Su, B.; Cao, Z. C.; Shi, Z. J., *Acc. Chem. Res.* **2015**, *48*, 886–896.

(29) Gandeepan, P.; Müller, T.; Zell, D.; Cera, G.; Warratz, S.; Ackermann, L., *Chem. Rev.* **2019**, *119*, 2192–2452.

(30) Zhu, X.; Su, J.; Du, C.; Wang, Z. L.; Ren, C.; Niu, J.; Song, M. P., *Org. Lett.* **2017**, *19*, 596–599.

(31) Song, W., Ackermann, L., *Angew. Chem. Int. Ed.* **2012**, *51*, 8251–8254.

(32) Punji, B.; Song, W.; Shevchenko, G. A.; Ackermann, L., *Chem. Eur. J.* **2013**, *19*, 10605–10610.

(33) Jagtap, R. A.; Soni, V.; Punji, B., *ChemSusChem* **2017**, *10*, 2242–2248.

(34) Pandey, D. K.; Vijaykumar, M.; Punji, B., *J. Org. Chem.* **2019**, *84*, 12800–12808.

(35) Sheldon, R. A., *Green Chem.* **2005**, *7*, 267–278.

(36) Anastas, P.; Eghbali, N., *Chem. Soc. Rev.* **2010**, *39*, 301–312.

(37) Prat, D.; Wells, A.; Hayler, J.; Sneddon, H.; McElroy, C. R.; Abou-Shehada, S.; Dunn, P. J., *Green Chem.* **2016**, *18*, 288–296.

(38) DeSimone, J. M., *Science* **2002**, *297*, 799–803.

(39) Hassan, J.; Sévignon, M.; Gozzi, C.; Schulz, E.; Lemaire, M., *Chem. Rev.* **2002**, *102*, 1359–1470.

(40) Jerphagnon, T.; Pizzuti, M. G.; Minnaard, A. J.; Feringa, B. L., *Chem. Soc. Rev.* **2009**, *38*, 1039–1075.

(41) Allen, S. E.; Walvoord, R. R.; Padilla-Salinas, R.; Kozlowski, M. C., *Chem. Rev.* **2013**, *113*, 6234–6458.

(42) McCann, S. D.; Stahl, S. S., *Acc. Chem. Res.* **2015**, *48*, 1756–1766.

(43) Zhu, X.; Chiba, S., *Chem. Soc. Rev.* **2016**, *45*, 4504–4523.

(44) Loup, J.; Dhawa, U.; Pesciaioli, F.; Wencel-Delord, J.; Ackermann, L., *Angew. Chem.*

Int. Ed. **2019**, *58*, 12803–12818.

(45) Do, H. Q.; Daugulis, O., *J. Am. Chem. Soc.* **2007**, *129*, 12404–12405.

(46) Zhao, D.; Wang, W.; Yang, F.; Lan, J.; Yang, L.; Gao, G.; You, J., *Angew. Chem. Int. Ed.* **2009**, *48*, 3296–3300.

(47) Phipps, R. J.; Gaunt, M. J., *Science* **2009**, *323*, 1593–1597.

(48) Ciana, C. L.; Phipps, R. J.; Brandt, J. R.; Meyer, F. M.; Gaunt, M. J., *Angew. Chem. Int. Ed.* **2011**, *50*,458–462.

(49) Cao, H.; Zhan, H.; Lin, Y.; Lin, X.; Du, Z.; Jiang, H., *Org. Lett.* **2012**, *14*, 1688–1691.

(50) Yang, Y.; Li, R.; Zhao, Y.; Zhao, D.; Shi, Z., *J. Am. Chem. Soc.* **2016**, *138*, 8734–8737.

(51) Ackermann, L.; K. Potukuchi, H.; Landsberg, D.; Vicente, R., *Org. Lett.* **2008**, *10*, 3081–3084.

(52) Jeyachandran, R.; Potukuchi, H. K.; Ackermann, L., *Beilstein J. Org. Chem.* **2012**, *8*, 1771–1777.

(53) Modha, S. G.; Greaney, M. F., *J. Am. Chem. Soc.* **2015**, *137*, 1416–1419.

(54) Vásquez-Céspedes, S.; Chepiga, K. M.; Möller, N.; Schäfer, A. H.; Glorius, F., *ACS Catal.* **2016**, *6*, 5954–5961.

(55) Yang, Y.; Gao, P.; Zhao, Y.; Shi, Z., *Angew. Chem. Int. Ed.* **2017**, *56*, 3966–3971.

(56) Phipps, R. J.; Grimster, N. P.; Gaunt, M. J., *J. Am. Chem. Soc.* **2008**, *130*, 8172–8174.

(57) Soni, V.; Jagtap, R. A.; Gonnade, R. G.; Punji, B., *ACS Catal.* **2016**, *6*, 5666–5672.

(58) Pandey, D. K.; Ankade, S. B.; Ali, A.; Vinod, C. P.; Punji, B., *Chem. Sci.* **2019**, *10*, 9493–9500.

(59) Rivara, S.; Lodola, A.; Mor, M.; Bedini, A.; Spadoni, G.; Lucini, V.; Pannacci, M.; Fraschini, F.; Scaglione, F.; Sanchez, R. O.; Gobbi, G.; Tarzia, G., *J. Med. Chem.* **2007**, *50*,6618–6626.

(60) Bedini, A.; Lucarini, S.; Spadoni, G.; Tarzia, G.; Scaglione, F.; Dugnani, S.; Pannacci, M.; Lucini, V.; Carmi, C.; Pala, D.; Rivara, S.; Mor, M., *J. Med. Chem.* **2011**, *54*, 8362–8372.

(61) Davies, D. L.; Al-Duaij, O.; Fawcett, J.; Giardiello, M.; S. T.; Hilton, Russell, D. R., *Dalton Trans.* **2003**, 4132–4138.

(62) Simmons, E. M.; Hartwig, J. F., *Angew. Chem. Int. Ed.* **2012**, *51*, 3066–3072.

(63) Qu, X.; Li, T.; Zhu, Y.; Sun, P.; Yang, H.; Mao, J., *Org. Biomol. Chem.* **2011**, *9*, 5043–5046.

(64) Jie, L.; Wang, L.; Xiong, D.; Yang, Z.; Zhao, D.; Cui, X., *J. Org. Chem.* **2018**, *83*, 10974–10984.

(65) Yu, T. Y.; Zheng, Z. J.; Bai, J. H.; Fang, H.; Wei, H., *Adv. Synth. Catal.* **2019**, *361*, 2020–2024.

(66) Zhang, L.; Chen, J.; Chen, J.; Jin, L.; Zheng, X.; Jiang, X.; Yu, C., *Tetrahedron Lett.* **2019**, *60*, 1053–1056.

(67) Shen, Z.; Pi, C.; Cui, X.; Wu, Y., *Chin. Chem. Lett.* **2019**, *30*, 1374–1378.

(68) Tiwari, V. K.; Kamal, N.; Kapur, M., *Org. Lett.* **2015**, *17*, 1766–1769.

(69) Turnu, F.; Luridiana, A.; Cocco, A.; Porcu, S.; Frongia, A.; Sarais, G.; Secci, F., *Org. Lett.* **2019**, *21*, 7329–7332.

(70) Benitez–Medina, G. E.; Ortiz–Soto, S.; Cabrera, A.; Amzquita–Valencia, M., *Eur. J. Org. Chem.* **2019**, 3763–3770.

(71) Wei, Y.; Deb, I.; Yoshikai, N., *J. Am. Chem. Soc.* **2012**, *134*, 9098–9101.

(72) Nenajdenko, V. G.; Zakurdaev, E. P.; Prusov, E. V.; Balenkova, E. S., *Tetrahedron* **2004**, *60*, 11719–11724.

Chapter-6

Summary and Outlook

6.1 SUMMARY

The thesis described our efforts in designing and developing pincer-based manganese complexes for chemoselective hydrogenation of unsaturated double bonds and indole's C(2)−H bond arylation under copper catalysis. We could successfully demonstrate the manganese(I) catalyzed chemoselective hydrogenation of conjugated ketones, aldehydes, epoxy ketones, imines, and α-ketoamides directly using lower hydrogen pressure at room temperature. Also, we developed an atom economical process for regioselective C(2)−H arylation of indoles directly using aryl iodides as coupling partners. Initial controlled experiments were conducted to determine the catalytic pathway of the arylation reaction, and detailed mechanistic investigations (experimental and DFT calculation) were completed to determine the reaction mechanism of the hydrogenation reaction.

In Chapter 1, We have provided an overview of recent advancements in the fields of regioselective C–H bond arylation of indoles and base metal-catalyzed chemoselective hydrogenation of different unsaturated moieties using molecular hydrogen. This chapter mainly highlights the recent developments on Mn, Fe, and Co-catalyzed chemoselective hydrogenation and C–H arylation of indoles under Cu, Co, and Ni-catalysis.

Chapter 2 describes the synthesis of $(R^2PN^3N^{Pyz})$Mn(I) manganese complexes and their applications in the chemoselective hydrogenation of α,β-unsaturated compound for C=C, C=O, and C=N double bonds at room temperature. Mechanistic studies of the catalytic pathway have been extensively conducted, probing the intricate interplay between metal and ligand for efficient H_2 activation.

Chapter 3 describes the implication of developed manganese catalysts for the synthesis of α-hydroxy epoxides *via* chemoselective C=O bond hydrogenation of α,β-epoxy ketones using molecular hydrogen. Applicability of this protocol was showcased by performing reactions on different α,β-epoxy ketones derived from biologically active molecules and aroma compounds.

Chapter 4 presents manganese-catalyzed direct hydrogenation of α-ketoamides to generate synthetically and biologically important α-hydroxy amides. This protocol was very selective for the hydrogenation of C=O bonds in the α-ketoamides leaving other reducible functional groups unreacted.

The development of a solvent-free, ligand-free method for the selective C-H bond arylation of indoles and similar heteroarenes with copper catalysts is covered in Chapter 5. The effective removal of the directing group and the synthesis of Tryptamine derivatives demonstrated the synthetic value of this procedure.

Lastly, Chapter 6 presents the overall summary of the thesis work, followed by the future direction related to the field.

6.2 OUTLOOK

Significant advancements have been made in the recent few decades in both the regioselective C-H bond arylation of indoles and related heteroarenes and the chemoselective hydrogenation of diverse unsaturated moieties. However, most of the protocols required higher reaction temperatures, strong activators, and different hydrogen sources for the hydrogenation which will result in the formation of multiple products in a single reaction and the generation of stoichiometric waste. Our study on developing mixed donor pincer manganese catalysts for the chemoselective hydrogenation of various polar and non-polar double bonds showed that designing the ligand backbone for these types of reactions is very important. Hence there is a need to develop numerous kinds of ligands and corresponding 3d metal-based catalysts for the chemoselective hydrogenation of discrete unsaturated functional groups using H_2.

Recent studies on the 3d metal-catalyzed regioselective C–H arylation of indoles showed significant advancement in this field. However, there is scope to develop more sustainable protocols using highly abundant and cheap metals such as iron-based catalysis. In addition, there is a need to broaden the applicability of base metal-catalyzed C–H bond arylation by developing a method without any directing group. The outlook of the thesis would help to find a better catalytic system for milder and selective hydrogenation and arylation of numerous unsaturated moieties and heteroarenes respectively.

www.ingramcontent.com/pod-product-compliance
Lightning Source LLC
LaVergne TN
LVHW082243150826
845677LV00009B/1519

* 9 7 9 8 2 3 0 7 4 0 4 5 2 *